Kinetic theory and entropy

C. H. Collie

Kinetic theory and entropy

Longman London and New York

Longman Group Limited
Longman House
Burnt Mill, Harlow, Essex, UK

Published in the United States of America
by Longman Inc., New York

First published 1982

British Library Cataloguing in Publication Data

Collie, C. H.
Kinetic theory and entropy.
1. Gases, Kinetic theory of
2. Chemical reaction, Rate of
I. Title
541.3′9 QD501

Library of Congress Cataloguing in Publication Data
Collie, C. H.
Kinetic theory and entropy.

Bibliography: P.
Includes index.
1. Gases, Kinetic theory of. 2. Entropy.
I. Title.
QC175.C58 530′.113 81-8332
ISBN 0-582-44368-7 AACR2

Printed in Singapore by

Selector Printing Co Pte Ltd

Contents

Preface

This book is intended for those who were so felicitously known to Clerk Maxwell as beginners. That is, those who have studied mathematics and the physical sciences to a good school level (A level), and intend to spend at least another two years studying physics. Owing to the very high standard of science teaching in schools, beginners can be in a quandary. They have an enviable knowledge of descriptive physics, which far exceeds their immediate capacity for mastering any branch of it. The discovery of the gap which on the whole is a mathematical one can be very frustrating to those whose main interest is in the experimental side of the subject. In fields like the kinetic theory in which the main work has been done, but of which all scientists must have some knowledge, a solution to the problem is to push on with the knowledge one has, which is essentially that summations can be carried out by means of the integral calculus. With guidance one can go a long way without going outside the range of mathematics taught in schools. The arguments may lack rigour and fall back on intuition, but they get somewhere.

In defence of this point of view one can say that most of the pioneers who have added to our ways of thinking by crossing the frontiers of knowledge did so on decidedly rickety structures. For this reason a glimpse of the historical development has been included.

It is a commonplace to say that we live in a scientific age, but students need reminding how rapidly this change has come about. It is also some comfort when struggling with a difficult argument to know that at some time it has attracted the attention of the masters. On the experimental side entropy has been treated as a simply measurable quantity. The extreme refinement both of measurement and notation used in physical chemistry have been avoided. It is really part of the experimental side of the subject and is apt to distract the attention of the student on a first reading from the central idea.

The examples have been composed either with the object of removing straightforward calculations from the text or of illustrating the application of kinetic theory reasoning to relatively modern situations. They are not intended to be examination questions and do not have the neatness of presentation and capacity for elegant solution in thirty minutes which make these so difficult to compose.

On the whole numerical values have not been given because facility in using tables of constants is an important part of a physicist's training and the issue of new tables of reasonable size has eased the student's difficulties in this respect. In conclusion, it is obvious that in a book of this kind all the credit goes to the authors of standard texts, Jeans, Kennard, Mayer and Mayer, Chapman and Cowley, Fowler, Schrödinger, Wilson and many others without which it could not have been written and to whom my debt is amply and I hope sufficiently acknowledged. In particular I must thank Dr. Handel Davies whose patience in dealing with my artless questions on integrating factors has, I hope, enabled me to explain the matter correctly.

Malta 1982

CHAPTER 1

Historical background

1.1 Introduction

This introductory chapter is a summary of the main ideas needed for the understanding of the subsequent chapters. Natural science is an experimental subject which for the most part has grown by a steady application of the scientific method usually referred to by philosophers as induction. This consists of observing some natural phenomenon and framing an explanation in terms of a crude model or working hypothesis. The scientific method works best when the explanation is limited and concrete and suggests further experimental work so that a further comparison with experiment enables the hypothesis to be modified and so account for a wider range of observations.

Scientists on the whole are insensitive to the objections of philosophers. Practitioners of the scientific method find no difficulty in assuming that matter occurs in what would otherwise be empty space and are easily satisfied with simple ideas about time as an ever-flowing stream. They regard a lack of rigour in their philosophical position as a small price to pay for the solid advances which the steady application of the scientific method has made in our knowledge of nature. A successful hypothesis which explains a wide range of observations is usually raised to the rank of a 'theory'. Even quite successful theories which incorporate the work of many generations of scientists can turn out to be quite wrong. In this case a new start has to be made. Well-known examples are the phlogiston theory of chemical change, the caloric theory of heat, and Newton's theory of universal constantly flowing time.

Physical theories are usually tested by a close numerical comparison of their predictions with the results of careful quantitative experimental work. It is at this stage that the scientist has to face two problems; the theoretical one of calculating what are the exact predictions of the theory and the experimental one of making accurate measurements of the predicted quantities, which may well not be the ones which are easy to measure. Neither of these problems are always easy to solve and individual scientists usually concentrate on either the experimental, or the theoretical aspects of the work. The number who achieve worthwhile work in both fields is very small indeed. In the kinetic theory of matter, which professes to explain the properties of real substances in terms of the motion of small particles or molecules, one must clearly be provided with a 'system of mechanics' which lays down how particles move. Until the turn of the century Newton's system based upon the three laws of motion was unrivalled and unquestioned. Since then it has been challenged by an alternative system usually referred to as 'quantum mechanics' due to Heisenberg and Schrödinger which is peculiarly suitable for predicting the motion of light particles like electrons. A description of atomic structure has to be cast entirely in the language of quantum mechanics. For molecules, which are many thousand times heavier than electrons, a description in terms of the classical or Newtonian mechanics of the school textbooks is usually adequate. As will be seen in later chapters, ideas (as opposed to detailed calculations) obtained from quantum mechanics are essential to an understanding of the more general parts of the subject.

1.2 Molecular theory

That matter as ordinarily understood, e.g. tables and chairs, water, air, natural gas and frogspawn, has a discontinuous structure is now universally conceded. For frogspawn this is obvious to the naked eye, the gelatinous mass being easily seen to be composed of cells which cannot be cut without destroying their fundamental property of growing into tadpoles. For wood this discontinuous structure is less obvious and for water, metals and gases it is not obvious at all but is deduced from the results of long series of experiments which are an intimate part of the growth of natural science. Direct evidence that the postulated molecules and atoms actually exist was not available until seventy years ago.

The last serious upholders of the contrary view, that matter is essentially continuous, were Descartes (1596–1629) and Leibniz (1664–1716). The term molecule to represent the small units from which matter is assembled was introduced by Avogadro (1811); the distinction between atoms (the word used by Democritus (410 B.C.) and meaning uncuttable) and molecules, which can be composite structures containing many tightly bound atoms, was clearly enunciated by Cannizzaro (1858). A fair idea of molecular size was available by 1815 (Young) and the general explanation of the three states of matter in molecular terms was understood after the work of

Andrews (1869) and van der Waals (1881). That the characteristic properties of the solid state are due to the regular arrangement of the molecules in a three-dimensional lattice was not experimentally verified until the discovery of X-ray diffraction by Friederich and Knipping (1911). The fundamental importance of the regular ordering of the molecules in distinguishing between solids and viscous liquids such as glass and fused silica is quite a modern idea.

Daniel Bernoulli (1738) first proposed that the pressure of a gas was due to the bombardment of the walls by more or less free molecules and Maxwell believed that the first proper demonstration that Boyle's law could be derived from these considerations was due to Joule. Early work was often based upon a very simplified picture in which the size of molecules and the forces between them were completely neglected. The attention of mathematical physicists was intermittently directed to the possibility of relaxing these extreme assumptions which naturally did not commend themselves to scientists who were beginning to find out something about the properties of real molecules by indirect chemical means.

Maxwell's theoretical work on the viscosity of gases (1860) directed the attention of experimental physicists to some unsuspected properties of simple gases. After its initial successes this way of finding out about molecules became very unrewarding. Mathematical difficulties confined mathematicians to molecular models with spherical symmetry, just at a time when chemists were thinking of simple molecules like H_2, N_2, O_2 as distinctly dumb-bell-like structures. Nevertheless, the good qualitative but poor numerical agreement between theory and experiment was for some time the best direct physical evidence for the existence of molecules.

This is no longer the situation. The main evidence for the existence of molecules comes from the diffraction of X-rays and electrons, mass spectrometry, radioactivity and a wealth of spectroscopic evidence undreamed of in 1900. The main problem of molecular physics is no longer to obtain evidence for the existence of molecules but to find out what is going on in a molecular system using all the methods available.

1.3 Kinetic theory of gases

The kinetic theory of gases is a fairly well defined branch of molecular theory. Its subject matter is gases at relatively low pressures in which the distance between the molecules is greater by a factor of ten or more than it is in the liquid or solid states. The term kinetic is used to indicate that it is the motion of the molecules and not the forces between them which dominates the theoretical explanation of the gaseous state. Experimentally it has been found that the equations representing the behaviour of gases at low pressure and well above their boiling points are mathematically simple. The explanation of these simple results in terms of the molecular theory is one of the landmarks in our understanding of the properties of matter.

The typical gas is the air with which we are all surrounded. A student asked to demonstrate the existence and fundamental physical properties of gases could not do better than submerge an inverted burette in a tank of water. The trapped air occupies the space defined by the walls of the burette and the pressure exerted by the air prevents the water entering the otherwise empty space. When the tap is opened, the air visibly flows out as a stream of bubbles and the water flows in to take its place.

This experiment was well described and correctly interpreted by Empedocles in 440 B.C.: quite in the modern style with a charming female operator (Russell, 1947). After this promising start the study of gases made no progress for two thousand years when van Helmont (1577–1644), a distinguished Belgian doctor, identified carbon dioxide and clearly stated that there were different sorts of gases just as there are different sorts of matter. It is to him that we owe the term gas. This lack of progress was mostly due to the mystical properties attributed to air as one of the four Aristotelian elements. Van Helmont was as mystical as any of his predecessors and it was left to the next generation of scientists to make the final step. The proper study of gases as physical substances begins with Boyle's investigation of the 'Spring of the Air' (1660–1669). An account of the same discoveries was published later by Mariotte (1676) and they are often reasonably attributed to him in French textbooks.

1.4 Molecular forces

The tensile strength of solids shows that there are strong attractive forces acting between molecules. That the two pieces of a broken solid do not join up when they are brought together shows that these attractive forces are of short range. The large forces required to compress a solid measured by the bulk modulus *K* are usually interpreted to mean that at still smaller distances molecular forces change sign and become repulsive. In a solid with free surfaces the molecules are supposed to oscillate about the equilibrium point at which the forces are zero. The forces between molecules can thus be roughly represented by a graph as in Fig. 1.1a. We know very little by direct experiment about the shape of this force distance graph. Theoretically these forces may be calculated from Schrödinger's equation and the electronic structure of the molecules if this is known. The attractive part of the force known as the 'molecular force' or 'van der Waals force' is an inverse seventh power force as shown in the diagram. It is exemplified by the attraction between a short bar magnet and a small soft iron sphere at large distances. Unlike the much larger (varying as $1/r^4$) force between bar magnets it is always attractive, since if the direction of the bar magnet is reversed so is the induced magnetic moment of the soft iron sphere.

The forces between molecules are of course not magnetic in origin but are due to the electrostatic forces between the constituent

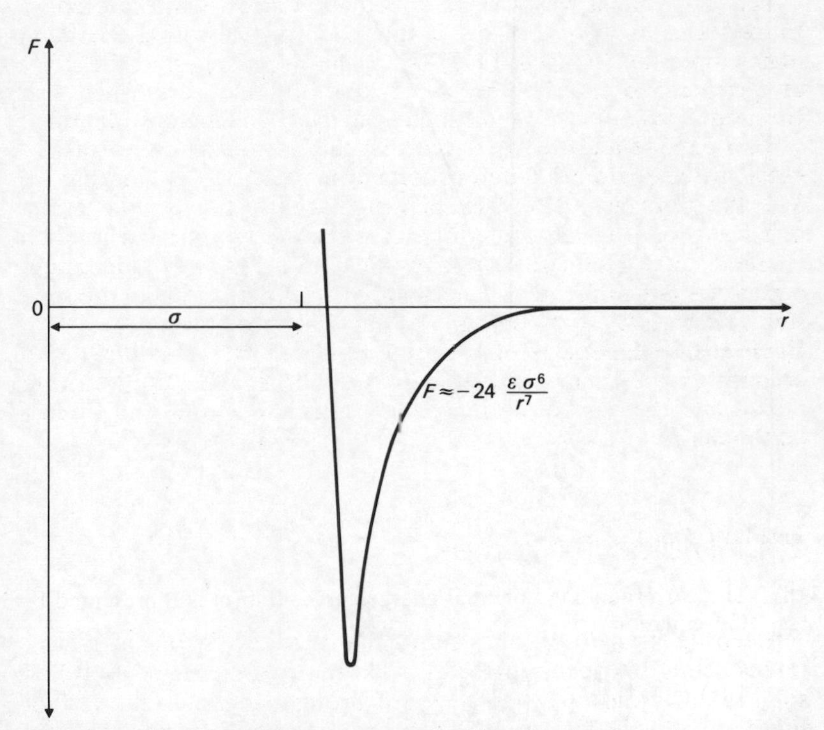

Fig. 1.1 The Lennard-Jones approximation for the force between two molecules.
(a) The attractive force F has a negative sign since it is opposite to the direction of r.

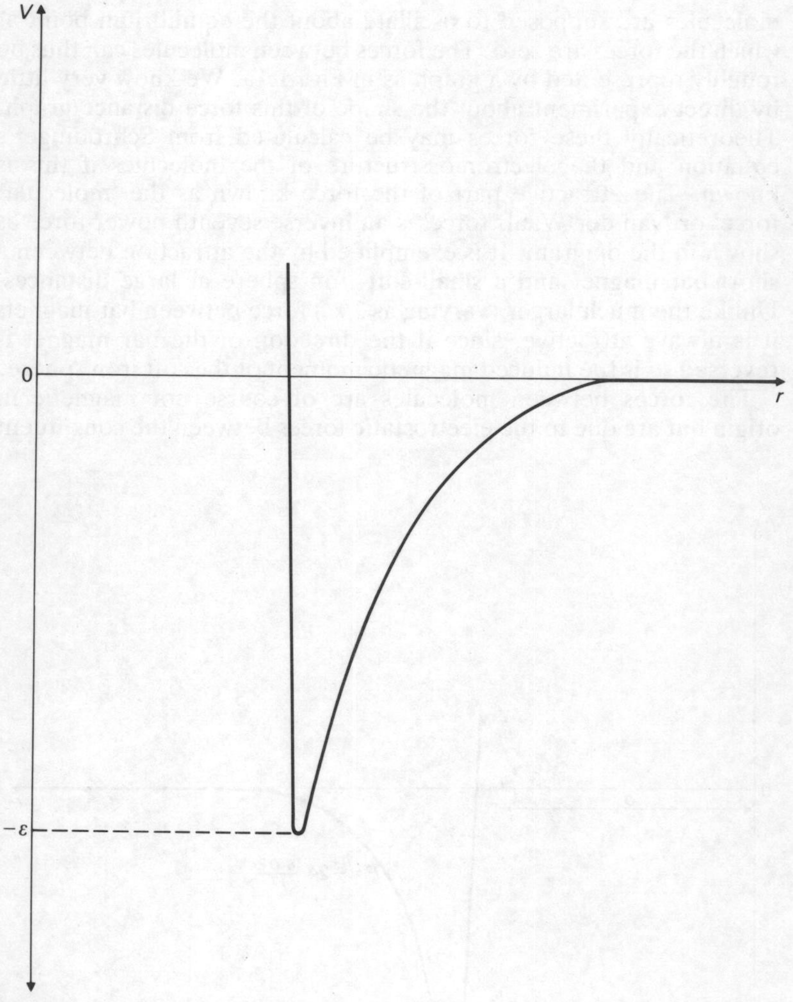

Fig. 1.1 (*Cont.*)

(b) The corresponding potential energy curve. If $V(r)$ is represented by the equation

$$V(r) = 4\varepsilon\left(\left(\frac{\sigma}{r}\right)^m - \left(\frac{\sigma}{r}\right)^6\right)$$

it is known as a Lennard-Jones 6 − m potential. The usual value of m is 12. Typical numerical values are $\varepsilon = 12 \times 10^{-22}$ J and $\sigma = 3.7 \times 10^{-10}$ m.

electrons and protons. Although each molecule is electrically neutral it exhibits a small residual electric field at large distances because the positions of the equal but opposite electric charges are not the same. Thus, there is always some force between neutral molecules at all distances. Since the hydrogen atom is represented by a neutral positive charge surrounded by a spherically symmetrical electron cloud whose density falls off with the radial distance r, as e^{-2r/a_0} it might be supposed that the force between two hydrogen atoms would also fall off as an exponential function of R the distance between them. Rather surprisingly direct calculation (London, 1930) shows that this is not so and that there is a residual attractive force $39e^2a_o{}^5/4\varepsilon_oR^7$ which is appreciable long after the exponential terms have become negligible. This is a general result (not confined to the special case of two hydrogen atoms) and may be explained in general terms in the following way. (A more detailed treatment is given by Margenau and Kester (Margenau, 1969).) The electron cloud in a molecule is not static and the electrons are always dodging about in a random way; it is only the time average of the electron density at a given place which is constant. Thus even molecules with a strictly spherical symmetry have a fluctuating electric dipole moment. The phase and magnitude of this dipole moment is random so that the dipole–dipole force between two molecules at large distances fluctuates rapidly and averages to zero over any time large enough for there to be any appreciable movement of the molecules. As the molecules approach each other the phases of the fluctuating dipoles are affected so that they are no longer independent of each other and the force between the molecules no longer averages to zero. This correlation of the phases can be described as a distortion of the electron clouds so that one can say that the dipole moment of each atom has induced a corresponding dipole moment in the other, thus giving rise to an inverse seventh power attractive force. This explanation agrees with the generally observed phenomenon that substances with high refractive indices and therefore large coefficients of polarisation also show large values of the constant a which represents the attractive forces between the molecules in van der Waal's equation (section 2.6). The forces due to further overlapping of the electron clouds are much more complicated and individual to the atoms and may even involve the formation of chemical compounds. Thus two hydrogen atoms form a hydrogen molecule; carbon atoms unite to form diamond or graphite. In this case it is the strongly directional nature of the forces which gives rise to the solid state with its regular arrangement of the atoms into a lattice.

The formation of graphite is an example of how unexpected the atomic interactions can be when there is considerable overlap of the electron clouds. Thus except at very high pressure diamond, the compact (density 3.52) giant organic molecule in which each carbon atom is linked to four others in a regular tetrahedral arrangement, is not the stable form. This is graphite (density 2.25), a much looser structure in which some of the electrons are so loosely bound that graphite is a moderate conductor of electricity. Correspondingly the solid and particularly the liquid states cannot be adequately described by simple general models and little progress can be made by studying them by elementary methods.

1.5 Conservation of energy

Fortunately for the development of the subject, an isolated system of interacting particles has an invariant energy which remains constant however much the individual positions and velocities of the particles may change with time.

The concept of energy is an important one. It can be approached by considering a simple system of two attracting particles. These two particles are considered first of all as two individual particles and then as a single system of two interacting particles. A single particle of mass m and velocity v is said to have kinetic energy $\frac{1}{2}mv^2$. This is a formal definition and is the meaning of the term kinetic energy. This kinetic energy will remain constant unless the particle is acted upon by a force F. During the motion the force will change its point of application and do work W. According to Newton's second law the work done is equal to the increase in kinetic energy for:

$$F\mathrm{d}x \quad = m\frac{\mathrm{d}v}{\mathrm{d}t}\,\mathrm{d}x \quad = m\frac{\mathrm{d}v}{\mathrm{d}t}\,v\,\mathrm{d}t \quad = mv\mathrm{d}v = \mathrm{d}\tfrac{1}{2}mv^2$$

work done   force = mass × acceleration $\mathrm{d}x = v\,\mathrm{d}t$

Thus for a particle of initial velocity v_0 and position vector $\mathbf{r}$

$$\tfrac{1}{2}mv^2 - \tfrac{1}{2}mv_0^2 = \int F \cdot \mathrm{d}\mathbf{r}$$

increase in the kinetic energy   work done by the force

This may be rearranged to give

$$\int \tfrac{1}{2}mv^2 - \int F \cdot d\mathbf{r} = \tfrac{1}{2}mv_0{}^2 = \text{constant depending on the initial conditions.}$$

Applying this result to two particles as shown in Fig. 1.2 one has

$$\tfrac{1}{2}m_1v_1^2 - \int F_1 \cdot d\mathbf{r}_1 = \text{constant for the first particle}$$

$$\tfrac{1}{2}m_2v_2^2 - \int F_2 \cdot d\mathbf{r}_2 = \text{constant for the second particle}$$

so that

$$\tfrac{1}{2}m_1v_1^2 + \tfrac{1}{2}m_2v_2^2 - \int F_1 \cdot d\mathbf{r}_1 - \int F_2 \cdot d\mathbf{r}_2 = \text{constant}$$

By Newton's third law $F_1 = -F_2 = F$ so that

$$\tfrac{1}{2}m_1v_1^2 + \tfrac{1}{2}m_2v_2^2 - \int F \cdot d\mathbf{r} = \text{constant}$$

In a simple system F can only depend upon the distance r between the particles and one can always introduce a potential function $V(r)$

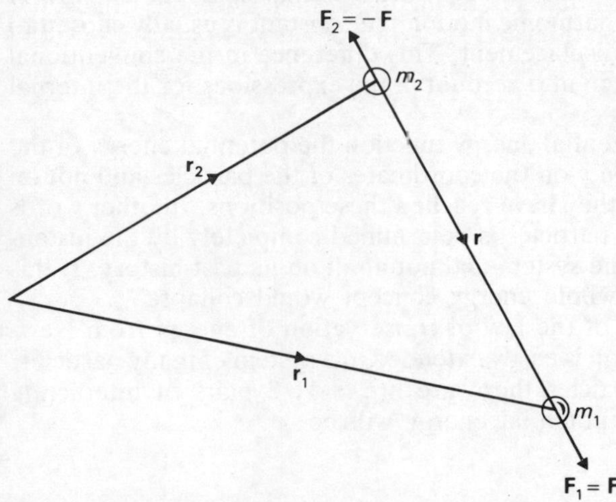

Fig. 1.2 Geometry of the interaction between two particles. Since $\mathbf{r}_2 + \mathbf{r} = \mathbf{r}_1$, $\mathbf{r} = \mathbf{r}_1 - \mathbf{r}_2$ and $d\mathbf{r} = d\mathbf{r}_1 - d\mathbf{r}_2$ also $F = F_1 = -F_2$.

such that

$$-\int F \cdot \mathbf{dr} = V(r) \text{ and one has}$$

$$\underbrace{\tfrac{1}{2}m_1v_1^2 + \tfrac{1}{2}m_2v_2^2}_{\text{kinetic energy}} + \underbrace{V(r)}_{\text{potential energy}} = \text{constant}$$

in which $V(r)$ is by definition the potential energy of the system. This is the law of conservation of energy. The step of introducing a potential function is not an undue limitation; all the ordinary forces between molecules have this property. The empirical Lennard–Jones potential function from which the forces between the molecules are often calculated is shown previously in Fig. 1.1b. The forces F_1 and F_2 have to be equal and opposite but need not act along the line of centres. The forces between dipoles are like this and are an example of tensor forces. A potential function (Stockmayer function) can still be established but it now depends upon the angles which the dipoles make with the lines of centres, as well as the distance between them. Alternatively, the system of the two dipoles can be treated as a system of four particles interacting through simple potential functions. By either treatment the kinetic energy of rotation of the dipoles will have to be included in the total energy of the system which is conserved. For free particles the arbitrary constant is chosen by convention so that the potential energy is zero when the particles are far enough apart for the forces between them to be zero. For particles oscillating about a centre of force with simple harmonic motion the constant is usually chosen to be zero for zero displacement. This difference in the conventional zero has to be taken into account in the expressions for the internal energies of solids.

If there is a potential energy function the potential energy of the system depends only on the coordinates of the particles and not on the way in which they have reached these positions. In other words the energy of the particles is determined completely by the instantaneous state of the system and not at all on its past history. If this were not so the whole energy concept would collapse.

The derivation of the law of conservation of energy from Newton's laws of motion is easily extended to a system of many particles. For n similar particles there are $n(n-1)/2$ pairs of interacting particles and the potential energy will be

$$\frac{1}{2}\sum_{\substack{p=1\\q=1}}^{\substack{p=n\\q=n}} V(r_{pq})$$

in which $V(r_{pq})$ is the potential energy of two particles p and q a distance r_{pq} apart in an obvious notation. In elementary work one views with horror and rejects any suggestion that $V(r_{pq})$ can be changed by the addition of a third body to the system. When $p = q$ the symbol $V(r_{pq})$ has no meaning and is omitted. This is sometimes indicated by writing the summation symbol as Σ'. The factor $\frac{1}{2}$ comes from the fact that each pair has been counted twice, once as $V(r_{pq})$ and once as $V(r_{qp})$. The restriction to forces derivable from a potential function is not necessary if one is considering external forces acting on the system. The $\int F \cdot d\mathbf{r}$, the work done by the external force, can then be evaluated or determined experimentally and will cause a corresponding increase in the energy of the system. The fact that the energy of the system depends upon its past history now causes no embarrassment and it is generally true that the increase in the total energy of a system of particles equals the external work done upon the system. A detailed example is discussed in (section 4.2).

The concept of energy was originally introduced by Lagrange (1758) as the basis of a new mathematical way of solving problems in celestial mechanics. The use of the potential functions was greatly extended by Laplace (1764): the name is due to Green (1828). The importance of energy as a fundamental concept of physics has steadily increased. The energy has turned out to be a fundamental property of a system composed of atoms and molecules and their subatomic constituents neutrons, protons and electrons. In quantum mechanics the energy of a system is quantised and can usually have only discrete values $E_1, E_2, E_3, E_4 \ldots$ often referred to as the energy levels of the system. The steps between the levels are so small that they usually only have to be taken into account when we are describing the properties of matter in terms of the individual atoms and molecules. Energy has become so important that the first objective of a scientist investigating the physics of a new molecular system is to establish both theoretically and experimentally the values of its energy levels. In quantum mechanics energy is conserved if the system is sufficiently long-lived for the energy levels to be well-defined. The energy levels of the simple systems to be described are given in Table 1.1. In the expression for the energy levels Planck's constant h is a fixed quantity 6.6×10^{-34} J s. Since the energy depends upon the mass of the particle, quantisation is much more obvious for light particles like electrons in an atom than it is for heavy particles like the earth moving in its orbit round the sun. For simple systems energy levels are additive. This means that the energy levels of a dumb-bell molecule in a box can be written down as the sum of the energy levels of a point particle with quantum numbers q, r, s and of a rotator with a quantum

Table 1.1

System	**Representation**	**Energy levels**
Particle of mass m in a cubical box of side a	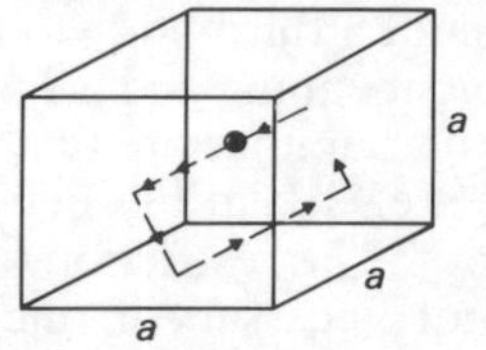	$(q^2 + r^2 + s^2)\dfrac{h^2}{8a^2}\mathrm{m}$ q, r, s are positive integers 1, 2, 3...
Dumb-bell molecule of moment of inertia I rotating in free space.	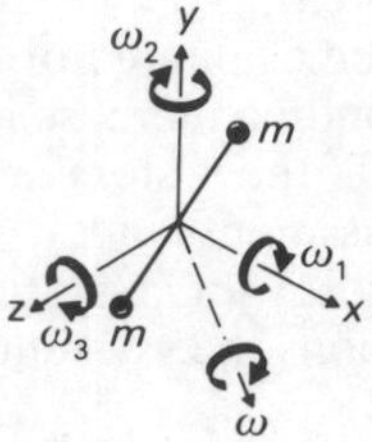	$\dfrac{K(K+1)h^2}{8\pi^2 I}$ K = positive integers 0, 1, 2, 3...
Simple harmonic oscillator in one dimension which if treated by Newton's laws has a frequency of oscillation $\nu = \omega/2\pi$ independent of the amplitude.	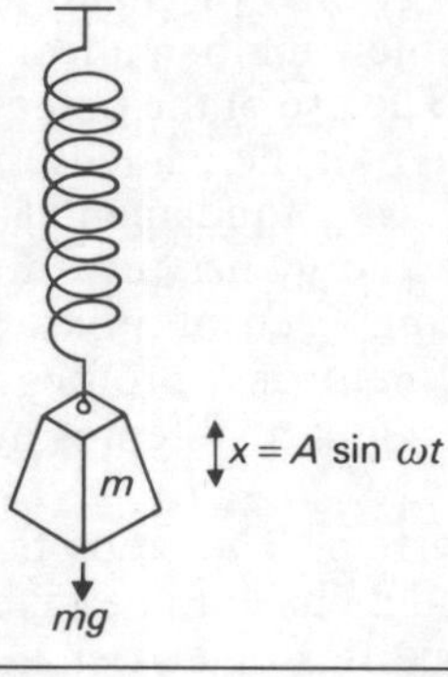	$(n + \frac{1}{2})h\nu$ n = positive integers 0, 1, 2, 3..

number K. A simple way of obtaining these energy levels is given in Appendix A.

Through Schrödinger's equation it is the energy which determines how the system changes with time. This contrasts with the Newtonian or classical view that changes are determined by the forces according to Newton's laws of motion. These two very different views of the way systems of particles change with time often give rise to similar predictions. This is one of the reasons quantum mechanics remained undiscovered for so long. In dealing as we shall be with molecules and their collisions we shall not on the whole be bothered with the complexities of quantum mechanics. The mere knowledge that energy levels exist is often sufficient to enable progress to be made. This is because molecular collisions can be described in terms of the energy, momentum and angular momentum of the particles, and all these quantities are conserved in both systems of mechanics.

1.6 Positive and negative energy

As already explained, potential energy has an arbitrary constant in its definition. If this is chosen so that the potential energy is zero when the molecules are at great distances from each other, energy content provides a convenient way of classifying simple molecular systems. Broadly speaking, for gases the total energy is positive, for solids it is negative. Between the two extremes lie the liquids. Common liquids such as can be poured from one vessel to another in the open have negative energy, but liquids near the critical temperature usually have a positive average energy for the molecules and a very small latent heat of evaporation. If the walls of the vessel are removed and the molecules are allowed to move further apart the attractive forces between them slow down the molecules and their kinetic energy is reduced; the potential energy correspondingly increases so that the total energy remains the same. If the potential energy is zero (by convention) at large distances, then a positive total energy means that at large distances apart the molecules still have kinetic energy and are free to move about and fill any space available to them. This property of filling any space available is the characteristic property of the gaseous state, and occurs when the total energy of the system is positive. If a system has negative total energy the attractive forces between the molecules are strong enough to bring all the molecules to rest before they have got so far apart that the attractive forces are negligible. The molecules thus cannot escape from each other's influence and so can occupy only part of a large space available to them.

1.7 Conservation of energy as an experimental law

The law of conservation of energy rests upon a double foundation. Firstly, as we have seen, it is demonstrably true under very general conditions for a system of particles obeying Newton's laws of motion (or Schrödinger's wave equation). Thus every verification of the predictions of mechanics, from the motions of the heavenly bodies to the spectrum of the hydrogen atom, can be looked upon as a confirmation of the principle of the conservation of energy. Secondly, conservation of energy is a principle firmly based on more direct experiments.

Early explanations of natural phenomena which were observed but not understood were often made in terms of an all penetrating fluid. Thus there was an electric fluid to account for electrical phenomena, a phlogiston or flame to account for combustion, a magnetic fluid, and a caloric fluid. Of these the only one to survive and play a part in modern physics is the electric fluid composed as we now know of electrons.

After a youthful Humphry Davy had claimed (Andrade, 1935) to have shown in 1799 that Count Rumford's (Benjamin Thompson), experiments on the production of heat by boring cannon could be extended to the melting of ice by friction it was no longer possible to maintain a caloric theory of heat. This statement, although correct, does not accurately present the downfall of the caloric theory which lingered on for a generation. The great Laplace based his famous correction (1816) to Newton's formula for the velocity of sound upon the caloric theory of heat; similarly Carnot's theorem (section 5.4) was at the time thought to be based upon the caloric theory. Nevertheless, Humphry Davy's crude experiment (he was a self-taught scientist of 19 at the time) was a crucial one. With great theoretical insight he saw that the caloric theory could not explain the melting of ice by friction since it was at that time generally accepted that water contained more caloric than ice. The mere suggestion of the experiment thus imposed the duty of showing that ice could not be melted by rubbing two pieces together, upon any one wishing to uphold the caloric theory. Needless to say, this challenge was simply ignored. The argument can, of course, be extended to any substance once Black's contemporary discovery of latent heat is taken into account. The work of Mayer, Joule, Thomson and Helmholz showed that the term heat or caloric content could be replaced by the total energy of the molecules forming the system. The experimental evidence favouring this view was supplied by Joule (1840–49). He took a large number of substances and arranged for measurable amounts of external work to be done on them in as many different ways as he could carry out. If matter is indeed a collection of molecules obeying the laws of mechanics, then equal quantities of external work must produce equal increases in the total internal energy irrespective of how the work was done. What Joule showed was that equal quantities of work always produced equal quantities of heat measured by conventional methods as (rise of temperature) $\times$ (heat capacity). Joule, who came from a brewing family, was an amateur scientist and an indifferent mathematician. The importance of his contribution to the theory of heat has not perhaps received the recognition it deserves. His name is usually associated with the much less theoretically important measurement of the exact value of the conversion factor between heat and work.

As the result of Joule's work, calorimetry was established as a convenient way of measuring the changes in energy of systems not otherwise amenable to direct mechanical measurement. For example, if carbon burns in oxygen the hot carbon dioxide formed should have the same internal energy as the original system: $C + O_2 = CO_2 + Q$ ($Q = 408.3$ J/mol)
The energy Q released, (heat capacity × rise in temperature), must be supposed to be the difference in the energy of the two electronic systems before and after the combustion. If the reaction takes place in two stages

$$2C + O_2 = 2CO + Q_1 \quad (Q_1 = 248 \text{ J})$$
$$2CO + O_2 = 2CO_2 + Q_2 \quad (Q_2 = 569 \text{ J})$$

since the initial and final states of the system are the same, the total energy must be the same. Thus apart from a minute correction for the external work done because of the change in volume one should find that

$$Q_1 + Q_2 = 2Q$$

This is seen to be so since $Q_1 + Q_2 = 817$ J and $2Q = 816$ J. The conservation of energy has been verified in this way for a surprisingly large number of chemical reactions.

The principle of the conservation of energy is also well verified experimentally by measurements in nuclear physics and high energy physics. In these fields in which the energy changes per particle are so large that they can be measured directly, the expression for the total energy contains no arbitrary constant but one has to use the relativistic expression

$$E = c\sqrt{p^2 + m_0{}^2c^2}$$

for the total energy E, in which c is the velocity of light, p is the momentum of the particle and m_0 is the rest mass which is characteristic of the individual particle, i.e. proton, neutron, electron, π meson, etc., under consideration. The potential energy of classical mechanics including the arbitrary constant has been swallowed up in the difference between the actual mass m and the rest mass m_0. This theoretically more satisfying but mathematically awkward expression for the energy need not be used in molecular theory, because of the relatively large molecular mass, and the low value ($\approx 10^3$ m/s) of molecular velocities compared with the velocity of light (3×10^8 m/s).

The molecular theory of heat shows that thermal phenomena can be accounted for in terms of the potential and kinetic energy of the molecules involved. For simple bodies at rest this is the same as the internal energy U. For a moving body mass M and velocity V, the mechanical energy of the centre of gravity $\frac{1}{2}MV^2$ is not usually counted in the internal energy because it can easily be converted into external work. The $\frac{1}{2}MV^2$ of kinetic energy can be converted into heat energy in inelastic collisions. The standard example of this is the head-on collision of equal spheres moving with equal and opposite velocities. If the collision is perfectly elastic the velocities are reversed and there is no conversion of kinetic energy $\frac{1}{2}MV^2$ into internal energy. If the collision is completely inelastic (two pieces of putty) kinetic energy $2 \times \frac{1}{2}MV^2$ disappears and an increase in internal energy takes place to allow energy to be conserved.

Finally, attention must be drawn to the notation now usual in this branch of physics. The symbol U or u or E is used for the energy of the molecules. This includes any internal electronic energy so that U is not strictly identical with the heat energy. The symbol E or ε is usually used for energy levels and in entirely theoretical work with individual molecules. For material in bulk U is usually used and unless otherwise stated it is taken to refer to the internal energy of the L molecules making up one mole of material: u is convenient in writing and as a general symbol not implying any particular quantity of matter. The symbol Q or q is used for the amount of heat energy which flows into a system. If a system changes from state 1 to state 2 conservation of energy requires that $Q = U_2 - U_1$ + external work done during the change.

1.8 Temperature

The identification of changes of heat or caloric content with changes in the total internal energy has solved the problem of the nature of heat. It has hardly touched on the related one of the nature of temperature. Everyone has an intuitive knowledge of temperature since differences in temperature affect the senses directly. An intelligence test which asked the candidates to underline the hotter of ice-cream and tap water would be looked upon as too easy; most children being able to answer the question correctly from their own experience before they have reached the examination age. Nevertheless, it is not immediately obvious what is the physical difference between the molecules, of say a cold piece of copper, an aluminium block at room temperature and a red-hot platinum wire which enables us unhesitatingly to arrange them in order of increasing temperature.

The first step in answering this question was taken by Fahrenheit, a London instrument maker (1714), when he invented an accurate thermometer and so enabled the question of temperature difference to be studied quantitatively. Fahrenheit based his thermometer on the assumption that mercury had a constant expansion coefficient independent of temperature. He chose 0 °F as the lowest temperature he could reach using a freezing mixture of ice and salt and human temperature as 96 °F. The advantages of using the freezing and boiling points of water as the fixed temperatures had already been pointed out by Newton (1701) and Hooke (1665). We are not here concerned with details of accurate thermometry but to find an answer on the molecular scale to the question 'What is the same about two bodies which have the same temperature?' The following facts are easily established once one has a dependable and sensitive thermometer:

(a) Bodies in good thermal contact reach the same temperature. The process by which this happens is described as the flow of heat energy from the hot to the cold body.
(b) A change in temperature is accompanied by a change in nearly all the measurable properties such as size, electrical resistance and vapour pressure.
 Bodies therefore act as their own thermometer and we can detect that there is no heat flow between two bodies by observing that their measurable properties such as size, electrical resistance and vapour pressure are not changing.
(c) If A and B have the same temperature and A and C have the same temperatures, then B and C have the same temperature and can be placed in contact without any heat flowing between them.

It should be noted that the bodies are not necessarily of the same material; A could be a copper cube, B a silver cube, and C a gold one. Without this reciprocal relationship the concept of temperature would collapse and its experimental verification has an important bearing on the nature of temperature. From it we know that temperature is a definite parameter which can be measured on an arbitrary scale and is as characteristic of the state of a simple piece of matter as say its volume. This idea is most easily developed when applied to real gases because the volume of a gas is so easily changed by changing either the temperature or the pressure.

This discussion has not answered the question 'What is the same on the molecular scale for substances at the same temperature?' posed at the beginning of the section. For perfect gases the answer is given in Chapter 2. The way in which this answer can be extended to real gases and matter in general forms the main material of Chapters 5 and 6.

1.9 Equations of state

If a gas, confined in a vessel so that its pressure p_1 and its volume v_1 can be varied, is placed in a thermostat, temperature T_1 as in Fig. 1.3, one can plot the relationship between p_1 and v_1. This can be represented by an equation $f_1(p_1, v_1) = k_1$. This relationship is unique and reproducible. By changing the temperature of the thermostat to T_2 one can obtain another relationship $f_2(p_2, v_2) = k_2$ and so on. The whole series of p, v relationships can be used to construct a surface in a p, v, T space whose equation is $f(p, v, T) = 0$. Each individual $f_m(p_m, v_m) = k_m$ graph is the line in which the plane $T = T_m$ cuts the surface as in Fig. 1.4. The

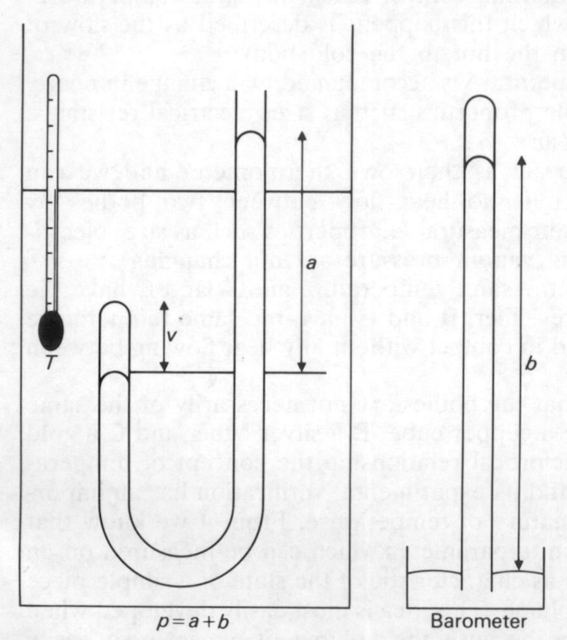

Fig. 1.3 Apparatus for measuring pv curves at any temperature T.

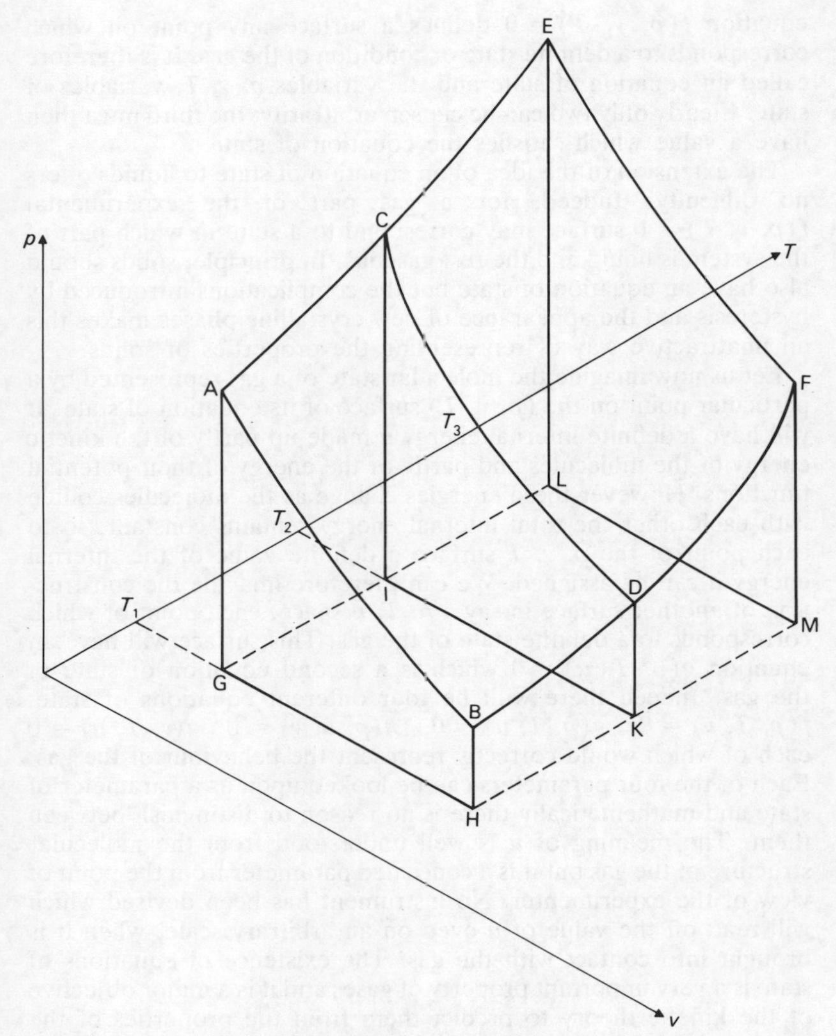

Fig. 1.4 The diagram illustrates the construction of a p, v, T surface. A p, v curve $f(p, v) = k$ is measured experimentally at a temperature T (in an apparatus shown diagrammatically in Fig. 1.3), and plotted on thin cardboard. This is then cut out to give the shape ABHG in which the profile AB represents the experimental data at temperature T_1. Similar profiles CDKI, EFML, etc. are prepared from data obtained at other temperatures T_2, T_3, etc. The cards are then stacked at the appropriate positions in a p, v, T coordinate frame. To get the p, v, T surface the space between the sections is filled with plaster of Paris and smoothed off with a straight edge as it is setting. The greater the number of profiles the more accurately the surface is delineated.

equation $f(p, v, T) = 0$ defines a surface any point on which corresponds to a definite state or condition of the gas. It is therefore called an equation of state and the variables p, v, T, variables of state. Clearly only two can be chosen arbitrarily; the third must then have a value which satisfies the equation of state.

The extension of the idea of an equation of state to liquids offers no difficulty. Indeed, for a gas part of the experimental $f(p, v, T) = 0$ surface may correspond to a state in which part of the system is liquid and the rest gaseous. In principle, solids should also have an equation of state but the complications introduced by hysteresis and the appearance of new crystalline phases makes this an unattractive way of representing the properties of solids.

Let us now imagine the molecular state of a gas represented by a particular point on the (p, v, T) surface of its equation of state. It will have a definite internal energy u made up partly of the kinetic energy of the molecules and partly of the energy of their potential functions. However these energies change as the molecules collide with each other the total internal energy remains constant, so to each point of the p, v, T surface a definite value of the internal energy u can be assigned. We can therefore imagine the construction of another surface in say a p, T, u space, each point of which corresponds to a definite state of the gas. This surface will have an equation $g(p, T, u) = 0$ which is a second equation of state of the gas. Indeed there willl be four different equations of state, $f(p, T, v) = 0$, $g(p, T, u) = 0$, $h(p, u, v) = 0$, $i(v, T, u) = 0$ each of which would correctly represent the behaviour of the gas. Each of the four parameters can be looked upon as a parameter of state and mathematically there is no reason to distinguish between them. The meaning of u is well understood from the molecular structure of the gas but it is a concealed parameter from the point of view of the experimenter. No instrument has been devised which will read off the value of u even on an arbitrary scale, when it is brought into contact with the gas. The existence of equations of state is a very important property of gases and it is a major objective of the kinetic theory to predict them from the properties of the molecules.

Problems

1 Leibniz used the following argument (amongst others) to show that matter is not made of molecules: if one picks two blades of grass from a lawn they are never identical. If grass were made of discrete molecules similar arrangements would occur and identical blades would be observed. Why do we not accept this conclusion? Show that Leibniz' argument leads to the correct conclusion when applied to atomic nuclei.

2 Lord Rayleigh once tried to measure the mass of a molecule by observing that the great drawing-room at Terling Place always smelt of musk kept in an open jar. He weighed the musk at the beginning and the end of the summer, argued that there must always be a molecule of musk in the nasal cavity if one could smell it and estimated the rate at which the air was changed in the drawing-room. The latter is a difficult measurement. How would you attempt it? Having found it, how would you calculate the mass of a molecule of musk.

3 Make a list of the experimental evidence you would use to defend the proposition that the hydrogen molecule contains two atoms.

4 Give an account of the simple methods of estimating molecular size available to Young in 1815.

5 It is said that when the Jesuits wished to open a Mission in China that amongst other advantages they offered to teach the Chinese about the great Natural Laws discovered in Europe, to which the Chinese objected that the Emperor made laws which were printed so that his subjects could read them and obey. It was absurd to suggest that sticks and stones which could not read could obey laws. How do you react to this argument?

6 Why do you think atoms contain electrons? Does your reasoning lead to the conclusion that radioactive nuclei contain α and β particles?

7 If gravity acted only on the rest mass of a particle the period of an ideal pendulum with a carbon bob would be different from the period of a similar pendulum with a uranium bob. Calculate the ratio of the two periods. Do you think the ratio is sufficiently large for it to be possible to obtain experimental evidence in this way that gravity acts on the total mass and not on the rest mass alone? (The masses can be obtained from a table of atomic masses and the assumption that nuclei are made up from neutrons and protons.)

8 How would you set about showing that there is no appreciable change in mass during a chemical change?
What limits would you set to the term appreciable? (The best investigated example is the reduction of silver nitrate by ferrous

sulphate in aqueous solution in which a decrease of mass was at one time observed.)

9 If electrons and protons have equal and opposite charges and neutrons have no charge, matter should be neutral.
A cylinder of argon 2 m long, 0.4 m diameter at 120 atmospheres was placed in a Faraday cage and discharged through an ion trap into the atmosphere. After all the argon had left the difference of potential between the cylinder and the cage was not more than 3 μV. If the gap between the isolated cylinder and the Faraday cage was $\sim$ 1 cm, calculate to what extent the charge on the electron and proton are the same. Describe the apparatus you would use in some detail and compare it with that described by **Hileas, A.M. and Cranshawe, T.E.** (1959) *Nature* **184**, 892.

10 Van Helmont planted a willow weighing 51 lb in 200 lb of dry soil and added nothing but water for five years. The willow then weighed 169 lb and the soil had changed little in weight. He concluded that willow trees were made of water and air in agreement with his mystical philosophy. Do we still agree with this conclusion? If not, what is the flaw in the argument?

11 An atom may crudely be represented by a point charge $+ Ze$ at the centre of a uniformly charged sphere of radius a and total charge $- Ze$. Show that in this representation the atoms will not interact when they are a distance $R > 2a$ apart and will repel each other if $R < a$.

12 A simple model of an atom is a central charge + Ze at the centre of a spherical distribution of negative charge of density = $A\mathrm{e}^{-r/a}$. Find the value of A if the atom is to be neutral and find the potential at any point distant r from the centre.

13 Bentley observed that BF_3 was absorbed on nickel in three layers, containing 1.8×10^{-8}, 1.8×10^{-8} and 3.1×10^{-8} mol/cm^2 respectively. Calculate the volume of BF_3 desorbed from 10 cm^2 of a nickel surface if measured at 25 °C and 1 Pa pressure. What explanation would you give of Bentley's observation in terms of molecular diameters? (**Bentley, P.G.** (1960) *J. Sci. Inst.,* **37**, 323).

14 The constants in the 6 – 12 Lennard–Jones potential for hydrogen are $\varepsilon = 45.9 \times 10^{-23}$ J, and $\sigma = 2.968 \times 10^{-10}$ m. Sketch the curve and give the values of r at which the potential is zero, the force is zero, the attractive force is a maximum, and the attractive force is 1 per cent of its maximum value.

15 Show that the principal moments of inertia of a molecule which has two point atoms of masses m_1 and m_2 a fixed distance r apart are

$$I_1 = I_2 = I = \frac{m_1 m_2 r^2}{m_1 + m_2} \text{ and } I_3 = 0.$$

If such a molecule is rotating about an axis perpendicular to the line of centres through the centre of gravity with angular velocity ω show that its k.e. of rotation is $\frac{1}{2} I \omega^2$.

Further reading

Brown, Sanborn C. (1964) *Count Rumford*, Heinemann Educational, London.

Goldstein H. (1964) *Classical Mechanics,* Addison-Wesley, Reading, Mass. Palo Alto; London.

Jeans, J.H. (1960) *The Growth of Physical Science,* CUP, London.

Jeans, J.H. (1942) *Physics and Philosophy,* CUP, London.

McRea, W.H. (1935) *Relativity Physics,* Methuen, London.

Milne, A.E. (1948) *Vectorial Mechanics,* Methuen, London.

Pauling L and Wilson E Bright Jr. (1935) *Introduction to Quantum Mechanics*, McGraw-Hill, N.Y.

Schiff, L.I. (1970) *Quantum Mechanics,* McGraw-Hill, N.Y.

Stebbing, Susan L. (1944) *Philosophy and the Physicists,* Penguin, Harmonsworth.

References

Andrade, E.N. da C. (1935) *Nature,* **135**, 359.

Margenau, H. and Kestner, V.R. (1969) *Theory of Molecular Forces,* Pergamon, Oxford.

Russell Bertrand (1947) *History of Western Philosophy,* Allen and Unwin, London.

CHAPTER 2

The gas laws and equations of state

2.1 Boyle's law

As the forces between the molecules play only a small part (section 1.3) in determining the pressure exerted by a gas, one expects all gases to behave quantitatively in much the same way on compression in contrast to the large differences in the modulus of compression shown by solids. This is so. The Honourable Robert Boyle, 'Seventh son of the Earl of Cork and the father of Chemistry', with the aid of Hooke (1662), observed that the pressure p and the volume v of air were related by the equation pv = constant if the temperature is constant.

All gases obey this law approximately at high temperatures and low pressures and it is a typical example of a relationship which is not a fundamental law like the inverse square law of electrostatic force, or Newton's laws of motion. These laws are supposed to govern without exception the behaviour of molecules. If a gas is a collection of molecules (assumed to be Newtonian particles), it should be possible to deduce Boyle's law from the laws of mechanics and the law of force between the particles. This can be done very simply on the assumption that the pressure is entirely due to the change in momentum suffered by a molecule when it collides with the walls, giving an explanation of Boyle's law in terms of the more fundamental laws obeyed by molecules.

Boyle's law can be rewritten as $p = k/v = k' \times$ density $= k'' \times$ number of molecules/unit volume. The last form is seen to be quite consistent with the assumption that each molecule in the gas moves independently of all the others and that the pressure due to N molecules is just N times the pressure due to one molecule. The latter is easily calculated. Consider a single particle of mass m and velocity c confined in a spherical cavity of radius r. The wall of the vessel is supposed to be hard and perfectly elastic so that the particle loses no energy when it collides with it. As shown in Fig. 2.1 let the particle leave the wall after a collision so that its path makes an

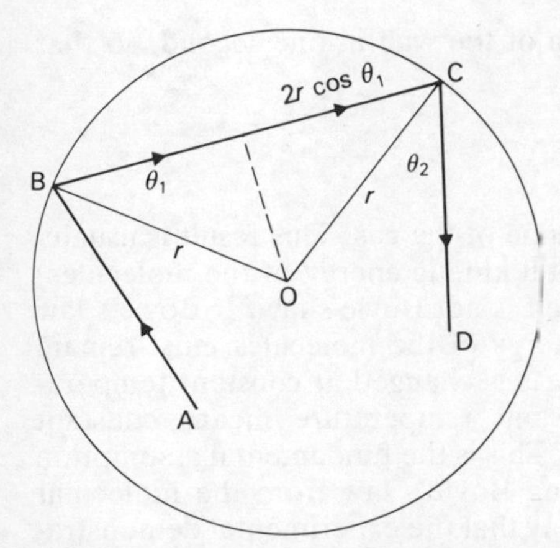

Fig. 2.1 The path of the molecule is ABCD: BC (two arbitrary points on the sphere) together with O (the centre of the sphere) define a diametral plane so that the radii OB, OC are always normal to the surface. The paths AB and CD are not necessarily in the plane BCO. BC is clearly $2r \cos \theta_1$.

angle θ_1 with the normal. The particle will again strike the wall of the vessel after travelling $2r \cos \theta_1$ in a time $t_1 = (2r \cos \theta_1/c)$ again making an angle θ_1 with the normal at the point of arrival. In this time it will have gained momentum $-mc \cos \theta_1$ along the outward normal on leaving the wall and brought up momentum $mc \cos \theta_1$ again normal to the wall on terminating its path. It then sets off again making a new angle θ_2 with the normal. Thus the total outward momentum transferred to the wall in a series of collisions is $2mc \cos \theta_1 + 2mc \cos \theta_2 + 2mc \cos \theta_3 \cdots$ in a time $t_1 + t_2 + t_3 \cdots$ $= 2r(\cos \theta_1 + \cos \theta_2 + \cos \theta_3 \cdots)/c$ and the total rate of change of momentum normal to the wall is:

$$\frac{\text{total change of momentum}}{\text{total time}} = \frac{2mc^2(\cos \theta_1 + \cos \theta_2 + \cos \theta_3 \cdots)}{2r(\cos \theta_1 + \cos \theta_2 + \cos \theta_3 \cdots)}$$

$$= \frac{mc^2}{r}$$

which is independent of θ and so of the direction of the molecular motion. Thus N molecules moving at random in the spherical vessel lose momentum normal to the wall at the rate of Nmc^2/r per second. If N and c are both large the succession of small impulses which the molecules give to the wall when they collide with it merge together and are equivalent to a steady outward pressure p. By Newton's second law p is equal to the total change of momentum of the molecules striking unit area of the wall in one second, so that

$$p = \frac{Nmc^2}{r} \cdot \frac{1}{4\pi r^2} = \frac{Nmc^2}{3v}$$

in which $v = 4\pi r^3/3$ is the volume of the gas. This result is usually written as $pv = \frac{1}{3}Nmc^2 = \frac{2}{3}$(total kinetic energy of the molecules) of a monatomic gas. This in itself is not Boyle's law. If Boyle's law is to be obeyed, the kinetic energy of the molecules must remain constant as the volume of the gas is changed at constant temperature, which implies that constant temperature means constant kinetic energy of the molecules. This is the fundamental assumption which must be made in deriving Boyle's law from the molecular theory. Alternatively, one can say that the experimental demonstration of Boyle's law shows that the temperature of an ideal gas is a direct measure of the molecular kinetic energy. The imaginary gas which obeys Boyle's law exactly is usually referred to as a perfect or ideal gas.

The simple result that for a monatomic gas $pv = \frac{1}{3}Nmc^2 = \frac{2}{3}$ (total kinetic energy of the molecules) is fundamental to the development of the kinetic theory. A great deal of the advanced part of the subject is devoted to justifying that the simple deductions about the nature of temperature which can be made from it remain true when the gas is a real gas for which molecular forces cannot be neglected. The reader coming upon it for the first time is sometimes a little surprised that so important a result should be derived from such artificial assumptions, and indeed, many objections can be made against the simple derivation which has been given... They are tabulated in the form of objection and answer below:

Objection	***Answer***
It cannot be right to assume that all the molecules have the same velocity c.	A valid objection which can easily be met by giving each molecule its own velocity c_1, c_2, c_3, etc. The result is to replace c^2 by $\overline{c^2}$ and the statement $pv = \frac{2}{3}$ total kinetic energy $= \frac{1}{3} Nm\overline{c^2}$ remains true.
The result is only derived for a spherical vessel.	A niggling objection. The simple derivation is easily extended to a cube or a box with a rectangular cross-section. The extension to a vessel of arbitrary shape requires

Objection	*Answer*
	sophisticated mathematics but introduces no new physical principle.
Neglecting the molecular collisions is such a ridiculous assumption that the whole demonstration is worthless.	The cogency of this serious objection depends upon the pressure of the gas. In a good vacuum in which there are still 10^{16} molecules per m^3 the molecules actually do cross from wall to wall without collision (vide section 3.2) and the proof as given stands. What is misleading is that, in the absence of a direct demonstration to the contrary, one naturally assumes that as soon as molecular collisions are appreciable, deviations from Boyle's law will appear and that as soon as the pressure is so high that molecular collisions are the rule rather than the exception, Boyle's law will break down completely. This is not so and the law continues to be obeyed quite accurately. This is shown by proofs in which molecular collisions are taken into account from the start such as the one which follows in section 2.2. The reason for Boyle's law continuing to be obeyed in the region when molecular collisions cannot be neglected is that the number of molecules hitting the wall is hardly changed by molecular collisions. A molecule prevented by collision from reaching the wall is replaced by another which only reaches the wall because it has suffered a collision. At moderate pressure this compensation is exact.

2.2. The virial

A derivation of Boyle's law which takes into account the collisions between the molecules can be given by applying Clausius' (1870) Virial Theorem to the motion of molecules in a gas. For a modern account of this theorem see Milne (1925). The Virial Theorem is proved as follows: If m is the mass of the molecule, u its velocity parallel to the x-axis and X the component of force acting on it parallel to this axis, we have by Newton's second law.

$$m\frac{\mathrm{d}u}{\mathrm{d}t} = X \text{ so that}$$

$$m\frac{\mathrm{d}}{\mathrm{d}t}(xu) = m\frac{\mathrm{d}x}{\mathrm{d}t}u + m\frac{\mathrm{d}u}{\mathrm{d}t}x = mu^2 + Xx$$

Similar equations apply to the y and z axes so that for a molecule of velocity c and velocity components u, v, w,

$$m\frac{\mathrm{d}}{\mathrm{d}t}(xu + yv + zw) = m(u^2 + v^2 + w^2) + Xx + Yy + Zz$$

Applying this relationship to N independent molecules one has

$$m\sum_{i=1}^{i=N}\frac{\mathrm{d}}{\mathrm{d}t}(x_iu_i + y_iv_i + z_iw_i) = \sum_1^N mc_i^2 + \sum_1^N(X_ix_i + Y_iy_i + Z_iz_i).$$

This is the Clausius virial relationship in which

$$-\frac{1}{2}\sum_i^N (X_ix_i + Y_iy_i + Z_iz_i)$$

is called the virial of the forces acting on the molecules. For a gas of uniform density devoid of mass motion the l.h.s. is zero since for each elementary volume it is proportional to the change in the flux of molecules in the direction of the origin.

The virial is made up of two parts: an internal part made up of the large number of terms due to the internal forces between the molecules, and an external part due to the forces exerted by the walls of the vessel in which the gas is confined. For simplicity, suppose this to be a cubical box of side a so arranged that one corner defines the axes of a cartesian system of coordinates x, y, z. Then, $\frac{1}{2}\Sigma mc_1^2$ = virial of the external forces + virial of the internal forces. The virial of the external forces is easily evaluated since the position of the walls is known. The contribution from the three faces passing through the origin (0, 0, 0) is identically zero. The remaining faces contribute $-a\Sigma X_1 - a\Sigma Y_1 - a\Sigma Z_1$ in which $\Sigma X_1, \Sigma Y_1, \Sigma Z_1$ are the total forces normal to the faces of the cube. For each face this is $-Pa^2$ in which P is the gas pressure. Thus we have $m/2\ \Sigma_1^N\ c_1^2 = 3Pa^3/2$ + virial of the internal forces or since $a^3 = V$ the volume of the gas, $PV = \frac{2}{3}$(total kinetic energy of the molecules) $-\frac{2}{3}$(virial of the internal forces).

If there are no internal forces, their virial is zero, and since there is then no potential energy we have for such a perfect monatomic gas $PV = \frac{1}{3}Nm\overline{c^2} = \frac{2}{3}$(total internal energy) = constant if the temperature is constant. This is the same result as obtained by the elementary method. A perfect gas can thus be described as one which obeys Boyle's law or one in which the forces between the molecules are so small as to be negligible. If the forces between the molecules have a very short range compared to the mean distance between the particles their virial will be small since only the small fraction of the molecules within this range will contribute and their individual contributions will also be small because they are so near together. The magnitude of the contribution of the repulsive forces to the virial of the internal forces can be estimated on the crude assumption that the molecules are hard spheres. If one takes a time average of the virial equation, it becomes, since PV and $m\overline{c^2}$ are constants $PV = \frac{1}{3}Nm\overline{c^2} - 2/3t \int$ (virial of the internal forces) dt. The integral represents the summation of a series of peaks contributed by the molecular collisions to the virial, separated by long periods when the molecules are not interacting with each other. It can readily be approximated to if the molecules are treated as hard elastic spheres of diameter σ. The molecular collisions are not all of the same type, varying from head-on collisions ($\theta = 0$) to grazing collisions ($\theta = \pi/2$): they can be described in terms of the angle θ which the line of centres at the moment of collision makes with the relative velocity V, as shown in Fig. 2.2. Consider first a head-on collision; if F is the unknown force between the molecules the impulse $\int F\mathrm{d}t$ reverses the momentum relative to the centre of gravity, so that $\int F\mathrm{d}t = mV$. The contribution to the integral of the virial is $-\frac{1}{2}\int \sigma F\mathrm{d}t = -\frac{1}{2}mV\sigma$, since throughout the collision the centres of the molecules are the sensibly constant distance σ apart. For molecules making glancing collisions the corresponding contribution is easily shown to be $-\frac{1}{2}mV\ \sigma\cos\ \theta$. A molecule moving with a velocity V relative to the other molecules will collide in time t

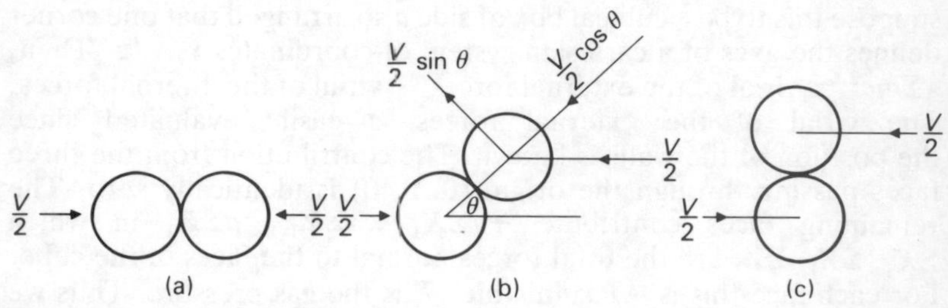

Fig. 2.2 (a) Head-on collision.
(b) Oblique collision making an angle θ with the line of centres.
(c) Glancing collision: $\theta = 90°$.

with $\pi\sigma^2 nVt$ (as shown in Fig. 3.2) molecules so that the total contribution of N molecules to the required integral is the mean value averaged over V and θ of

$$\frac{N}{2} \times \pi\sigma^2 nVt \times -\tfrac{1}{2}mV\sigma\cos\theta = -\frac{Nmn\pi\sigma^3 t}{4}V^2\cos\theta$$

in which n is the number of molecules per unit volume and the factor $\frac{1}{2}$ allows for each molecule having been counted twice in the averaging. The fraction of the collisions which occur between θ and $\theta + d\theta$ is $2\sin\theta\cos\theta\,d\theta$ and a simple integration gives that the average contribution to the virial of the internal forces is $-Nmn\pi\sigma^3\overline{V^2}/6$. Substituting the known value (section 3.2) of $\overline{V^2} = 2\overline{c^2}$ gives at once that

$$PV = \tfrac{1}{3}Nm\overline{c^2} + \frac{Nmn\overline{c^2}}{6} \cdot \frac{4\pi}{3}\sigma^3$$

A more elegant (but harder) derivation of this result is given by Lorentz (1881). Elimination of $nm\overline{c^2}$ in terms of the perfect gas pressure assumed, for the purpose of calculating a correction, to be the same as the real gas pressure P gives $PV = \frac{1}{3}Nm\overline{c^2} + Pb$ usually written for one mole of gas as $P(V - b) = RT$ in which b is $\frac{2}{3} \cdot N_A\pi\sigma^3 =$ four times the volume of the $N_A =$ L molecules. For permanent gases (N_2, O_2, H_2) at room temperatures b is about 0.1% − 0.2 per cent of the volume occupied by the gas. The virial of the attractive forces ($-\mu/r^7$) is easily calculated provided one is prepared to assume that the molecules are evenly distributed throughout the available space in spite of the attractive forces between them. This is a reasonable approximation for a gas at low pressures and high temperatures. It is clearly a very bad one at low temperatures when the molecules are very unevenly distributed throughout the available volume and are nearly all to be found in the small fraction of the volume occupied by the liquid state. On this simplifying assumption the virial of the attractive forces is given by

$$\frac{N}{2} \times -\frac{1}{2}\int_{\sigma} \frac{-\mu}{r^7} \cdot r \quad 4\pi n r^2 \mathrm{d}r$$

N particles counted once only	Virial of a pair of molecules *r* apart	Number of molecules between *r* and *r* + d*r*

$$= \frac{Nn\pi\mu}{3\sigma^3} = \frac{KN^2}{V}$$

Combining the two independent contributions to the virial gives an approximate equation of state

$$\left(P + K\left(\frac{N}{V}\right)^2\right)V = RT + bP = \left(P + \frac{a}{V^2}\right)V = RT + bP$$

in which b represents the repulsive forces and a the attractive forces acting between the molecules. This is as far as it is profitable to push this elementary method of attacking the problem. A similar equation obtained in a different way is discussed in section 2.5. The correction to Boyle's law due to the finite size of the molecules can also be estimated by analogy. If a Boyle's law experiment is carried out in a vessel of volume V to which a volume v of sand has been added there is no doubt that the observed result will be $P(V - v) = RT$; the sand could be said to have excluded vm^3 from the Vm^3 previously available to the gas. By analogy one could say that the volume of the molecules must be excluded from the volume V apparently occupied by the gas. Since when two spherical molecules collide their centres are 2 radii apart the volume excluded by each pair of molecules is eight times the volume of one molecule and one should write $P(V - b) = \frac{2}{3}U$ in which b is four times the volume of the molecules. These results give a theoretical interpretation of the empirical relationship found by Amagat that $pv = K + bp$ + higher terms. It enables the pv behaviour of a perfect gas to be predicted by extrapolating pv for a real gas to zero pressure.

In the relationship $pv = \frac{1}{3}Nm\overline{c^2}$ since Nm is the total mass of the gas $\overline{c^2}$, the mean square velocity of the molecules can be determined accurately without having to know Avogadro's number. This was the first completely new result to flow from the mathematical study of molecular theory.

The equivalent velocity $\sqrt{c^2}$ is of the same order as the velocity of sound in the gas. Some numerical values at 0 °C are given below:

Gas	H_2	N_2	CO_2	Hg
r.m.s. velocity m/s	1840	493	393	185

The high value of the molecular velocity is quite large enough to justify the usual neglect of gravitational forces on the motion. Hence, this accounts for the fact that the density of a gas is sensibly constant throughout the small volumes (~1 l) used in laboratory experiments. (see problem 2.10).

A knowledge of the average kinetic energy of the molecules does not tell us how this energy is distributed amongst the very large number of molecules of which the gas is composed.

The relationship

$$\frac{dN}{N} = \frac{4}{\sqrt{\pi}} \cdot \left(\frac{m}{2kT}\right)^{3/2} \cdot e^{-mc^2/2kT} \cdot c^2 dc$$

which gives the fraction dN/N of the molecules which have velocities between c and $c + dc$ was discovered by Maxwell (cf. section 8.3) and is universally known as the Maxwell distribution function. It is a sharply peaked function and explains why calculations based upon the simplifying assumption that all the molecules have the same velocity frequently lead to nearly correct results. The value of the velocity at the peak of the distribution is easily shown by differentiation to be $\sqrt{(2kT/m)}$ and is usually referred to as the most probable velocity since there are more molecules in a fixed velocity range Δc centred on this velocity than at any other.

The mean value of any function of the velocity $f(c)$ is by definition given by

$$\overline{f(c)} = \frac{1}{N}\int f(c) dN$$

Worked example

Find the average value of the nth power of the velocity c^n for a Maxwell distribution. By definition

$$\overline{c^n} = \int_0^\infty c^n dn = \frac{4}{\sqrt{\pi}}\left(\frac{m}{2kT}\right)^{3/2} \int_0^\infty c^{n+2} e^{-mc^2/2kT} \, dc = \frac{4}{\sqrt{\pi}}\left(\frac{m}{2kT}\right)^{3/2} I_{n+2}$$

The integral can be simplified by integrating by parts as explained in section 8.2 and is tabulated in Table 8.1.

The most frequently required mean values are $\bar{c}$ the mean velocity ($n = 1$) and $\overline{c^2}$ the mean square velocity ($n = 2$).
From Table 8.1 one obtains

$$\bar{c} = \frac{4}{\sqrt{\pi}}\left(\frac{m}{2kT}\right)^{3/2} I_3 = \frac{2}{\sqrt{\pi}}\left(\frac{m}{2kT}\right)^{3/2}\left(\frac{2kT}{m}\right)^2 = \frac{4}{\sqrt{\pi}}\sqrt{\frac{kT}{2m}}$$

$$\overline{c^2} = \frac{4}{\sqrt{\pi}}\left(\frac{m}{2kT}\right)^{3/2} I_4 = \frac{4}{\sqrt{\pi}}\left(\frac{m}{2kT}\right)^{3/2} 3\sqrt{\frac{\pi k^5 T^5}{2m^5}} = \frac{3kT}{m}$$

from which one easily obtains the frequently used relationship

$$\overline{c^2} = \frac{3\pi}{8}(\bar{c})^2$$

Worked example

Show that neutrons with a Maxwellian velocity distribution (such as are obtained when they are diffusing through graphite) have a most probable velocity of 2200 m/s when their temperature is 20.4 °C.

From the text the most probable velocity is $\sqrt{(2kT/m)}$ so that

$$T = \frac{2.2^2 \times 10^6 \times m_n}{2k}$$

Taking the standard values $m_n = 1.6749 \times 10^{-27}$ kg, $k = 1.3806 \times 10^{-23}$ J K^{-1} from the tables gives

$$T = \frac{4.84 \times 1.674 \times 10^2}{1.3806 \times 2} = 293.58 \text{ K} = 20.4(2)\,°\text{C}.$$

Maxwell's relationship has been tested experimentally (section 8.3) and it is known that it is substantially correct although the precision of the experiments is not high, nor extended over a large velocity range.

2.3 Charles' law

A less arbitrary way of measuring temperature than Fahrenheit's can be based upon the experiments of Charles (1787). Citizen Charles, a pioneer balloonist, showed that equal volumes of nearly perfect gases at constant pressure expanded by the same amount when their temperature was increased by the fixed interval between the ice point and the boiling point of water. This result is true for any fixed interval and is in marked contrast with the behaviour of liquids which expand by different amounts. Charles' result suggests that gases are more likely to have uniform expansion coefficients than liquids and approximately obey the law

$$p_t v_t = p_0 v_0 \left(1 + \frac{t}{273}\right) = p_0 v_0 \left(\frac{t + 273}{273}\right)$$

in which t is the Celsius temperature obtained by dividing the change in pv between the ice point and the boiling point of water into 100 equal parts. The figure 273 is derived from experiment and is subject to the usual margin of experimental error. This assumption can be further refined to define an absolute perfect gas temperature T such that

$$T = \frac{273[p_T v_T]}{p_0 v_0} = \frac{[p_T v_T]}{K}$$

in which $[p_T v_T]$ is the value of $p_T v_T$ obtained by extrapolating the pv values for a real gas to zero pressure.

The constant K is adjusted so that $T = 273.16$ at the triple point of water. At the triple point, water, ice and water vapour are in equilibrium at a unique temperature which is 0.01 K above the ice point. The figure 273.16 gives the best agreement between the perfect gas scale (which as shown in section 5.3 is the same as the Kelvin absolute scale) and the International Practical Temperature Scale, 1968. This scale t_{68} is based on a Platinum resistance scale in which the triple point is 0.01 t_{68}/°C and the boiling point of water is 100 t_{68}/°C. Such a perfect gas temperature is absolute in the sense that it does not depend on which real gas was used to supply the data for extrapolation to zero pressure. Theoretically the difference between perfect gas temperatures and temperatures measured with a gas thermometer is important. In fact the differences are quite small and it requires very accurate experimental work to detect them at all. The difference between the reading of a constant pressure gas thermometer assumed to obey the law $pv = rt$ and the higher perfect gas temperature obtained by using $[pv]$ the value of pv extrapolated to zero pressure is shown in the following table:

Table 2.1 Corrections to the constant pressure gas thermometer found by Holborn and Otto.

Temperature K	723	523	273	173	90
Helium thermometer	0.012	0.004	0	0.005	0.029
Nitrogen thermometer	0.67	0.225	0	0.399	

The pressure of the gas thermometers was 1 mHg.

From the point of view of molecular physics Boyle's law and Charles' law are two aspects of the theoretical relationship $pv = \frac{1}{3} N m \overline{c^2} = \frac{2}{3} u'$ in which u' is that part of the internal energy of the gas which is due to the motion of the centre of gravity of the molecules. The remaining internal energy $u - u'$ due to the rotation and internal vibration of polyatomic molecules (i.e. H_2, N_2, O_2) plays no part in determining the behaviour of the gas in this context.

Boyle's law tells us that $T = f(u')$ and Charles' law shows that $f(u')$ can have the simple form constant $\times\, u'$ so that $pv =$ constant $\times\, T =$ constant $\times\, u'$. This is the equation of state obeyed by a perfect gas and is usually written in the form:

$$PV = \tilde{n}\, RT = \text{constant} \times \tfrac{1}{2} \tilde{n} L m \overline{c^2}$$

in which $\tilde{n}$ is the number of moles of gas under consideration, $L = N_A =$ the number of molecules in a mole, and R is the gas constant for 1 mole. Its value is 8.3167 J/K. The numerical value of R must be determined experimentally. It depends on the arbitrary choice of the gramme as the unit of mass in defining the mole and 273.16 as the fixed point on the absolute scale. In theoretical work it is usual to write the same equation in the form

$$pv = NkT \quad \text{or} \quad p = nkT$$

in which N is the total number of molecules in the gas and n the number of molecules per unit volume. The constant $k = R/L$ is the gas constant per molecule. Its value is 1.380×10^{-23} J/K and it is usually referred to as Boltzmann's constant, in honour of the great Viennese physicist whose work we shall explore in Chapter 6.

If $pv = NkT$ from the definition of perfect gas temperature and $pv = \frac{1}{3} N m \overline{c^2}$ from molecular theory it follows that $T = \frac{2}{3} k \left(\frac{1}{2} m \overline{c^2}\right)$ and when two perfect gases have the same perfect gas temperature they have the same value for the mean kinetic energy of translation $\frac{1}{2} m c^2$ of their molecules. This is an important result and answers the question 'What is the same about two perfect gases when their temperature is the same?' It is the mean value of the kinetic energy of translation of their molecules. This is, of course, the equivalent of Avogadro's hypothesis on which the whole elementary theory of chemistry was built up: 'Equal volumes of gases at the same temperature and pressure contain equal numbers of molecules'. If two gases of unequal temperature are in contact, molecular collision provides the mechanism by which their temperatures become the same. If two molecules of mass m_1 and m_2 and velocity u_1 and u_2 collide, it is easy to calculate the exchange of energy between them. It is nevertheless surprisingly difficult to show that on the average this exchange ceases when the mean kinetic energies of the two types of molecule are the same. The difficulty lies in averaging over all types of collision because this averaging must be carried out over both space and time. The rate at which a given type of collision is taking place depends upon the relative velocity of the molecules and the time average cannot be found until the velocity distribution function is known. The way to arrive at the important result that when the two gases are mixed equilibrium is reached when $m_1\overline{c_1^2} = m_2\overline{c_2^2}$ and the mean kinetic energies are equal was found by Maxwell who showed that it is only true if both gases have a Maxwellian energy distribution.

2.4 Kinetic energy and temperature

That molecules of perfect gases in thermal equilibriums have the same mean kinetic energy is a result that can be extended to molecular systems in general but not by elementary methods. Thus, the short answer to the question posed in section 1.8 – 'What is the same about the molecules of substances which have the same temperature?' – is that they have the same average kinetic energy. The simplest way of establishing this result is to rely upon the theorem of the equipartition of energy. This theorem is itself a deduction from a more general result due to Boltzmann about the way energy is shared amongst the degrees of freedom (dynamical variables) of an isolated system of molecules interacting through a mutual potential energy function. A simple account of the way in which Boltzmann's relationship is derived is given in Chapter 6.

The theorem of the equipartition of energy is restricted to those degrees of freedom which occur as squared terms in the expression for the energy of the molecule. With this restriction the theorem states that the average energy of a molecule is $kT/2$ for each degree of freedom. The term degree of freedom as employed in this context needs explaining.

In classical mechanics each point particle requires six independent quantities to specify its mechanical state at a given time. The six independent quantities are called the degrees of freedom of the particle. They are the three cartesian coordinates x, y, z and the three components of velocity $\dot{x} = u$, $\dot{y} = v$, $\dot{z} = w$. It is fundamental to the whole method of representing mechanical systems that the number (six) of degrees of freedom is invariant and cannot be changed by inventing a new coordinate system. For example in polar coordinates the three independent space coordinates are r, θ, ϕ and the three independent velocities $\dot{r}$, $r\dot{\theta}$ and

r Sin $\theta\dot{\phi}$. A system of two particles has $2 \times 6 = 12$ degrees of freedom and so on. If the two particles (1 and 2) are linked to form a rigid dumb-bell molecule with the two atoms a fixed distance R apart, there are two relationships

$$(x_1 - x_2)^2 + (y_1 - y_2)^2 + (z_1 - z_2)^2 = R^2 \quad \text{(atoms always } R \text{ apart).}$$

$$\frac{dR^2}{dt} = 0 \quad \text{(atoms must move so that they remain } R \text{ apart)}$$

between the coordinates and the number of independent degrees of freedom is reduced from 12 to 10. These ten degrees of freedom can be rearranged in the following way. The three cartesian coordinates x, y, z of the centre of gravity of the molecule (three degrees of freedom), the direction of the line of centres, $\theta\phi$ (two degrees of freedom), the velocity of the centre of gravity $\dot{x}$, $\dot{y}$, $\dot{z}$ (three degrees of freedom) and the angular velocity of the molecule which is a vector with three components ω_1, ω_2, ω_3 (two degrees of freedom) of which only two are independent making $3 + 2 + 3 + 2 = 10$ degrees of freedom in all. The three components ω_1, ω_2, ω_3 of the angular momentum vector are not independent because they must be chosen so that there is no angular momentum about the axis of the molecule; this component of the angular momentum must be zero because the two particles forming the molecule have been assumed to be points, not small spheres. The energy of a particle depends upon the coordinates and so upon the degrees of freedom. Thus the energy of a point particle moving in the earth's gravitational field is given by the expression $E = \frac{1}{2}mu^2 + \frac{1}{2}mv^2 + \frac{1}{2}mw^2 + mgz$ in a cartesian coordinate system in which z is measured upwards. The six degrees of freedom (dynamical coordinates) enter differently into this expression; u, v, w enter as squared terms, z as a linear term and x, y are not represented. The theorem of the equipartition of energy is a theorem about that part of the energy which is due to the squared terms and makes no pronouncement about the energy due to the other terms. The kinetic energy always depends upon the square of the dynamical coordinates u, v, w through the relationship k.e. $= \frac{1}{2}m\,(u^2 + v^2 + w^2)$; it is therefore in a special position and is always $3kT/2$ whatever the potential energy may be. The restriction to coordinates 'whose squares represent energy' is merely a mathematically convenient way of quoting the theorem. The mean energy associated with any degree of freedom can be calculated (provided the integrations can be carried out) but it will depend upon the details of the particular system being considered. The potential energy will only have the equipartition value if the molecules are moving with simple harmonic motion for which the potential energy is proportional to the square of the displacement. The outstanding example of such a system is the idealised monatomic solid in which the atoms are supposed to oscillate with s.h.m. about fixed lattice points. The mean energy of an atom is accordingly $kT/2$ for each degree of freedom $= 6 \times kT/2$ corresponding to an atomic heat $C_v = 3Lk = 3R$ J/mole.

It is an instructive exercise to calculate that part of the energy of the atmosphere (treated as a perfect gas) which is in the form of potential energy. It is easily found to be $2NkT/2$ which is twice the equipartition value. This is, of course, another way of calculating the value of $C_p - C_v = R = Lk$. To see this one has only to imagine any section of the gas to be confined between vertical walls so that all the expansion is upwards. The work done by any section of the gas between z and $z + \mathrm{d}z$ raises the gas above it so that all the external work done in the expansion i.e. $(C_\mathrm{p} - C_v) \times$ number of moles, goes into raising the centre of gravity of the gas. The effect of associating potential energy mgz with one of the spatial coordinates has been to add R not $R/2$ to the molar heat. If the potential energy had been mgz^2 then the theorem of equipartition would have applied and the extra molecular heat would have been $R/2$.

Worked example

Find the potential energy of a perfect gas in a gravitational field. Let gravity act along the z axis so that the potential energy of a particle is mgz for each of the n molecules per unit volume. For an elementary cylinder of gas of unit cross sectional area and height $\mathrm{d}z$ the downward force due to gravity is $mgn\mathrm{d}z$. This is balanced by the difference in the pressure $p + \mathrm{d}p$ acting downwards on the top of the cylinder and p acting upwards on the bottom of the cylinder. Putting the net force equal to zero gives

$$nmg\mathrm{d}z + p + \mathrm{d}p - p = 0 \quad \text{or} \quad \frac{\mathrm{d}p}{\mathrm{d}z} = -nmg$$

but for a perfect gas $p = nkT$ so that

$$\frac{\mathrm{d}n}{\mathrm{d}z} = -\frac{nmg}{kT} \text{ and } n = n_0\mathrm{e}^{-mgz/kT}$$

The potential energy of a molecule is mgz so the average p.e. of the molecules in a cylinder of unit cross sectional area is

$$\frac{\text{Total p.e. of the molecules}}{\text{Total number of molecules}} = \frac{\displaystyle\int_0^\infty mgz\, n_0\mathrm{e}^{-mgz/kT}\mathrm{d}z}{\displaystyle\int_0^\infty n_0\mathrm{e}^{-mgz/kT}\mathrm{d}z} = kT$$

The equipartition theorem depends upon Newtonian mechanics and is not true in those fields in which classical and quantum mechanics give different results. Notable examples are the kinetic energy of electrons in atoms and metals which are very much higher than $3kT/2$, and the kinetic energy of the molecules of solids at low temperature which can be much lower than $3kT/2$.

2.5 Laplace's law of adiabatic compression

From the point of view of the molecular theory the compression of a thermally isolated monatomic gas is theoretically simpler than the compression of a gas at constant temperature since the relationship between kinetic energy and temperature does not have to be known. Suppose the compression is carried out by means of a piston of area A. At each collision with the piston the molecules will have the velocity of the piston added to their own and will so continually increase their kinetic energy. For a small motion of the piston dx the work done on the gas = force × distance = $-F\,dx \equiv -PA\,dx = -PdV$ = increase in the kinetic energy = dU. Conservation of energy requires that $PdV + dU = 0$. But at any stage of the compression $PV = \frac{2}{3}U$ for a monatomic gas so that

$$d(PV) = PdV + VdP = \tfrac{2}{3}dU$$

eliminating dU gives at once that

$$5PdV + 3VdP = 0 \text{ and } PV^{5/3} = \text{constant} = PV^{\gamma}\left(\gamma = \frac{c_p}{c_v} = \frac{5}{3}\right)$$

By combining the adiabatic equation $PV^{5/3} = K$ with the equation of state $PV = RT$ one obtains the rise in temperature on compression in the form

$$\frac{T_1}{T_2} = \left(\frac{V_2}{V_1}\right)^{2/3} = \left(\frac{P_1}{P_2}\right)^{2/5}$$

For diatomic gases U = kinetic energy of translation + energy of rotation of the molecule. It is the first term only which determines PV. As a first approximation the molecule is looked upon as a dumb-bell-like structure made from two point masses m_1 and m_2

separated by a fixed inter-molecular distance r. The principal moments of inertia are

$$I_1 = I_2 = \frac{m_1 m_2 r^2}{m_1 + m_2}, \; I_3 = 0$$

and its kinetic energy of rotation is $\frac{1}{2}I_1\omega_1^2 + \frac{1}{2}I_2\omega_2^2$ in which ω_1 and ω_2 are the angular velocities about the principal axes. These are dynamical coordinates so that the equipartition theorem applies and the kinetic energy of rotation is $2 \times kT/2$ and U for a diatomic gas is $5RT/2$. Using this relationship leads at once to the law

$$PV^{7/5} = K = PV^{\gamma}\left(\gamma = \frac{c_p}{cv} = \frac{7}{5}\right)$$

for the adiabatic expansion of a perfect diatomic gas.

The relationship $PV^{\gamma} = K$ was discovered by Laplace (1816) to hold for the compression of an isolated gas and used by him in his justly celebrated correction to Newton's formula for the velocity of sound. It is well obeyed as is shown by the following experimental results.

Gas	He	Ne	A	Hg	Theory
Experimental value of γ	1.6(6)	1.64	1.67	1.67	1.666
Gas	H_2	N_2	O_2	NO	
Experimental value of γ	1.41	1.41	1.40	1.40	1.40

It is of historical interest to note that the simple derivation just given of the law obeyed by a gas when it is compressed adiabatically is not that given by Laplace. He first of all deduced the existence of the law from the discrepancy between the observed and calculated velocity of sound in gases, and then after a considerable interval was able to deduce it from the caloric theory of heat. This is an example of how slowly the idea of conservation of energy displaced the caloric theory. It is also a good example of the dictum which scientists forget at their peril, that agreement between theory and experiment is no guarantee that the premises of the theory are correct. Nor is it a bad thing occasionally to remember that the incomparable Newton, and Laplace, sometimes called the French Newton, who was indeed a great mathematical physicist, both made theoretical mistakes in this field, for which 'A' level candidates would be severely marked down if not failed.

2.6 Van der Waals' equation

The first successful attempt to formulate an equation of state which took into account both the attractive and the repulsive forces between the molecules was made by Van der Waals (1873). He distinguished between the internal and the external pressure of a gas. In the equation $P(V - b) = RT$ (section 2.2) P may be said to refer to the internal pressure, a pressure calculated from the collision rate with an imaginary area dS placed inside the gas. The constant b takes into account the repulsive forces and it can be argued that in calculating the internal pressure it is legitimate to neglect the longer range attractive forces because each molecule in the body of the gas is surrounded by attracting molecules arranged at random so that on the average their attractions cancel out. The external pressure as measured by a manometer outside the gas is less than the internal pressure because the molecules have to leave the body of the gas against the attractive forces of all the molecules in the range of the attractive forces. The difference between the internal and external pressures will vary as the square of the density of the gas. The density ρ enters twice into the correction. Once because it effects every molecule and so is a small fraction of the pressure; secondly, because the effect on each molecule is proportional to the number of molecules in the range of attractive forces, a quantity proportional to the density. Thus one can write

$$P_{\text{external}} = P_{\text{internal}} - K\rho^2$$

in which K is the constant representing the attractive forces. The relationship

$$P_{\text{internal}} \cdot (V - b) = RT$$

becomes

$$(P_{\text{external}} + K\rho^2)(V - b) = RT$$

usually written as

$$\left(P + \frac{a}{V^2}\right)(V - b) = RT$$

in which a represents the attractive part of the molecular forces and b the strong repulsion when the spheres collide. By adjusting the two arbitrary constants a and b the equation can be made to reproduce the deviations of a real gas from the ideal Boyle–Charles law $PV = RT$ over a wide range of pressures and temperatures with moderate accuracy. It differs from the equation of state derived from the virial in section 2.2 only in the second order term ab/V^2. This is an important difference and indicates that van der Waal's equation has, in a rudimentary way, taken into account the possibility of three molecules interacting together. It is because of this term that van der Waals' equation has implications which transcend the simple approximations used in deriving it. It can be rewritten as a cubic equation in v

$$pv^3 - v^2(RT + pb) + av - ab = 0$$

The value of the independent variables p and T determining whether it has three real roots or two imaginary and one real. A set of pv curves is shown in Fig. 2.3. The upper curves represent the deviations of a real gas from the behaviour of a perfect gas and corresponds to the cubic equation having one real and two imaginary roots. By suitable choice of a and b the behaviour of real gases is reasonably reproduced. The lower curves (corresponding to a

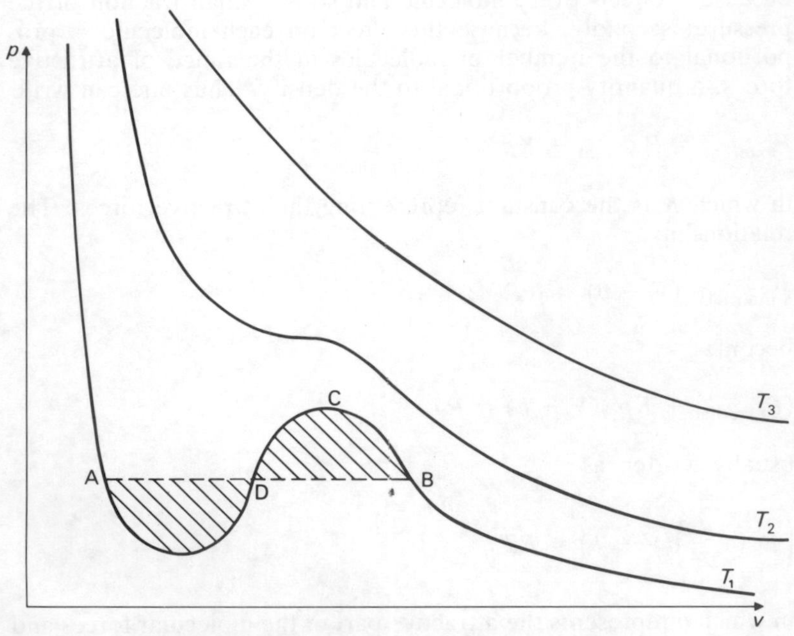

Fig. 2.3. Schematic pv diagram for a gas obeying van der Waals' equation.

cubic equation in v with three real roots) do not represent the behaviour of any known substance. This is not surprising since it represents conditions far from those assumed in the simple approximate derivation of van der Waals' equation. This part of the diagram can however be interpreted by drawing a constant pressure line AB so that the shaded areas are equal. This horizontal portion is to represent the condensation of the gas to a liquid state with a constant vapour pressure. With this subsidiary assumption van der Waals' equation becomes a rudimentary equation of state covering at least qualitatively the pv relationship between the liquid and gaseous state. It gives no hint that a solid state also exists. The rule for drawing the line AB is to a certain extent arbitrary. Since it is a well established experimental result that the saturated vapour pressure of a liquid depends upon the temperature only, all equations of state which claim to represent the liquid-vapour equilibrium must include a horizontal line in their representation. The choice of the pressure at which the line is to be drawn is equivalent to predicting the vapour pressure of the liquid and this is beyond the scope of simple equations of state of the van der Waals'

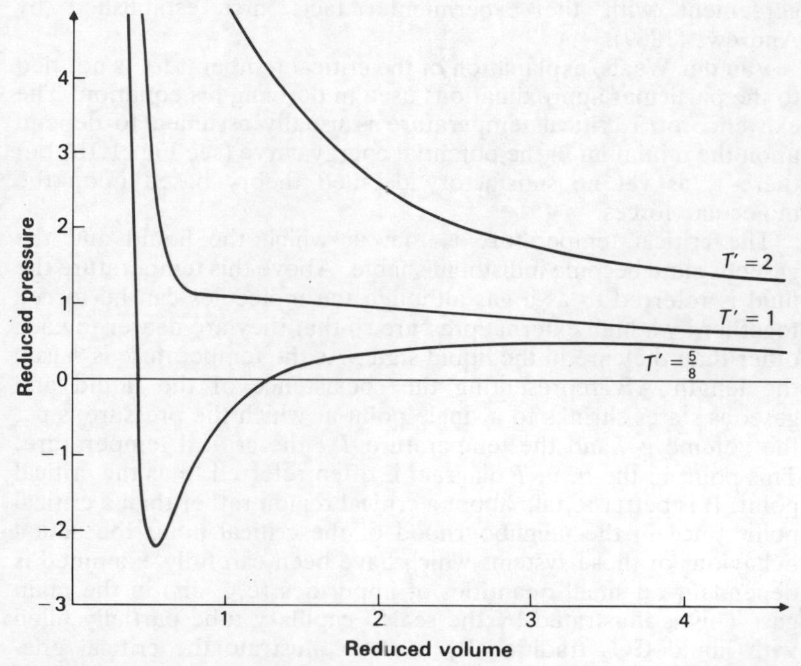

Fig. 2.4 *PV* diagram for a gas obeying the reduced van der Waals' equation.

type. The equal area rule (usually attributed to Maxwell) is not quite arbitrary and is based upon the consideration that $\oint p\,dv$ round a closed isothermal reversible cycle is zero in accordance with thermodynamic reasoning.

If this is applied to the imaginary cycle ABCDA (Fig. 2.3) the equal area rule follows immediately. This demonstration however carries no great weight since it is only valid if all stages of the cycle pass through states of equilibrium. This condition is manifestly not fulfilled for the humped part of the cycle which in fact cannot be realised and would be unstable if it were realised since it contains sections in which dp/dv is positive. Positive (dp/dv) corresponds to a negative compressibility and chance changes in density create pressure gradients which promote the fluctuations so that they continue to grow; correspondingly a negative (dp/dv) creates pressure gradients which reduce the fluctuations and they decay.

In equations of state of the van der Waals' type it is the numerical value of the temperature which determines whether there are three real roots, or only one. Only for three real roots can the liquid state be described and there is a critical temperature above which no liquid state is possible however great the pressure. This is in agreement with the experimental facts first established by Andrews (1869).

Van der Waals' explanation of the critical temperature is not tied to the particular approximations used in deriving his equation. The existence of a critical temperature is usually assumed to depend upon the minimum in the potential energy curve (see Fig. 1.1b) but there is as yet no satisfactory detailed theory based upon the molecular forces.

The critical temperature is that at which the liquid and the gaseous state become indistinguishable. Above this temperature the fluid is referred to as a gas although the molecules can be forced together by a high external pressure so that they are nearer to each other than they are in the liquid state. As the temperature is raised the length AB representing the coexistence of the liquid and gaseous states shrinks to a single point at which the pressure is p_c, the volume v_c, and the temperature T_c, the critical temperature. This point in the p, v, T diagram is often referred to as the critical point. It is better to talk about a critical region rather than a critical point since in the neighbourhood of the critical point the actual behaviour of those systems which have been carefully examined is dependent on small quantities of impurities (e.g. air) in the main gas. This is illustrated by the sealed capillary tube partially filled with liquid CO_2 traditionally used to illustrate the critical phenomena. On its being heated in a current of warm air the liquid is observed to expand, the meniscus to flatten and the surface to disappear suddenly in a flurry of opalescence near the 'critical temperature'. A little consideration shows that if there were a sharp critical point the amount of carbon dioxide in the tube would have to be very accurately predetermined before the tube was sealed, whereas in fact the tubes can be made without much difficulty. The behaviour of a fluid in the critical region is both complicated and difficult to investigate; it is certainly not to be explained in terms of simple equations derived in the first instance to account for relatively small deviations from the perfect gas laws. In the first place the behaviour in this region sometimes depends critically on the purity of the sample. Callendar (1928) found that small quantities of dissolved air very noticeably affected the behaviour of water in this region. For air free water he was able to show that the latent heat of evaporation had not fallen to zero when the meniscus vanished at 374 °C but was 3.013×10^4 J/kg and that distinct density curves for water and steam could be followed up to about 380 °C as shown in Fig. 2.5. Secondly, later work has amply confirmed that the vanishing of the meniscus is not a sufficiently precise way of recognising the identity of the liquid and gaseous phases; differences of density ($\sim$10%) can be detected after the meniscus has vanished. The compressibility of the liquid is also 30–50 times the already very large value predicted by van der Waals' equation so that gravitational effects cannot be neglected even in quite small experimental tubes.

The fluctuations in density produced in this way can be removed by stirring but they reappear in a few hours. Finally, the simple kinetic theory relationship between heat conductivity and viscosity is widely departed from. The viscosity (Fig. 2.6) changes smoothly in passing through the critical region but the heat conductivity shows a large anomaly as shown in Fig. 2.7.

Worked example

Show that for a van der Waals' gas in the critical region (critical pressure p_c, critical volume v_c) that an isothermal increase in pressure Δp produces an increase in volume Δv given by

$$\frac{\Delta p}{p_c} = -\frac{3}{2}\left(\frac{\Delta v}{v_c}\right)^3.$$

Since at the critical point

$$\left(\frac{\partial p}{\partial v}\right) = \left(\frac{\partial^2 p}{\partial v^2}\right) = 0$$

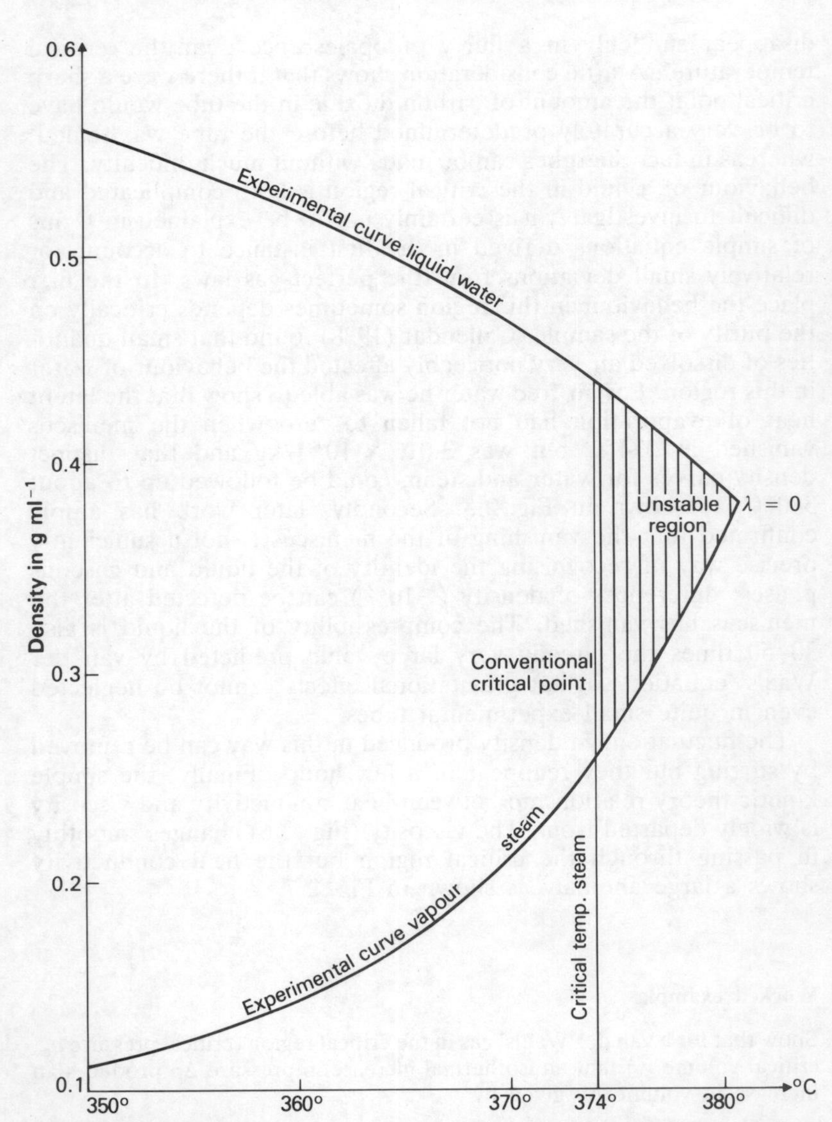

Fig. 2.5 Critical data for water. Callendar's (1928) results for air-free water showing that the densities of water and steam are not the same above the conventional critical temperature 374 °C.

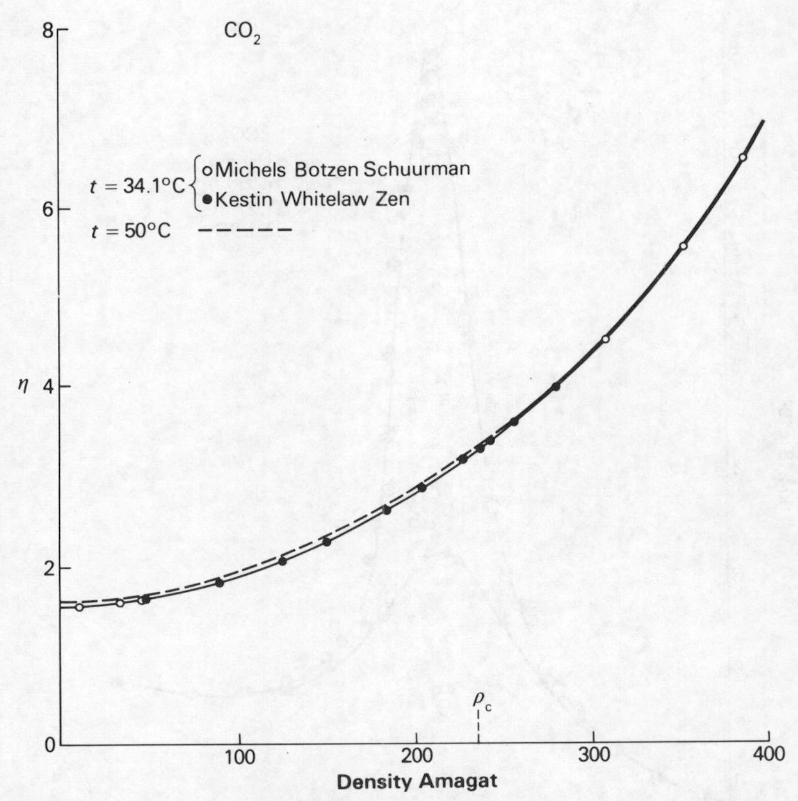

Fig. 2.6 Viscosity of CO_2 in the critical region. (Sangers 1960).

the change in pressure Δp accompanying a change in volume Δv is never a linear function of Δv, and must be calculated from the third term of a Taylor theorem expansion.

Thus

$$\Delta p = p - p_c = \frac{(\Delta v)^3}{3!}\left(\frac{\partial^3 p}{\partial v^3}\right)_c$$

in which $(\partial^3 p/\partial v^3)_c$ is the numerical value of $(\partial^3 p/\partial v^3)$ calculated from the equation of state at the critical point p_c, v_c, T_c.

Direct differentiation of van der Waal's equation gives that

$$\left(\frac{\partial^3 p}{\partial v^3}\right)_T (v - b) + 3\left(\frac{\partial^2 p}{\partial v^2}\right)_T = \frac{6a}{v^4} - \frac{24ab}{v^5},$$

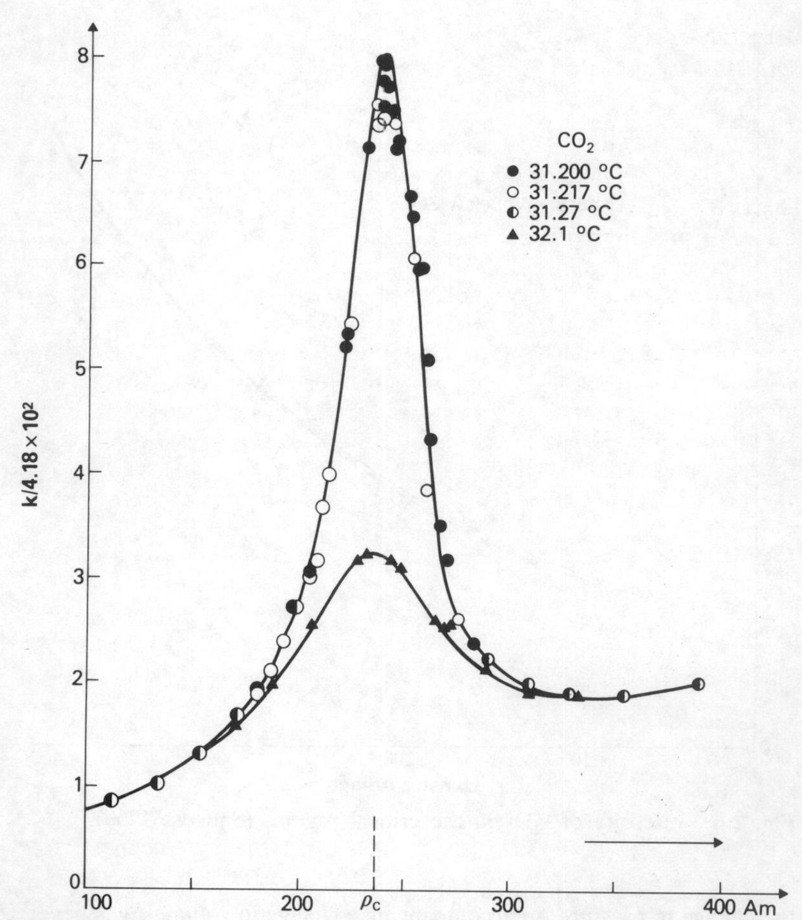

Fig. 2.7 Heat conductivity of CO_2 in the critical region (Sangers 1960).

so that

$$\left(\frac{\partial^3 p}{\partial v^3}\right)_{T_c} = \frac{6a(v-4b)}{v^5(v-b)}.$$

Eliminating a and b in terms of the critical constants gives at once that

$$\frac{\Delta p}{p_c} = -\frac{3}{2}\left(\frac{\Delta v}{v_c}\right)^3.$$

Many attempts have been made to apply van der Waals' ideas in a different way to give a modified equation of state. The effect of the attractive force between the molecules can be assumed to depend upon the temperature to give equations of the type

$$\left(p + \frac{a'}{(v + c)^2 T}\right)(v - b) = RT.$$

This is Clausius' equation, or with $c = 0$, Berthelot's. Alternatively, the relationship between the internal pressure and the external pressure can be modified. If one adopts the relationship which obtains in an ideal liquid whose vapour pressure is given by the relationship $p = p_0 e^{-\lambda/RT}$ one easily finds that $p(v - b) = RT$ $e^{-a'/RTv}$ an equation of state first given by Dieterici. These equations are useful semi-empirical equations for representing the behaviour of gases over limited ranges of the variables.

Van der Waal's equation may be compared with experiment through the values of the critical constants p_c, v_c, and T_c. At the critical point

$$\left(\frac{\partial p}{\partial v}\right)_T = \left(\frac{\partial^2 p}{\partial v^2}\right)_T = 0$$

and these two equations can be used to determine a and b in terms of the critical constants. The result is

$$a = \frac{27 R^2 T_c{}^2}{64 p_c} \text{ and } b = \frac{RT_c}{8 p_c}.$$

The value of b obtained in this way is some 20–50 per cent lower than that obtained from the gas laws and other kinetic theory data. A similar degree of agreement is obtained between the theoretical value for the dimensionless constant $(p_c v_c / RT_c) = \frac{3}{8}$ and its experimental value which usually lies between 0.3 and 0.75. In general, van der Waal's equation tends to smooth out the differences between individual substances and predicts that they behave more uniformly than they do. This appears from the 'reduced equation'. If a and b are eliminated from the equation in terms of the critical constants one obtains the 'reduced equation of state'

$$\left(\mathfrak{p} + \frac{3}{\mathfrak{v}^2}\right)(3\mathfrak{v} - 1) = 8\mathfrak{t}$$

in which

$$\mathfrak{p} = \frac{p}{p_c}, \quad \mathfrak{v} = \frac{v}{v_c}, \quad \text{and} \quad \mathfrak{t} = \frac{T}{T_c}$$

are the reduced values of the pressure, volume and temperature. This is a universal equation and should be obeyed by all substances.

The isotherm corresponding to the boiling point T_B of a normal liquid ($T_B/T_c \approx 0.6$) is shown in Fig. 2.5. A rough quadrature shows that the predicted ratio (density of liquid/density of vapour) is much smaller than the experimental value which is usually 500–1000.

Worked example

How does the saturated vapour pressure of a fluid obeying van der Waals' equation depend upon the temperature in the critical region?

The vapour pressure p_{vap} is that fixed pressure which complies with the Maxwell equal area condition; this may be written as

$$p_{vap}(v_g - v_l) = \int_{v_l}^{v_g} p\,dv$$

in which v_l and v_g are the volumes of the liquid and vapour respectively at the points A and B in Fig. 2.3 and the integral is taken along the path AEDCB. It is convenient to change to the reduced variables $\mathfrak{p}$, $\mathfrak{v}$, $\mathfrak{t}$ in which the reduced equation of state can be written as

$$\mathfrak{p} = \frac{8\mathfrak{t}}{3\mathfrak{v} - 1} - \frac{3}{\mathfrak{v}^2}.$$

The Maxwell condition remains the same in the new variables and

$$\mathfrak{p}_{vap}\,(\mathfrak{v}_g - \mathfrak{v}_l) = \int_{\mathfrak{v}_l}^{\mathfrak{v}_g} \mathfrak{p}\,d\mathfrak{v} = \frac{8\mathfrak{t}}{3}\log\frac{3\mathfrak{v}_g - 1}{3\mathfrak{v}_l - 1} + 3\left[\frac{1}{\mathfrak{v}_g} - \frac{1}{\mathfrak{v}_l}\right]$$

and

$$\mathfrak{p}_{vap} = \frac{8\mathfrak{t}}{3(\mathfrak{v}_g - \mathfrak{v}_l)}\log\frac{3\mathfrak{v}_g - 1}{3\mathfrak{v}_l - 1} - \frac{3}{\mathfrak{v}_g\mathfrak{v}_l}$$

This expression is not much use outside the critical region because $\mathfrak{v}_g$ and $\mathfrak{v}_l$ are the real roots of a cubic equation for which there is no simple algebraic expression. (Cardan's irreducible case: *see* Loney 1907 or other text). In the critical region write $\mathfrak{v}_g = 1 + \mathfrak{v}_g$, $\mathfrak{v}_l = 1 + \mathfrak{v}'_l$ in which $\mathfrak{v}'_g$ and $\mathfrak{v}'_l$ are small quantities of opposite sign. The expression for the vapour pressure then becomes

$$\mathfrak{p}_{vap} = \frac{8\mathfrak{t}}{3(\mathfrak{v}'_g - \mathfrak{v}'_l)}\log\frac{1 + \frac{3}{2}\mathfrak{v}'_g}{1 + \frac{3}{2}\mathfrak{v}'_l} - \frac{3}{(1 + \mathfrak{v}'_g)\,(1 + \mathfrak{v}'_l)}$$

which to the first order in $(\mathfrak{v}'_g - \mathfrak{v}'_l)$ is

$$\mathfrak{p}_{vap} = 4\mathfrak{t} - 3 + 3(\mathfrak{v}'_g + \mathfrak{v}'_l)(1 - \mathfrak{t}).$$

Now $\mathfrak{v}'_g$ and $\mathfrak{v}'_l$ are small, of opposite sign and nearly equal in magnitude (as may be shown by solving the appropriate cubic) so that $\mathfrak{p}_{vap} = 4\mathfrak{t} - 3$.

Modern experimental work on gases is usually expressed in the form

$$PV = RT\left(1 + \frac{B_v}{V} + \frac{C_v}{V^2} + \frac{D_v}{V^3} \cdots\right) = RT(1 + B_pP + C_pP^2 \cdots)$$

The coefficients are known as virial coefficients. This is the form most easily compared with calculations based upon an assumed potential function for the interaction between the molecules. The first approximation gives

$$B_v = 2\pi N \int_0^\infty (1 - e^{-V(r)/kT})r^2 dr$$

in which $V(r)$ is usually taken to be a Lennard–Jones potential.

$$V(r) = 4\varepsilon\left(\left(\frac{\sigma}{r}\right)^{12} - \left(\frac{\sigma}{r}\right)^{6}\right)$$

In retrospect, looking back with the benefit of hindsight over the work of the last hundred years, one sees that the chief defect of the rather general van der Waals' approach is that it makes insufficient provision for the molecules to form temporary clusters as the liquid state is approached.

Problems

1 Show that in calculating the momentum transferred to the walls in elementary derivation of Boyle's law, it is reasonable to omit the attractive forces between the walls and the molecules. (Apply Newton's third law).

2 The B_4C control rod of a nuclear reactor can capture about 10^{21} neutrons/cm^3 before it has to be replaced. If each neutron absorbed releases an atom of helium, calculate the pressure exerted by the helium at 0 °C on the assumption that

(a) it occupies a space equal to the volume of the control rod.

(b) it diffuses into the interstitial space of the rod looked upon as a collection of small closely packed spheres.

3 Calculate the value of b in terms of the liquid density and molecular weight (for hard spherical molecules) in van der Waals' equation on the assumption the molecules in a liquid at low temperatures

(a) are as closely packed as possible

(b) are arranged on a cubic lattice.

4 Making use of the relationship

$$\left(\frac{\partial^2 p}{\partial v^2}\right)_T = \left(\frac{\partial p}{\partial v}\right)_T = 0$$

at the critical point (or otherwise) derive the expressions for a and b and $P_c V_c / RT_c$ quoted in the text for van der Waals' equation. Make a similar calculation for Berthelot's equation, and Dieterici's equation.

5 Derive Dieterici's equation in the way suggested in the text. What simplifying assumption do you make?

6 The stopcock of a helium cylinder (120 atm) is accidentally opened so the gas escapes approximately adiabatically. Calculate the lowest temperature reached by the gas remaining in the cylinder, on this assumption. How well would you expect your simple calculation to agree with experiment? Make a similar calculation starting at 30 atm and 20 K. In this case it is legitimate to neglect the heat capacity of the container.

7 A helium compressor compresses 1 mole/s from 1 to 100 atm at room temperature (300 K).

Assuming that helium obeys the perfect gas laws calculate the minimum horse power required:

(a) if the compression is isothermal

(b) if the compression is adiabatic.

If the compression is in two adiabatic stages with an interstage cooler between stages, find the pressure ratio of the stages for minimum power consumption.

8 Find the value of the virial coefficients A_v and B_v in

$$pv = RT\left(1 + \frac{A_v}{v} + \frac{B_v}{v^2}\right)$$

for a gas which obeys van der Waals' equation.

9 Show that a constant volume gas thermometer which obeys van der Waals' equation will read perfect gas temperatures.

10 The mean energy of the particles in a beam of ultra cold neutrons is 10^{-7} eV. How far can they rise against gravity? (*Scientific American, June* 1979).

11 Two bulbs of volume V and v containing argon and helium respectively at 1 atmosphere pressure and 300 K are connected by a capillary tube of negligible volume.
If the large bulb is placed in a bath at 280 K and the small one is in a bath at 600 K, what will be the final pressure, and what will be the proportions of the two gases in the bulbs.
(a) before there is mixing by diffusion?
(b) when mixing by diffusion is complete?

12 Sketch the p, v, T surface for a liquid-gas system which obeys van der Waals' equation.

13 Explain how to calculate $\gamma = C_p/C_v$ for a mixture of a perfect monatomic gas ($\gamma = \frac{5}{3}$ and a perfect diatomic gas ($\gamma = \frac{7}{5}$). An equimolecular mixture of argon and nitrogen at 300 K is compressed adiabatically from a pressure of $\frac{1}{50}$ atm to 1 atm. Calculate the rise in temperature on the assumption that they are perfect gases.

Further reading

Kestin, J. (1966) *A Course of Thermodynamics*, Blaisdell, London.

Roberts, J.K. and **Miller, A.R.** (1960) *Heat and Thermodynamics*, Ch. IV, Blackie, London.

Temperley, H.N.V. (1956) *Changes of State*, Cleaver Hume Press, London.

References

Callendar, H.L. (1928) *Proc. Roy. Soc. A*, **120**, 460.

Lorentz, A.H. (1881) *Annalen der Physik* (Wiederman's) **12**, 127.

Milne, A.E. (1925) *Phil. Mag.*, **L**, 409.

Sangers, J.V. (1960) *Bureau of Standards Miscellaneous Publications* 273, 165, Washington D.C.

CHAPTER 3

Molecular collisions, diffusion, effusion and molecular flow

3.1 Specific properties of real gases

So far we have considered the forces between the molecules as the cause of deviations from ideal gas laws obeyed by all gases if the pressure is sufficiently low and the collisions between the molecules can be neglected. Gases also have properties which depend upon the forces between the molecules and have a numerical value which varies from gas to gas. These properties are represented by the coefficients of viscosity, heat conductivity and diffusion and the numerical values of these coefficients do not tend to a common value at low gas pressures. If one wishes to study these properties theoretically one must clearly introduce some mathematically simple model of the molecule in which the numerical values of the parameters distinguish one molecule from another. Newton wrote in his *Opticks* 'It seems probable to me that God in the beginning formed matter in solid massy impenetrable moveable particles . . .'. It is from this beginning that the 'hard sphere' model gradually evolved, in which different molecules are distinguished by a characteristic mass m and a characteristic molecular diameter σ. This model has been very successful in accounting for the viscosity of gases and also for their heat conductivity. The spheres are usually assumed to be smooth so that there are no tangential forces. The merit of this model is its mathematical simplicity. The collision between smooth hard spheres is not only a completely soluble problem but one whose solution can be very easily stated in general terms. For collisions between hard smooth spheres the angular distribution after collision is symmetrical with respect to the centre of gravity and can be expressed in terms of the ratio of the velocities of the colliding particles. For other collisions the absolute values of the velocities must be known. These are mildly surprising results which greatly simplify any mathematical treatment of colliding particles. The alternative molecular model is due to Boscovitch

(1711–1787), a talented but by repute egotistical expositor of the Newtonian system. He represented molecules by a massive point surrounded by a field of force. As a practical device it was not much use in his day. It failed because the law of force was quite unknown and because even with a known law of force scattering calculations were difficult. Nowadays all models are essentially Boscovitchian; the electron, the proton, the neutron, all enter our calculations as point particles influencing each other through complicated fields. Both molecular models are unsatisfactory; the hard sphere because it is too simple and the centre of force model because it is too complicated and theoretical.

A particularly simple Boscovitchian model in which the molecules repel each other according to an inverse fifth power law and the potential function $V(r) = \mu/r^4$ was used by Maxwell because it was mathematically convenient. For a uniform gas the velocity distribution is supposed to be Maxwellian. For a non-uniform gas (e.g. one in which there is a temperature gradient) this distribution is distorted by an amount which it is difficult to calculate. For an inverse fifth power law the averaged results of collisions can still be expressed in terms of average values so that to a first approximation deviations from Maxwell's law can be neglected. The special mathematical position of an inverse fifth power law of force can be seen by considering the head on collision of molecules of velocities $\pm V/2$ interacting through a potential function μ/r^n. Conservation of energy shows that they will come to rest and start moving apart again when they are a distance

$$r = \left(\frac{4\mu}{m}\right)^{1/n} V^{-2/n}$$

apart. In this respect they behave like spheres of diameter σ and collision cross section

$$\pi\sigma^2 = \pi\left(\frac{4\mu}{m}\right)^{2/n} V^{-4/n}.$$

If the molecular diameter is σ the collision cross section is $\pi\sigma^2$ not $\pi\sigma^2/4$, as explained in Fig. 3.2).

Now the collision rate between molecules (section 3.4) is proportional to (collision cross section) × (relative velocity) = constant × $V^{-4/n}$ × V = constant × $V^{(n-4/n)}$. Accordingly for an inverse fifth power law of force ($n = 4$) the collision rate is independent of the relative velocity for head on collisions. For more general collisions (Jeans 1925, section 284) a similar simplification occurs when $n = 4$ which enables the distorted distribution function to be calculated relatively easily. For other values of n the calculation is much more difficult. There is, of course, no physical justification for the fifth power law. It is not surprising that results obtained in this way were looked upon with great suspicion at the turn of the century owing to the artificial nature of the models used. Indeed a hard sphere model seemed unsatisfactory for representing molecules like nitrogen or hydrogen which are undoubtedly dumb-bell-like structures with axial rather than spherical symmetry, and capable of inelastic collisions in which energy is absorbed into both internuclear vibrations and rotation about an axis perpendicular to the line joining the nuclei. The discovery of quantum mechanics has reduced the significance of this objection to hard sphere models. The wave function representing the scattered molecule always has a component corresponding to hard sphere scattering and for low velocities it can be the dominant one.

3.2 Mean free path approximation

The difficulties encountered if one works in terms of collisions between molecules calculated from a molecular model are so great that they make any elementary exposition of the subject almost impossible. They can be avoided by working in terms of a free path l. This is the distance a molecule travels between collisions. The average of all the paths of all the molecules is the mean free path λ. This procedure presupposes that the term collision has a definite meaning so that one can say whether a molecule is undergoing a collision or not. Since two molecules 'are in collision' over a distance which is comparable to the range of the forces between them the mean free path approximation can only be usefully employed when the gas pressure is so low that the distance l between collisions is very large, compared with the range of the molecular forces. It turns out that this condition is sufficiently fulfilled for gases at or below the density corresponding to s.t.p. The lower the pressure the greater the validity of the mean free path approximation. At low pressures λ is almost independent of the molecular model used and the molecule can always be represented by a hard sphere model whose diameter σ is adjusted to give the correct collision rate. If greater refinement is required the angular distribution of the molecular velocities after collision and the dependence of λ on the relative velocities of the colliding molecules can be introduced as a correction.

The survival of the elementary kinetic theory as a subject depends upon the fact that calculations based upon a simple mean free path model have wide applications. Moreover, owing to the peaking of the Maxwell distribution function, even simple calculations in which all the molecules are assumed to have the same velocity and free path are valuable and yield results which seldom differ from the correct one by more than 50 per cent. There is nothing fundamental about the fact that the properties of gases were first studied at atmospheric pressure when their density is about $\frac{1}{1000}$th of that in the liquid state. A very modest vacuum technology enables the physicist to experiment and make measurements at about 1 millionth of the density corresponding to atmospheric pressure. With present day technology another factor of 1 million is available, and even at this low pressure there are still 3.10^{13} molecules per cubic metre; very far from the ideal of empty space. In hydrogen at a few atmospheres pressure the state of affairs is roughly as represented in Fig. 3.1. The dark core represents the region of strong repulsive forces; it is not quite spherical since it corresponds to the electron distribution

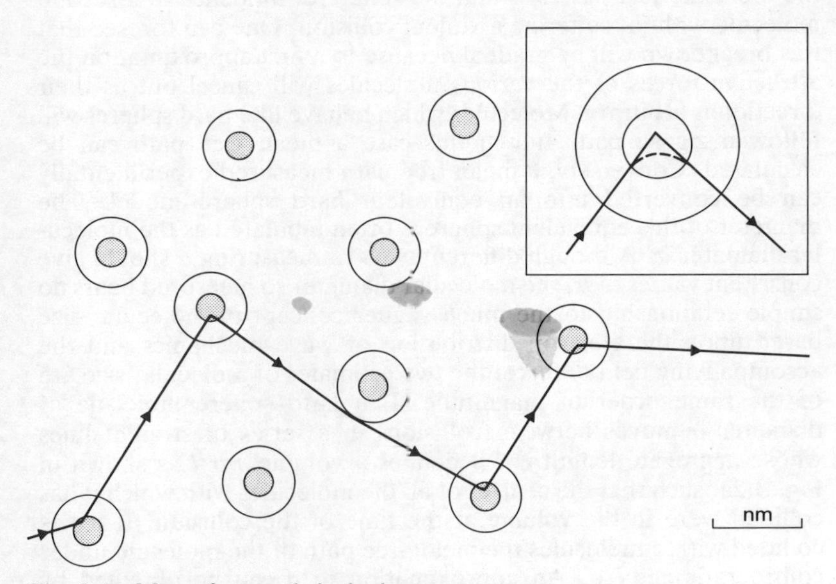

Fig. 3.1 Mean free path in hydrogen.
The point molecules are represented by an inner core (stippled) about 0.33 nm diameter inside which the molecular forces are strongly repulsive and a larger, about 0.8 nm diameter, sphere outside which the attractive forces have become negligible. Although the forces have nearly spherical symmetry the mass distribution is not spherical and is to a first approximation to be represented by two point masses (the protons) 0.074 nm apart.

round two protons. The outer sphere represents the region outside which the weak attractive forces, falling off as the inverse seventh power of the distance, can be neglected. An imaginary random path for one of the molecules is indicated.

It will be observed that for the most part it moves in a field free region and therefore in a straight line. The linear sections are joined by sharply curved sections. The mean free path approximation consists of supposing that the real paths can be replaced by the zigzag path obtained by extending the straight portion as shown in Fig. 3.1b. The motion of the molecules is thus supposed to consist of a series of 'free paths' of unequal length. The average of all the free paths of all the molecules is called the mean free path. The mean free path is a way of expressing the average interaction of the molecules with each other. The concept is almost independent of the exact law of the strong repulsive forces, but will break down as soon as there are no sections in which the path is straight owing to the influence of the much weaker attractive forces of longer range. Thus, in Fig. 3.1*a* the second leg of the zigzag is not quite straight as the molecule just passes within the sphere of influence of a second molecule without suffering a violent collision. One can foresee that this breakdown will be gradual because to a first approximation the attractive forces of the various molecules will cancel out as their direction is arbitrary. Molecules which behave like hard spheres will follow a zigzag path and in this case a mean free path can be calculated. Conversely, a mean free path measured experimentally can be converted into an equivalent hard sphere model. The diameter of this equivalent sphere is often tabulated as the molecular diameter σ. Although different ways of measuring σ should give consistent values of σ, the molecular diameter so measured bears no simple relationship to the much vaguer concept of molecular size based upon the electron distribution of wave mechanics and the accompanying field of force: the two estimates of molecular size are of the same order of magnitude. If a hard sphere molecule of diameter σ moves between collisions in a series of straight lines whose aggregate length is l it defines a volume $\pi\sigma^2 l$ as shown in Fig. 3.2a, such that the centres of all the molecules with which it has collided were in the volume at the time of the collision. If it has collided with q molecules the mean free path of the molecule under consideration is l/q. An approximation to q can be obtained by calculating the number of molecules in the critical volume $\pi\sigma^2 l$ averaged over a long time. If there are n molecules per unit volume of gas this gives at once that

$$\lambda = \frac{1}{n\pi\sigma^2}.$$

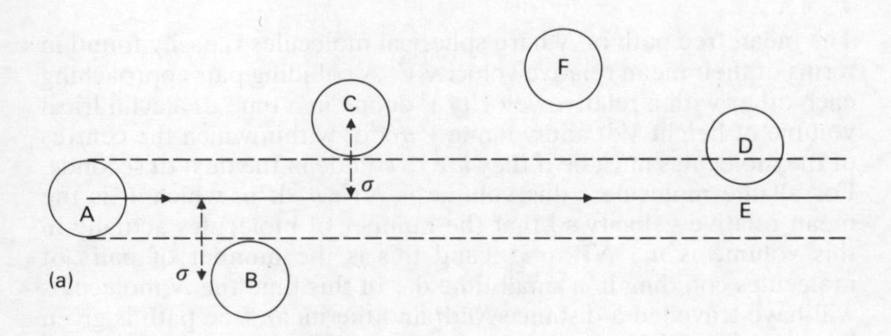

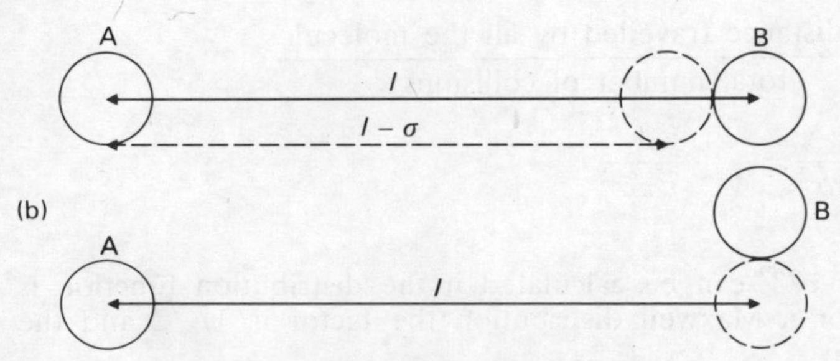

Fig. 3.2 Free path of hard spherical molecules.
(a) The molecule A represented by a hard sphere of diameter σ moves amongst stationary molecules B, C, D, F. It just avoids collision with molecules B and C whose centres lie outside the cylinder of radius σ and axis AE but collides with molecule D whose centre lies within the cylinder. If B, C, D, F are moving their motion cannot even on the average be neglected since it is their motion relative to A and not their absolute motion which determines whether they wander into the critical volume $\pi\sigma^2 l$ at the appropriate time.
(b) The distance l travelled between collisions is not a well defined quantity and lies between the extreme values, l and $l - \sigma$, as the angle between the line of centres at the time of collision varies from $0° - 90°$.

This is the mean free path of a very fast molecule which completes the length l in a time so short that no molecules have entered or left the critical volume $\pi\sigma^2 l$. The true value of λ can be calculated if the distribution function of the velocities of the molecules is known. It never exceeds

$$\frac{1}{n\pi\sigma^2} \quad \text{so that} \quad \lambda < \frac{1}{n\pi\sigma^2}$$

The mean free path of N hard spherical molecules is easily found in terms of their mean relative velocity $\bar{V}$. A colliding pair approaching each other with a relative velocity V define in a time $\mathrm{d}t$ a cylindrical volume of height $V\mathrm{d}t$ and volume $V\pi\sigma^2\mathrm{d}t$ within which the centres of the molecules must lie if they are to collide in the next $\mathrm{d}t$ seconds. For all the molecules, this volume is $N\bar{V}\pi\sigma^2\mathrm{d}t$ in which $\bar{V}$ is the mean relative velocity so that the number of molecules actually in this volume is $n \cdot N\bar{V}\pi\sigma^2\mathrm{d}t$, and this is the number of pairs of molecules colliding in a small time $\mathrm{d}t$. In this time the N molecules will have travelled a distance $N\bar{c}\mathrm{d}t$ and the mean free path is given by

$$\lambda = \frac{\text{total distance travelled by all the molecules}}{\text{total number of collisions}}$$

$$= \frac{N\bar{c}}{Nn\bar{V}\pi\sigma^2} = \frac{\bar{c}}{\bar{V}} \cdot \frac{1}{n\pi\sigma^2}$$

The factor $\bar{c}/\bar{V}$ can be calculated if the distribution function is known. For a Maxwell distribution the factor is $1/\sqrt{2}$ and the relationship

$$\lambda = \frac{1}{\sqrt{2}n\pi\sigma^2}$$

is usually used to calculate molecular diameters from measured mean free paths. If all the molecules are supposed to have the same velocity the factor is $\frac{3}{4}$ and

$$\lambda = \frac{3}{4n\pi\sigma^2}. \quad \text{(Clausius)}$$

Worked example

Find the mean value $\bar{V}$ and the mean squared value $\overline{V^2}$ of the relative velocity V of the molecules in a gas.

As it is obvious that, for an isotropic gas, the mean value of V is always zero it is usual to understand by $\bar{V}$ the mean value of all the negative values of V, i.e. relative velocity of approaching molecules and to omit the value of V for receding molecules. As this is a problem of considerable complexity it is best to start by considering the collisions between molecules of velocities c_1 and c_2 ($c_1 > c_2$). Although the molecules are moving in three dimensions the problem is really a problem in two dimensions since the velocities of molecules about to collide must be in a plane and form a triangle as shown in Fig. 3.3.

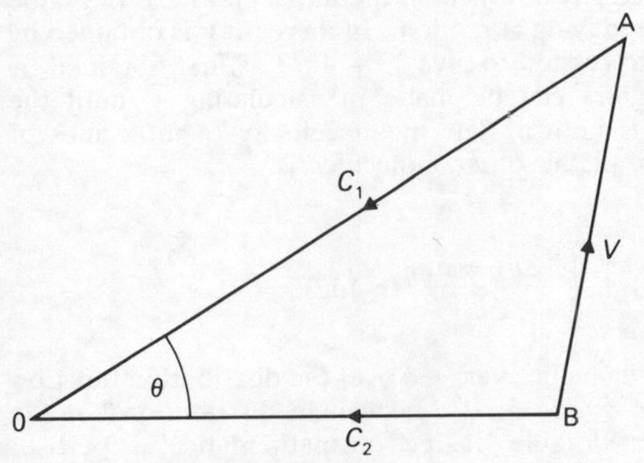

Fig. 3.3 Relative velocity.
Molecules at A and B will collide at 0. If $c_1 > c_2$, an observer on A will observe B approaching with a velocity V whose limits lie between $c_1 - c_2$ when $\theta = 0$ and $c_1 + c_2$ if $\theta = \pi$.

Thus $V_{12}^2 = c_1^2 + c_2^2 - 2c_1c_2 \cos\theta$ and

$$\overline{V_{12}^2} = c_1^2 + c_2^c - 2c_1c_2 \overline{\cos\theta} = c_1^2 + c_2^2$$

since $\cos\theta = 0$ when averaged over all directions in space. As this result holds for all pairs of molecules one has at once that $\overline{V^2}$ for a gas $= \overline{c_1^2} + \overline{c_2^2} = 2\overline{c^2}$ since c_1 and c_2 have been chosen at random from pairs of colliding molecules. This tacitly assumes that there is no correlation between velocity and direction of motion. The mean value of V_{12} can be calculated by making use of the fact that the number of molecules with velocity c_2 in a direction making an angle θ with c_1 is proportional to the solid angle $d\omega$ between θ and $\theta + d\theta = 2\pi \sin\theta \, d\theta$

$$\text{Thus, } \overline{V_{12}} = \int_0^\pi \frac{\sqrt{c_1^2 + c_2^2 - 2c_1c_2\cos\theta} \cdot \sin\theta \, d\theta}{\int_0^\pi \sin\theta \, d\theta}$$

$$= \frac{1}{2}\int_0^\pi \sqrt{c_1^2 + c_2^2 - 2c_1c_2\cos\theta} \cdot \sin\theta \, d\theta$$

$$\text{and} \quad \overline{V_{12}} = \frac{3c_1^2 + c_2^2}{3c_1}$$

This surprisingly unsymmetrical form is due to the assumption that $c_1 > c_2$ and that V must always correspond to approaching molecules.

If $c_1 < c_2$ then $\bar{V} = \dfrac{3c_2^2 + c_1^2}{3c^2}$

The further averaging over all possible pairs c_1, c_2 in the gas is only easily carried out for the simplified gas in which all the molecules have the same velocity c but are otherwise moving at random. In this case $\bar{V}$ is obtained by putting $c_1 = c_2$ in either expression to give $\bar{V} = 4c/3$ as first obtained by Clausius. No further progress can be made in calculating $\bar{V}$ until the distribution function $f(c)$ is known. The mean velocity of molecules of velocity c_1 with respect to all the other molecules is

$$\bar{V}_{c_1} = \int_0^{c_1} \frac{3c_1^2 + c_2^2}{3c_1} f(c_2)\mathrm{d}c_2 + \int_{c_1}^{\infty} \frac{3c_2^2 + c_1^2}{3c_2} f(c_2)\mathrm{d}c_2$$

To obtain $\bar{V}$ the result must again be averaged over the distribution function $f(c_1)$. The result is that $\bar{V} = \sqrt{2}\bar{c}$ for a Maxwellian distribution but the integration requires a considerable degree of mathematical expertise. Maxwell quoted the factor $\sqrt{2}$ but never published the details of the calculation. Some assistance is given in Kennard (1938, p. 169). For an alternative method see Jeans (1925 section 33).

3.3 Probability of a path l

The distribution of the actual path lengths about the mean path is not symmetrical but can readily be calculated. Imagine a beam of N particles per second per unit area launched into the gas parallel to the x axis. Consider a section of gas between x and $x + \Delta x$ and let the probability of a collision in the section be $\mu\Delta x$ and the probability that a molecule reaches x without collision be $F(x)$. To reach $x + \Delta x$ without collision for which the probability is $F(x) + F'(x)\Delta x$ a molecule must first arrive at x and then proceed a further Δx without collision for which the probability is $(1 - \mu\Delta x)$ since the sum of the probabilities of the only two possible events, namely free passage or collision must be unity from the definition of probability, (see section 6.2).

Thus $F(x) + F'(x)\Delta x = F(x)(1 - \mu\Delta x)$ and $F'(x) = -\mu F(x)$ so that $F(x) = A\mathrm{e}^{-\mu x}$

Since none of the particles have made a collision at the origin $A = N$. The number of particles making a collision between x and $x + \Delta x$ is $N\mathrm{e}^{-\mu x}\,\mu\Delta x$ so that the total path of all the particles is given by

$$N\mu \int_0^\infty x e^{-\mu x} dx = \frac{N}{\mu} = N\lambda$$

from the definition of λ the mean free path. Thus, the probability that a molecule has a free path greater than x is $e^{-x/\lambda}$. The mean value of

$$l^2 = \overline{l^2} = \int_0^\infty \mu x^2 e^{-\mu x} dx = 2\lambda^2.$$

It should be noted that these results are only correct if the conditions implied by using the arguments of the integral calculus can be justified from the physical conditions of the gas. The small quantity Δx cannot rationally be reduced below a molecular diameter. It is however part of the argument that Δx is so small that only a single collision can occur in crossing the elementary section. This limitation effectively means that the mean free path formulae will start to break down at a few atmospheres pressure. In deriving these results it has also been assumed that all the target molecules are at rest. If they are moving this can be taken into account by always working in terms of the relative velocity of the colliding molecules. If, as is usual in the kinetic theory, calculations are carried out in terms of the absolute velocities c of the molecules, then the mean free path of a molecule of velocity c in a gas of moving molecules is a function of its velocity. The limits between which λ_c lies are clearly zero (a molecule does not avoid collisions by being at rest) and $1/n\pi\sigma^2 = \sqrt{2}\lambda$ the value for a molecule moving so fast that all the others may be considered to be stationary. The intermediate values have been calculated by Meyer and are shown in Table 3.1.

Table 3.1

$c/\bar{c}$	0	0.25	0.25	0.627	0.886	1.0	1.535	2.0	4.0	6.0	∞
λ_c/λ	0	0.34	0.64	0.72	0.96	1.03	1.21	1.29	1.38	1.40	1.41

3.4 Molecular collision rate

In a long finite time t, N molecules travel a distance $N\bar{c}t$ (by definition of $\bar{c}$) and so make $N\bar{c}t/\lambda$ collisions, so that the collision rate for N molecules is $N\bar{c}/2\lambda$ per second since each collision has been counted twice, once as the incident molecule and once as the struck molecule. It is instructive to calculate the collision rate in another way. Consider the molecules which collide in Δt seconds. Each pair defines a volume $\pi\sigma^2 V\Delta t$ in which lie two molecules. Thus, for N molecules the volume is $N\pi\sigma^2\bar{V}\Delta t$ which contains $Nn\pi\sigma^2\bar{V}\Delta t = 2$ molecules. The

$$\text{collision rate} = \frac{1}{\Delta t} = \frac{Nn\pi\sigma^2\bar{V}}{2} = \frac{N\sqrt{2}n\pi\sigma^2\bar{c}}{2} = \frac{N\bar{c}}{2\lambda}$$

since for a Maxwell distribution $\bar{V} = \sqrt{2}\bar{c}$. Each collision provides two molecules whose directions are on the average uniformly distributed in space. Thus, an elementary volume $\mathrm{d}\tau$ of a gas containing n molecules per unit volume provides

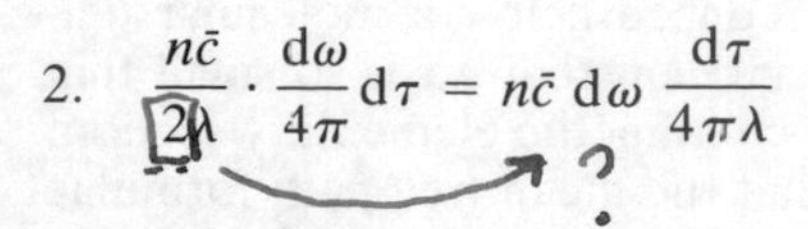

$$2.\ \frac{n\bar{c}}{2\lambda}\cdot\frac{\mathrm{d}\omega}{4\pi}\,\mathrm{d}\tau = n\bar{c}\,\mathrm{d}\omega\,\frac{\mathrm{d}\tau}{4\pi\lambda}$$

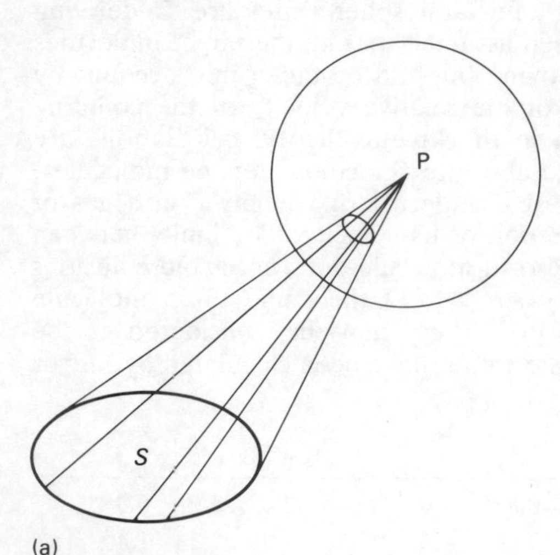

Fig. 3.4 Solid angle.
(a) To find the solid angle ω subtended by S at P, the points on the perimeter of S are joined to P to form a cone. This cone is the solid angle and its measure in steradians is the magnitude of the area defined by the cone when it cuts a sphere of unit radius centred on P. This definition ensures that if P is a source emitting N particles per second uniformly in all directions the flux of particles through S is $N\omega/4\pi$.

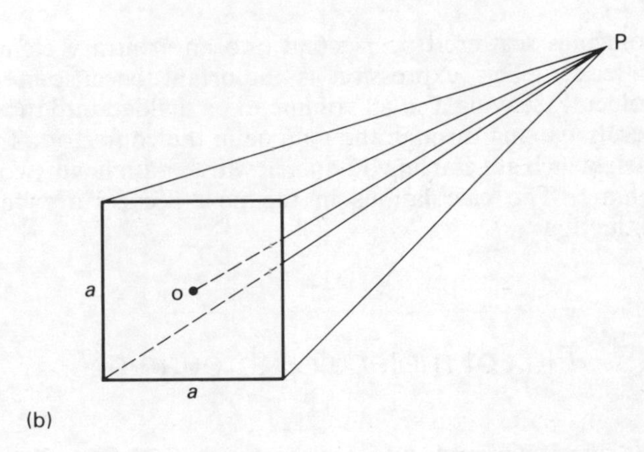

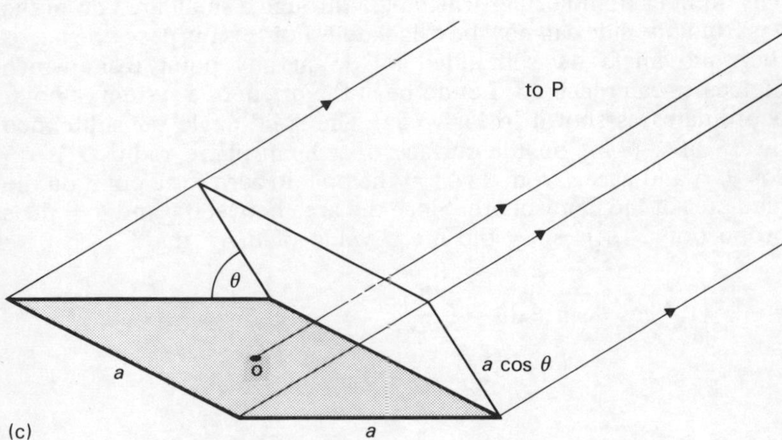

(b) To calculate ω, S must be split up into small squares of side a. If as in the diagram the square is normal to OP it subtends an elementary solid angle $d\omega = a^2/OP^2$ at P. The side of the square must be small enough for the angle a/OP to be negligible. This can always be done because

$$\cos\left(\frac{a}{OP}\right) = 1 - \frac{1}{2}\left(\frac{a}{OP}\right)^2 + \cdots$$

and contains no term in a/OP.

(c) If the normal to the elementary area makes an angle θ with OP one sees from the diagram that $d\omega$ is $a^2 \cos\theta/OP^2$. This maintains the flux relationship $N d\omega/4\pi$. The value of ω for a finite area is then obtained by summation so that $\omega = \Sigma d\omega$. In general these summations are difficult. A straightforward treatment is given by Richardson (1928) and Milne (1929).

molecules scattered per second into an arbitrary element of solid angle $d\omega$. This expression is important because it enables the molecules leaving a small volume to be divided into those which are merely passing through the volume in the course of a free path and those which are starting off on a new free path having collided in the volume. The calculations in the next section are based on this distinction.

3.5 Flux of molecules through dS

The number of molecules which pass through a small area dS in the gas from one side can now be calculated. For this purpose one needs the solid angle $d\omega$ subtended by dS at any point from which molecules can reach dS. Let dS be at the origin of a system of polar coordinates as shown in Fig. 3.5a. The solid angle $d\omega$ subtended by dS at a point on the surface of a hemisphere radius r is $dS \cos\theta/r^2$ and varies from dS/r^2 at the pole to zero for a point on the diameter of the hemisphere. Since the area between θ and $\theta + d\theta$ is proportional to $r^2 \sin\theta$ the mean value of $d\omega = \overline{d\omega}$

$$= \frac{dS}{r^2}\int_0^{\pi/2} \cos\theta \sin\theta \, d\theta = \frac{1}{2}\frac{dS}{r^2}.$$

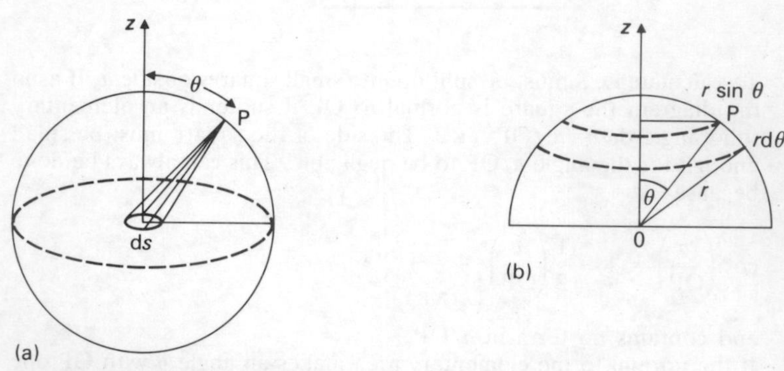

Fig. 3.5 Solid angle in polar coordinates.
(a) The solid angle $d\omega$ subtended by the small area dS at the origin by a point P is by definition $dS \cos\theta/r^2$.
(b) dS subtends the same solid angle $dS \cos\theta/r^2$ at all the points in a spherical belt lying between θ and $\theta + d\theta$ whose area is $2\pi r^2 \sin\theta \, d\theta$.

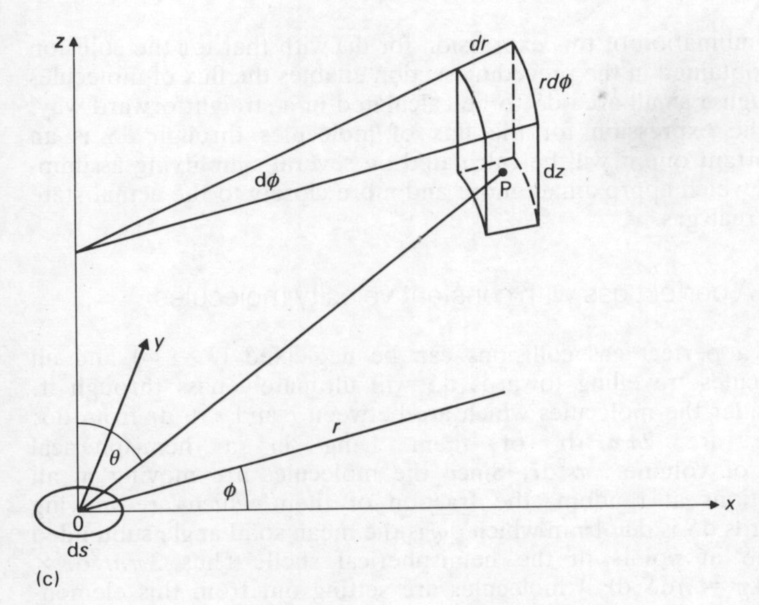

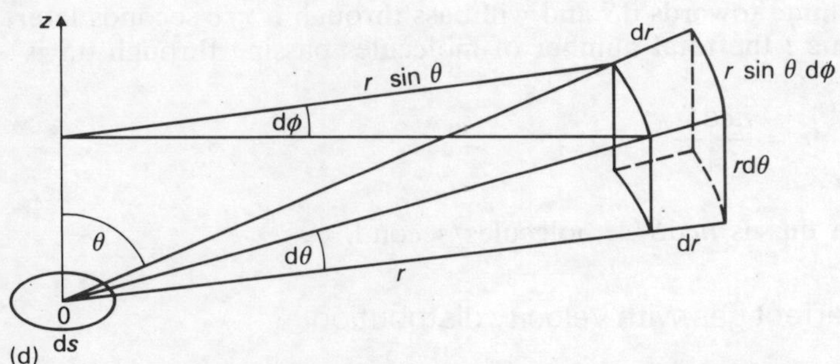

(c) If the position of P is described in cylindrical coordinates r, ϕ, z, the element of volume $\mathrm{d}\tau$ containing P is a rectangular block $\mathrm{d}r$ thick, $r\mathrm{d}\phi$ wide, and $\mathrm{d}z$ deep, of volume $\mathrm{d}\tau = r\mathrm{d}r\,\mathrm{d}\phi\,\mathrm{d}z$. The small area $\mathrm{d}S$ at the origin subtends the solid angle

$$\frac{\mathrm{d}S\cos\theta}{\mathrm{OP}^2} = \frac{z\mathrm{d}S}{(r^2+z^2)^{3/2}} \quad \text{at all the points in } \mathrm{d}\tau.$$

(d) If the position of P is described in polar coordinates, r, θ, ϕ, the element of volume $\mathrm{d}\tau$ is a small rectangular block, $\mathrm{d}r$ thick, $r\sin\theta\,\mathrm{d}\phi$ wide, and $r\mathrm{d}\theta$ deep of volume $r^2\sin\theta\,\mathrm{d}r\,\mathrm{d}\theta\,\mathrm{d}\phi$. The area $\mathrm{d}S$ at the origin subtends a solid angle

$$\frac{\mathrm{d}S\cos\theta}{r^2} = \frac{z\mathrm{d}S}{r^3} \quad \text{at all points in } \mathrm{d}\tau.$$

A combination of this expression for $\overline{d\omega}$ with that for the collision rate obtained in the preceding section enables the flux of molecules through a small area dS to be calculated in a straightforward way. As the expression for the flux of molecules through dS is an important one it will be calculated on several simplifying assumptions which approximate more and more closely to the actual state in a real gas.

(a) A perfect gas with constant velocity molecules

In a perfect gas collisions can be neglected ($\lambda \rightarrow \infty$) and all molecules travelling towards dS will ultimately pass through it. Consider the molecules which are between r and $r + dr$ from dS: there are $2\pi nr^2dr$ of them lying in a hemispherical shell of volume $2\pi r^2dr$. Since the molecules are moving in all directions at random, the fraction of them which are moving towards dS is $\overline{d\omega}/4\pi$ in which $\overline{d\omega}$ is the mean solid angle subtended by dS at points in the hemispherical shell. Thus $2\pi nr^2dr \times \overline{d\omega}/4\pi = ndS\ dr/4$ molecules are setting out from this elementary volume towards dS and will pass through it r/c seconds later. In a time t the total number of molecules passing through dS is

$$\frac{ndS}{4}\int_0^{ct} dr = \frac{ncdSt}{4}$$

and the flux is $ncdS/4$ molecules/second.

(b) Perfect gas with velocity distribution

If the molecules of the gas have different velocities so that n_1 have velocity c_1, n_2 velocity c_2, etc., each group will behave independently and the total flux is $dS(n_1c_1 + n_2c_2 \ldots)/4 = n\bar{c}dS/4$ in which $n = \Sigma n_1$ and $\bar{c}$ the mean velocity is

$$\frac{n_1c_1 + n_2c_2 \ldots}{n_1 + n_2 \ldots}$$

by definition.

(c) Nearly perfect gas with a constant mean free path

The molecules in the elementary volume $d\tau$ of Fig. 3.5d are divided into two parts; those merely passing through it and those colliding in the volume and thus ending one free path and commencing another. Each volume dτ thus acts as a source emitting molecules uniformly in all directions at a rate equal to the rate at which molecules are colliding in it. A fraction of these molecules set off towards dS. Not all the molecules which set off towards dS from a particular volume dτ will reach it since some of them will collide with other molecules on the way but the fraction which reach dS can be calculated if the mean free path and so the attenuation coefficient is known. In this way the rate at which molecules are reaching dS after collision in the element of volume dτ can be calculated and by summing the contributions from all the elementary volumes the rate at which molecules reach dS from the whole of the gas is obtained. In a steady state for every molecule with velocity c which enters into a collision another molecule with velocity c must be created on the average; otherwise collisions would be changing the distribution function. If this Maxwellian distribution is written as $\mathrm{d}n = nf(c)\mathrm{d}c$ the flux of molecules of velocity c through dS from the element of volume $2\pi r^2\mathrm{d}r$ is

$$2\pi r^2\mathrm{d}r \cdot \frac{\bar{c}\mathrm{d}n}{\lambda} \cdot \overline{\frac{\mathrm{d}\omega}{4\pi}} \cdot \mathrm{e}^{-r/\lambda}$$

element of volume	rate of production of molecules moving towards dS	attenuation due to molecular collisions

$$= \frac{\bar{c}\mathrm{d}n}{4\lambda}\,\mathrm{e}^{-r/\lambda}\mathrm{d}r\mathrm{d}S$$

The flux of molecules, velocity c, from the whole of the gas on one side of dS is

$$\frac{\bar{c}\mathrm{d}n\mathrm{d}S}{4\lambda}\int_0^\infty \mathrm{e}^{-r/\lambda}\mathrm{d}r = \frac{\bar{c}\mathrm{d}n\mathrm{d}S}{4\lambda}\left[-\lambda\mathrm{e}^{-r/\lambda}\right]_0^\infty = \frac{\bar{c}\mathrm{d}n\mathrm{d}S}{4}$$

and the total number of all velocities is

$$\frac{\bar{c}\mathrm{d}S}{4}\int_0^n \mathrm{d}n = \frac{n\bar{c}\mathrm{d}S}{4}$$

This expression for the flux is not changed if the variation of λ with velocity is taken into account since one can reasonably assume from their definitions that

$$\int_0^n \lambda_c \mathrm{d}n = n\lambda \qquad \text{and the flux through } \mathrm{d}S \text{ remains } n\bar{c}\mathrm{d}S/4.$$

This is an important result. It is independent of λ and so of the molecular size: so to a high degree of approximation it is true for real as well as ideal gases. It has important applications which will be discussed at the end of the chapter.

If $\mathrm{d}S$ instead of being an imaginary area in the body of the gas is an element of the wall of the containing vessel, the reversal of the momentum normal to $\mathrm{d}S$ brought up per second by the $n\bar{c}\mathrm{d}S/4$ molecules is the origin of the pressure exerted by the gas. Thus, the Boyle's law relationship $p = \frac{1}{3}nm\overline{c^2}$ already obtained is not disturbed by introducing molecular collisions as long as the density is low enough for the mean free path approximation, a conclusion confirmed by a direct calculation.

Worked example

Show that the flux of momentum through

$$\mathrm{d}S = \frac{mn\overline{c^2}}{6}.$$

Let the velocity distribution function be $\mathrm{d}n = nf(c)\mathrm{d}c$. Consider the molecules of velocity c created by collision in an elementary volume $\mathrm{d}\tau$ at the point r, θ, ϕ. On arriving at $\mathrm{d}S$ each molecule contributes a normal component of momentum $= mc \cos\theta$. The flux of momentum contributed by these molecules is thus the (solid angle subtended by $\mathrm{d}S$) $\times$ (rate of creation by collision per unit solid angle) $\times$ $mc\cos\theta$ $\times$ (chance of arriving at $\mathrm{d}S$ without further collision) $= \mathrm{d}S\mathrm{d}nmc^2\cos^2\theta\,\mathrm{d}\tau/4\pi\lambda \cdot \mathrm{e}^{-r/\lambda}$.

The total flux of momentum due to molecules of velocity c is

$$\frac{\mathrm{d}S\mathrm{d}nmc^2}{4\pi\lambda}\underbrace{\int_0^\infty \mathrm{e}^{-r/\lambda}\mathrm{d}r}_{\lambda} \times \underbrace{\int_0^{\pi/2}\cos^2\theta\sin\theta\,\mathrm{d}\theta}_{\frac{1}{3}} \times \underbrace{\int_0^{2\pi}\mathrm{d}\phi}_{2\pi} = \frac{mc^2\mathrm{d}n\mathrm{d}S}{6}$$

and the total flux of momentum due to all molecules on one side of $\mathrm{d}S$ is

$$\frac{m\mathrm{d}S}{6}\int c^2\mathrm{d}n = \frac{mn\mathrm{d}S}{6}\int_0^\infty c^2 f(c)\mathrm{d}c = \frac{mn\overline{c^2}\mathrm{d}S}{6}$$

Worked example

Find the flux of energy falling on one side of an area in a perfect gas all of whose molecules have the same velocity c. If as the result of collisions the gas acquires a Maxwellian distribution, show that the flux of energy goes up by a factor 4/3.
For the gas all of whose molecules have the same velocity c the flux of molecules is $nc/4$. Each molecule has energy $\frac{1}{2}mc^2$ so that the flux of energy is

$$\frac{nc}{4} \cdot \frac{1}{2} mc^2 = \frac{nc}{4} \cdot \frac{3}{2} kT.$$

If dn molecules have a velocity between c and c + dc they will contribute

$$\frac{\mathrm{d}n \cdot c}{4} \cdot \frac{1}{2} mc^2 = \frac{m\mathrm{d}n}{8} c^3$$

to the energy flux and the total flux will be $nm/8\Sigma c^3 \mathrm{d}n = nm/8$ (mean value of c^3) for the gas. For a gas with a Maxwellian velocity distribution $\overline{c^3}$ is found from table 8.1 to be

$$\overline{c^3} = 8\sqrt{\frac{2}{\pi}} \times \left(\frac{kT}{m}\right)^{3/2} = \frac{4kT\bar{c}}{m}$$

and the energy flux is

$$\frac{nm}{8}\overline{c^3} = \frac{n\bar{c}}{4} \cdot 2kT$$

and the ratio of the two fluxes is 4/3 as stated.

An estimate of the density at which the mean free path approximation will break down can be made, using a hard sphere model, in the following way. The gas laws will be obeyed as long as the number of molecules striking an element of surface dS is independent of the molecular forces, and is given by $n\bar{c}/4$ dS. This expression does not depend upon λ because of the cancellation of λ from the collision rate with a λ which comes from integrating $\mathrm{e}^{-r/\lambda}$ in the free path formula. These two expressions, while correct at low pressure when $\lambda \gg \sigma$ the molecular diameter, behave differently as the density increases. The value of λ in the collision formula is not very well defined. If the free path between centres of a pair of molecules is l as shown in Fig. 3.2b the actual distance travelled by the molecules before they collide lies between l and $l - \sigma$. The average value is $l - 2\sigma/3$ and this should be used instead of l in calculating the collision rate. So

$$\frac{\text{p(real gas)}}{\text{p(ideal gas)}} = \frac{\text{(collision rate) corrected}}{\text{(collision rate) uncorrected}} = \frac{l}{l - \dfrac{2\sigma}{3}}$$

$$= 1 + \frac{2\sigma}{3l}$$

Using $ln\pi\sigma^2 = 1$ to elminate l, one readily finds that $PV_{(\text{real})} = PV_{(\text{perfect gas})} + Pb$. The effect of molecular size calculated in this way is thus the same as the estimate given by the method of the excluded volume.

3.6 Diffusion

If a single foreign molecule, say hydrogen, is introduced into a large volume of gas, say xenon, it will wander away from the point at which it was released. This motion may be described in terms of a mean free path λ and a mean velocity $\bar{c}$. Let R be the distance from the origin after time t. Then from the diagram (Fig. 3.6) it follows that the change in R^2 produced by one free path l is given by $(R = \Delta R)^2 = R^2 + l^2 - 2Rl \cos\theta$. Averaged over a very large number of trials the last term vanishes and $\overline{(R + \Delta R)^2} - R^2 = \overline{\Delta R^2} = \bar{l}^2$ per trial. This change in R^2 takes place in a

$$\Delta t = \frac{\lambda}{\bar{c}}$$

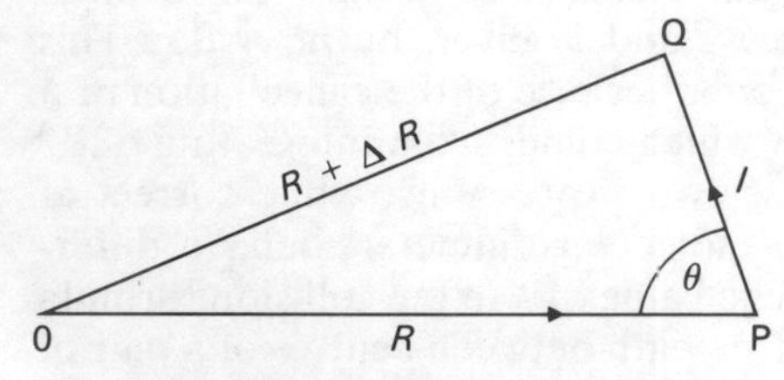

Fig. 3.6 Diffusion.
The molecule at P a distance R from the origin collides and travels a free path l to Q where it makes a second collision.

so that

$$\overline{\Delta R^2} = \frac{\bar{l}^2\bar{c}\Delta t}{\lambda} = 2\lambda\bar{c}\Delta t$$

since the mean value of l^2 is 2λ as shown in section 3.3. Thus on the average the mean value of R^2, the mean squared distance the molecule has wandered from the origin, increases linearly with the time and

$$\overline{R^2} = 2\lambda\bar{c}t.$$

Since $R^2 = x^2 + y^2 + z^2$ and the value of $\overline{R^2}$ is quite independent of the cartesian system in which it is represented it follows that

$$\overline{x^2} = \overline{y^2} = \overline{z^2} = \frac{\overline{R^2}}{3} = \frac{2\lambda\bar{c}t}{3}$$

The assumption that $\overline{\cos\theta} = 0$ on which this simple derivation is based requires comment. Formally, it is correct since as stated $\overline{\cos\theta}$ is an average over a large number of trials of the same experiment in which the initial value of $\cos\theta$ is arbitrary and so $\overline{\cos\theta} = 0$ by definition. However, in the course of the argument the meaning of $\cos\theta$ is allowed to change and the average is taken over the successive values of $\cos\theta$ which occur as the hydrogen molecule wanders about in the xenon. For the example chosen $\overline{\cos\theta}$ determined in this way is still very nearly zero (≈ 0.01) because the mass of the hydrogen molecule (2) is so much smaller than the mass of the xenon (132). If one had been considering the diffusion of a single xenon atom in hydrogen $\overline{\cos\theta}$ would not have been zero because the heavy xenon atom always comes out of any collision with a hydrogen molecule with some of its momentum in the initial direction and the successive values of $\cos\theta$ are not unrelated to each other. However, after many (q) collisions the bias in favour of the initial direction is smeared out and θ again becomes random. It is evident that the true average motion of the diffusing atom is not very different from the motion of an imaginary atom for which $\overline{\cos\theta} = 0$, with a fictitious mean free path $l' = \lambda$ chosen so that after q collisions the imaginary atom has wandered on the average the same distance as the real atom. If λ is interpreted in this sense the simple theory can be used even when the mass ratio is not small enough to justify putting $\overline{\cos\theta} = 0$ when averaged over the first q collisions.

It is instructive to calculate the actual rate of diffusion of mercury atoms as this is one of the factors involved in assessing the danger of mercury poisoning in laboratories. In air at room temperature $\lambda \approx 47$ nm, $\bar{c} = 1.7 \times 10^2$ m/s and if $R = 3$ m is taken to represent the dimensions of a small room, the time t for the mercury vapour from a drop spilled on the floor to diffuse throughout the room will be given by

$$t = \frac{R^2}{2\lambda\bar{c}} = \frac{9}{2 \times 4.7 \times 10^{-8} \times 1.7 \times 10^2} 5.6 \times 10^5 \text{ s}$$

$$\approx 6\tfrac{1}{2} \text{ days}$$

Thus, only in laboratories in which the windows are very well fitting and never opened is there much danger of mercury vapour poisoning from this cause. While the preceding calculation shows the need for reasonable ventilation in laboratories it only deals with one aspect of the problem. Mercury vapour is a dangerous substance to which no one should be innocently exposed. The tentative safe limit set by the Air Pollution Technical Information Centre in 1971 is 1 μg/m^3. The present limit of detection is about one fifth of this (Hadeishi 1975).

In a helium discharge tube in which the gaseous pressure is about 1/100th atm the mean free path is $\sim 10^{-5}$ m and a mercury atom will be able to diffuse 1 cm and reach the walls is about 2.5×10^{-2} seconds. It should be noted that in order to diffuse a direct distance $n\lambda$ it will take a time $= (n\lambda)^2/2\lambda\bar{c} = n^2\lambda/2\bar{c}$. In this time it will make $n^2\lambda/2\bar{c} \times \bar{c}/\lambda = n^2/2$ collisions. This important result explains why the electrical properties of helium discharge tubes depend so critically on the purity of the gas. During an electrical discharge in helium many of the helium atoms are excited into metastable states by electron collision, with an excitation energy of 19.73 eV. These cannot radiate but can lose their energy by colliding with a mercury atom with an ionisation potential of only 10.39 eV. In reaching the walls which under ordinary discharge conditions are some hundred mean free paths from the centre of the discharge tube the metastable atoms make some tens of thousands of collisions. So they have a high probability of colliding with and ionising a mercury atom even though this is present to only one part in ten thousand and the gas is 99.99 per cent pure. It is well known that in these circumstances the excited gas shows a strong mercury spectrum and weak helium lines in spite of the very great preponderance of helium in the discharge gas. The same argument applies to the sodium lamps used in street lighting. These are easily observed to show the neon spectrum on starting up and only the sodium spectrum when they have warmed up enough for the vapour pressure of sodium to be appreciable.

3.7 Diffusion coefficient

Diffusion problems are usually treated mathematically in terms of a diffusion coefficient D defined according to Fick's law so that the flux of diffusing molecules is equal to $D \times -$ (concentration gradient). A relationship between D and the molecular constants of the gas can be obtained by the following primitive argument. Consider a gas, say a very small amount of mercury vapour, diffusing along a straight cylindrical tube full of nitrogen. If the mercury vapour has a concentration gradient k the number of molecules per m^3 will be $n_0 - kx$ in which x is measured along the axis of the tube as in Fig. 3.7. Suppose in time t the mercury atoms have diffused through a distance $\pm l$ along the x axis. Then half the atoms in section A will have diffused into section B and half the atoms in section B will have

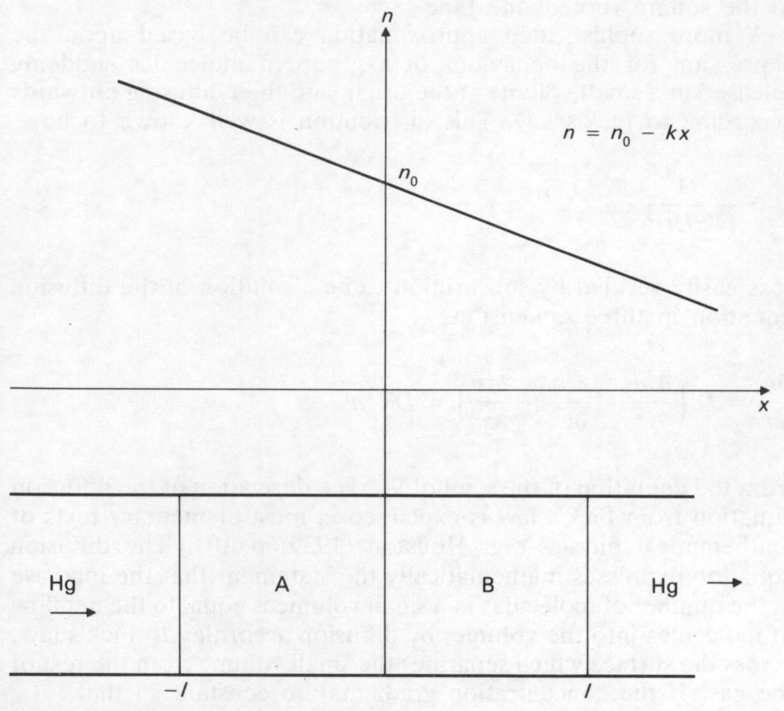

Fig. 3.7 Diffusion in a linear gradient.
The mercury vapour is diffusing along the tube in the direction A to B. The upper graph shows how the concentration of the mercury atoms (molecules/per unit volume) varies along the length of the tube.

wandered into Section A. The flux of molecules is thus

$$\frac{1}{2t}\text{(Hg molecules in Section A–Hg molecules in Section B)}$$

$$= \frac{1}{2t}\left(n_0 l + \frac{kl^2}{2} - \left(n_0 l - \frac{kl^2}{2}\right)\right) = \frac{kl^2}{2t}$$

But l^2 is the mean squared distance the molecules have diffused along the x axis in time t, which is $2\lambda\bar{c}/3t$. Thus, the flux of diffusing mercury atoms is $k/2 \cdot 2\lambda\bar{c}/3 = kD$ and D is $\lambda\bar{c}/3$. While this simple argument is open to the criticism that it assumes that all the molecules diffuse the same distance it has the merit of giving a picture of the way in which random molecular wandering produces a net flux proportional to the concentration gradient of the diffusing matter although the distance diffused by a molecule is proportional to the square root of the time.

A more sophisticated approximation can be based upon the expression for the behaviour of a group of molecules suddenly released in a small volume at the origin and then diffusing outwards according to Fick's law. This distribution is well known to be

$$n = \frac{1}{(4\pi Dt)^{3/2}}\,\mathrm{e}^{-r^2/4Dt}$$

It is easily verified by substitution to be a solution of the diffusion equation in three dimensions

$$\frac{\partial n}{\partial t} = D\left(\frac{\partial^2 n}{\partial x^2} + \frac{\partial^2 n}{\partial y^2} + \frac{\partial^2 n}{\partial z^2}\right) = D\nabla^2 n$$

from the definition of the symbol ∇^2. The derivation of the diffusion equation from Fick's law is explained in most elementary texts of mathematical physics e.g. Houston (1929) p. 101. The diffusion equation expresses mathematically the statement that the increase in the number of molecules in a small volume is equal to the net flow of molecules into the volume, by diffusion according to Fick's law, across the surface which separates the small volume from the rest of the gas. If the concentration gradients are constant so that

$$\frac{\partial}{\partial x}\left(\frac{\partial n}{\partial x}\right) = \frac{\partial^2 n}{\partial x^2} = 0 \text{ etc}, \quad \frac{\partial n}{\partial t} = 0,$$

there is a steady state and all the molecules diffusing in from one side of the small volume diffuse out again on the opposite side. The mean value of $\overline{r^2} = \overline{R^2}$ for the molecules in the diffusing group is

$$\frac{\int_0^\infty 4\pi n r^4 \mathrm{d}r}{\int_0^\infty 4\pi n r^2 \mathrm{d}r} = 6Dt$$

as can be found by using the table of integrals in Table 8.1. Equating the value of $\overline{R^2}$ found in this way with the value $2\lambda \bar{c} t$ found in section 3.3 for the wandering of a simple molecule gives $6Dt = 2\lambda \bar{c} t$ and $D = \frac{1}{3}\lambda \bar{c}$.

A comparison between this approximate result and experiment shows that the corrected value of λ which occurs in it is about 30 per cent higher than the uncorrected kinetic theory value when the masses of the two types of molecules are nearly the same.

A definite expression for the diffusion coefficient has been obtained because by implication the discussion has been confined to the random wandering of a single molecule: in these circumstances the mean free path λ and the mean velocity $\bar{c}$ have a definite meaning. As soon as one considers the diffusion of one gas (A) into another gas (B) the problem becomes more complicated. There are now three types of collision AA, AB, and BB, three mean free paths λ_{AA}, λ_{AB}, λ_{BB} and two mean velocities $\bar{c}_A$ and $\bar{c}_B$. While in general one can hope that a diffusion equation using some hybrid diffusion coefficient $D = f(\lambda_{AA}, \lambda_{AB}, \lambda_{BB}, \bar{c}_A, \bar{c}_B)$ will be obeyed no satisfactory progress can be made without introducing different mathematical methods. Nevertheless diffusion is an important physical process which controls the rate of many chemical changes especially those carried out on a large scale. It has been extensively studied not only for this reason but also for its academic interest. It is a field in which the ability to make accurate calculations is highly desirable because experimental work on a large scale is both expensive and time consuming. This need for detailed understanding is particularly great in designing plant for isotope separation because large plants have to be designed using data obtained from very small scale laboratory work. The elementary theory can be applied with confidence however to the diffusion of neutrons through matter since neutron-neutron collisions can legimately be neglected and the condition that the linear part of the mean free path is long compared with that part in which the 'collision' is actively in progress is more amply fulfilled than it usually is with gases in laboratory apparatus. Unfortunately, calculations on neutron diffusion are often complicated by absorption, an uncertain velocity spectrum and a mean free path which is a rapidly varying function of the velocity.

3.8 Neutron diffusion

Neutrons having no electric charge do not interact with the electron cloud which surrounds an atomic nucleus and gaseous matter presents itself to neutrons as a collection of nearly spherical nuclei distributed at random in empty space. The diameter of these spheres ($\sigma \sim 10^{-14}$ m) is so small that the conditions for a mean free path treatment ($\sigma \ll$ distance between the particles) is easily fulfilled even for liquids and solids. The problem of neutron diffusion in gases hardly arises: to neutrons most gases look like 'a good vacuum' with the path between collisions very much greater than the size of the containing vessel. The walls which contain the gas do not contain the neutrons which leave the system without interacting with the gas. A simple mean free path treatment of neutron diffusion in liquids and solids cannot usually be justified because a classical treatment of the neutron collisions is quite wrong and a full quantum treatment is necessary. There is however a class of substances known as moderators, in which the diffusion of thermal neutrons ($\frac{1}{2}m\bar{c}^2 \approx 3kT/2$) can be treated in terms of a diffusion coefficient $D = \lambda\bar{c}/3$. The moderator containing N nuclei per unit volume is said to moderate high energy incoming neutrons; they are slowed down by collisions until they approach a Maxwellian velocity distribution and have the same temperature as the moderating medium. It is the subsequent diffusion of the thermal neutrons which can be simply treated. As the neutrons diffuse through the moderator two independent processes are taking place. Firstly, the neutrons collide with the atomic nuclei and are scattered as if they had collided elastically with a hard sphere of diameter σ_t. This scattering process can be described in terms of a mean free path

$$\lambda_{\text{transport}} = \lambda_t = \frac{1}{N\pi\sigma_t^2}$$

and a corresponding diffusion coefficient $D = \lambda_t\bar{c}/3$. The nuclei are so nearly at rest that there is no Maxwellian factor $1/\sqrt{2}$.

Secondly, in a small proportion of collisions the neutron is captured and removed from the free neutron system. The rate of capture can be described in terms of a mean free path

$$\lambda_a = \frac{1}{N\pi\sigma_a^2}$$

The mean free path λ_a is the mean distance the thermalised neutron travels before capture and the mean time to capture is $\lambda_a/\bar{c}$. While the distance travelled by the neutron is λ_a this is not motion in a straight line and the neutron usually makes many elastic collisions before it is captured. The simplification which makes this particular problem amenable to a mean free path treatment is that while σ_t is a constant characteristic of the moderating medium, σ_a depends upon the velocity c of the neutron at the moment of capture so that one can write $\sigma_a^2 = K/c$ in which K is a constant characteristic of the medium. The relationship $\sigma_a^2 = K/c = K/v$ (with v the velocity of the neutron), the so called '$1/v$ law' is not an arbitrary assumption. It is the natural consequence of a wave mechanical treatment of neutron capture by nuclei with a special pattern of excited states peculiar to those light elements (^2_1D, ^9_4Be, $^{12}_6C$) which can be used as moderators. The capture rate of neutrons of velocity c and density n_c per unit volume is (using the methods used in calculating molecular collisions) $Nn_c c\pi\sigma_a^2 = Nn_c\pi K = \text{constant} \times n_c = \kappa^2 n_c$ in which κ^2 is independent of c. Accordingly the probability of capture is the same for all neutrons and once can write $dn/dt = -\kappa^2 n$ in which n is the total number of neutrons in the system and κ^2 is a constant and does not depend upon the velocity distribution or the spatial distribution. It is interesting to note that the '$1/v$ law' has introduced the same sort of simplification that Maxwell achieved by introducing the μ/r^5 law into molecular scattering theory. The approximate equation governing the behaviour of thermal neutrons in a moderator can now be written down as

$$\frac{\partial n}{\partial t} = D\nabla^2 n - \kappa^2 n$$

$\frac{\partial n}{\partial t}$	$D\nabla^2 n$	$\kappa^2 n$
rate of increase of neutrons in an elementary volume	net gain due to diffusion	loss due to capture

If $\kappa^2 = 0$ (no capture $\lambda_a = \infty$) the ordinary diffusion equation with a solution $n = f(x, y, z, t)$ describes the behaviour of the neutrons. If $\nabla^2 n = 0$ (uniform neutron distribution) $dn/dt = -\kappa^2 n$, $n = n_0 e^{-\kappa^2 t}$ and the neutrons disappear exponentially with time throughout the volume of the moderator. The two contributions to dn/dt are independent of each other and it is not difficult to show (by substitution) that the solution of the thermal diffusion equation is the product of the two extreme solutions just described so that $n =$ (pure diffusion solution) $\times$ (exponential decay). It should be noted that the pure diffusion solution also contains the time but in such a way that the total number of neutrons does not change with time, they merely diffuse from one part of the system to another.

By comparing the results of experiment with the theoretical solution the two mean free paths λ_t and λ_a can be measured. Two types of experiment are usually used.

Dynamic

A pulse of neutrons a few microseconds long is released in a tank of water (or other moderator) and their behaviour over the next 1400 μs is observed. Results of such an experiment are shown in Fig. 3.8 in which the separate effects of capture and diffusion can be observed.

Static

A continuous point source emitting Q neutrons per second is placed in a large volume of moderator and a series of measurements is made of the way in which the neutron density depends upon r the distance from the source. Owing to the symmetry of the experiment $n = f(r)$ only and the diffusion equation becomes:

$$D\left(\frac{\mathrm{d}^2 n}{\mathrm{d}r^2} + \frac{2}{r}\frac{\mathrm{d}n}{\mathrm{d}r}\right) - \kappa^2 n = \frac{\mathrm{d}n}{\mathrm{d}t} = 0 = \frac{\mathrm{d}^2 n}{\mathrm{d}r^2} + \frac{2}{r}\frac{\mathrm{d}n}{\mathrm{d}r} - \frac{n}{L^2} = 0$$

in which $\kappa^2/D = 1/L^2$ and the length L is referred to as the diffusion length. The solution of this time independent equation is $n = A/r \cdot \mathrm{e}^{-r/L}$ as is readily verified by substitution. Equating the total rate of absorption of neutrons $= \int_0^\infty \kappa^2 n \, 4\pi r^2 \, \mathrm{d}r$ to the rate Q at which they are being emitted into the moderator shows that $A = Q/L^2$. Other exercises in integration show that the mean distance $\bar{r}$ at which a neutron is captured is $2L$ and the mean square distance $\overline{r^2}$ is $6L^2$. The mean time for capture is $\lambda_a/\bar{c}$. This time can also be approximately calculated from the pure diffusion equation without capture (section 3.6), as $\bar{r}^2/2\lambda_t\bar{c}$. Equating the two times gives the approximate result $\lambda_a/\bar{c} = \overline{r^2}/\lambda_t\bar{c}$ and $\lambda_a \lambda_t \approx 3L^2$. A much more detailed analysis (Davison (1958) or Beckurts and Wirtz (1964)) shows that this approximate result is correct as long as $\lambda_a \gg \lambda_t$ as it is in all useful moderators. Some numerical values are given in Table 3.2 which shows the accuracy which can be achieved in this type of measurement.

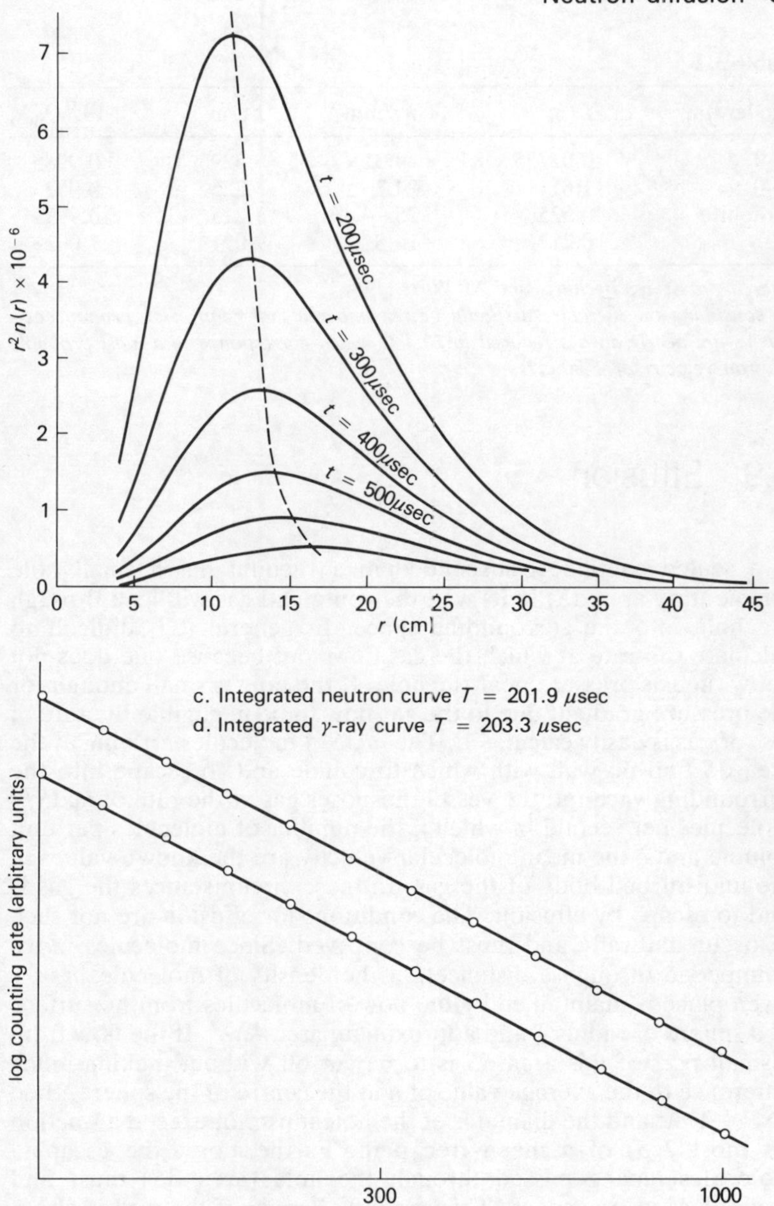

Fig. 3.8 (a) Shows the change in density of neutrons, r from the origin, in a tank of water at 100 μs intervals. Note the decay due to neutron capture and the outward shift of the maximum due to diffusion.
(b) The measured decay of the total number of neutrons in the tank; by integrating the neutron distribution given in fig. 3.8(a) the effect of diffusion is eliminated and the decay due to capture is exponential. (Taken from Meads (1956) Copyright The Institute of Physics).

Table 3.2

Moderator	L/m	λ_t/mm	λ_a/m	$3L^2/\lambda_a\lambda_t$
H_2O	0.02755	4.31	0.5291	0.9985
D_2O	1.61	24.3	3.226	0.992
Graphite	0.525	25.74	3.215	0.999
Be	0.212	14.8	9.091	1.002

(Based on K. H. Beckurts and K. Wirtz, 1964.)
In consulting the literature it should be remembered that neutron absorption coefficients are nearly always reduced to 20.4°C which corresponds to a most probable neutron velocity of 2200 m/s.

3.9 Effusion

If a sealed container, suspended in a vacuum has a small hole (elementary area $\mathrm{d}S$) in its wall the contained gas will leak through the hole into the surrounding space. In general it is difficult to calculate the rate at which the gas flows out because one does not know the gas pressure near the hole. If the hole is small enough for the pressure gradient due to the gas flow to be negligible the rate of loss of gas is easily calculated. The $n\bar{c}\mathrm{d}S/4$ molecules arriving at the area $\mathrm{d}S$ find no wall with which to collide and so escape into the surrounding vacuum; the vessel thus loses gas at the rate of $n\bar{c}\mathrm{d}S/4$ molecules per second in which n the number of molecules per unit volume and $\bar{c}$ the mean molecular velocity are the known values in the undisturbed body of the gas. In these circumstances the gas is said to escape by effusion. The conditions for effusion are not such as occur naturally and must be contrived. Since molecules move unimpeded through a distance $\sim\lambda$ the density of molecules n at a given place is maintained by the flow of molecules fromthe surface of a sphere of radius λ and approximate area $4\pi\lambda^2$. If the flow from a small part of this area $\mathrm{d}S$ is to be cut off without making much difference to the average value of n at the centre of the sphere, then $\mathrm{d}S \ll 4\pi\lambda^2$ and the diameter of the hole must not exceed a fraction (at most 2/5) of a mean free path. Furthermore, the escaping molecules having passed through the hole (area $\mathrm{d}S$) must find themselves in free space. This can only happen if the wall is also a fraction of a mean free path thick otherwise the missing area of wall $\mathrm{d}S$ would not be a hole but the entrance to a capillary tube in which conditions of flow are quite different (cf. section 3.2). Since practical considerations limit the thickness of foils to 10 μm and so λ to 1 mm or more effusion is essentially a low pressure phenomenon.

Taking the mean free path of gases at s.t.p. to be about 10^{-7} m, $\lambda \sim 1$ mm when the pressure is 10^{-4} atmospheres = 10 Pa. It is below this pressure that gas escapes by effusion through pin holes in a thin foil. The calculation of the rate of flow is shown in the following example.

Worked example

The mouth of a litre flask containing carbon dioxide at 0 °C is closed by a piece of 10 μm aluminium foil pierced with a hole 100 μm in diameter. The gas escaping by effusion is pumped away by a fast pump. Estimate the upper limit of the pressure in the flask for effusion not viscous flow to occur through the hole and the time for the pressure to drop to half its initial value (λ for CO_2 at s.t.p. = 39 nm).

Working in absolute units, it is sufficiently accurate to take one atmosphere = 10^5 Pa and 1 m^3 = 1000 l. The condition for effusion from a hole of diameter d is that $\pi d^2/4 = dS \ll 4\pi\lambda^2$. Interpreting this as $d^2 < 16\lambda^2 \cdot 10^{-2}$ gives $\lambda > 2.5d = 2.5 \cdot 10^{-4}$ m. Now λ is inversely proportion to the pressure so that for carbon dioxide $p_{atm} = 39 \cdot 10^{-9}/\lambda$ and effusion will be the main process when $p < 39 \cdot 10^{-9}/2.5 \cdot 10^{-4} = 15.6 \cdot 10^{-5}$ atm = 15.7 Pa. The rate of effusion is $n\bar{c}dS/4 = \pi n\bar{c} \cdot 10^{-8}/16$ mol/second. If there are N molecules in the litre flask $N = n/1000$ and $dN = -1000N\bar{c}\pi \cdot 10^{-8}$ $dt/16 = -\alpha N dt$ $(\alpha = \pi\bar{c} \cdot 10^{-6}/1.6)$; $dN/N = -\alpha dt$ so that $N = N_0\,e^{-\alpha t}$ and $N/N_0 = \frac{1}{2}$ when $\alpha t = 0.6932$ and $t = 0.6932/\alpha$.

Calculation of $\bar{c}$: For 1 mole $PV = RT = \frac{1}{3}M\overline{c^2}$ J and

$$\overline{c^2} = \frac{3RT}{M} = \frac{3 \times 8.314 \times 273.15}{44} \times 1000$$

$$= 1.548 \times 10^{+5}\ m^2/s^2$$

but $(\bar{c})^2 = 8/3\pi\,\overline{c^2}$ as shown in section 2.2 and $\bar{c} = 362.4$ m/s so that

$$\alpha = \frac{\pi\bar{c} \times 10^{-6}}{1.6} = 7.11 \cdot 10^{-4}$$

and

$$t = \frac{0.6932}{\alpha} = 974\ s.$$

3.10 Measurement of small vapour pressures

Very small vapour pressures can conveniently be measured by observing the rate at which the vapour escapes by effusion from a vessel in which a constant vapour pressure is maintained. These measurements are not merely an academic exercise. As will be shown in Chapter 6 the kinetic theory of a perfect gas cannot be developed without linking it with quantum mechanics. The resulting theory enables a calculation to be made of the absolute value of the vapour pressure of a monatomic solid in the temperature range for which the vapour can be treated as a perfect gas. The experimental verification of this vapour pressure equation is thus of fundamental importance. It is also of immediate practical importance since the quantum mechanical ideas on which the vapour pressure equation is based can be applied to the prediction of gaseous chemical equilibria, and so have a great influence on present day chemical technology.

For substances like sodium and potassium which have a negligible vapour pressure at room temperatures the effusion method can be applied directly. A vacuum furnace is constructed with a hole a few millimetres in diameter in the wall and the rate of effusion into the vacuum as a function of temperature is directly measured. This can conveniently be done by collecting the evaporated material on a cooled surface and weighing it directly or determining it chemically. The vapour pressure is so low that there is no difficulty in meeting the conditions necessary for effusion with holes of a size that can be accurately measured and walls that can be fabricated.

Worked example

According to Edmondson and Egerton the vapour pressure of potassium is given by the relationship (p in mm of mercury)

$$\log_{10}p \text{ mm} = \frac{-4507}{T} + 7.3447$$

Calculate the mass of potassium leaving in one hour a hole 2 mm diameter in the side of a furnace containing the metal at 177.54 °C. If the potassium is collected on a cooled plate and subsequently dissolved in water what volume of N/100 HCl will be required to neutralise the solution? Make an estimate of the uncertainties in such a measurement.

From the formula $\log_{10}p = \bar{3}.3447$ and $p = 2.212 \times 10^{-3}$ mm Hg $= 2.212 \times 1.333 \times 10^{-1} = 0.295$ Pa. The number of molecules leaving in time t is

$$\frac{n\bar{c}}{4}\,\mathrm{d}S \cdot t = \frac{p}{RT}\sqrt{\frac{RT}{2\pi M}} \cdot \mathrm{d}St$$

moles. This is equivalent to

$$\frac{p\,\mathrm{d}St \times 10^5}{\sqrt{2\pi RTM}} \text{ ml of N/100 HCl}$$

$$= \frac{0.295 \times \pi \cdot 10^{-6} \times 3.6 \cdot 10^3 \times 10^5}{(2\pi \times 8.314 \times 450.7 \times 39.10 \times 10^{-3})^{1/2}} = 11.0 \text{ ml N/100 HCl}$$

Errors: The diameter of the hole can be measured to 1 μm = 0.1 per cent in area. The temperature can be maintained with good thermostatic control to 0.01 per cent. The titration can be made to 0.01 ml = 0.1 per cent. The errors of measurement are thus quite small. The main source of uncertainty is the presence of K_2 molecules. Neuman and Volker (1932) have estimated about 0.8 per cent at 700 °C, and Miller and Kusch (1935) 0.25 per cent at 500 °C. The error due to the isotopic composition of the metal (6.6 per cent of ${}^{41}_{19}K$) can be calculated and is small.

The furnace from which the vapour is escaping suffers a reaction and is acted upon by a force acting along the inward normal to dS. This force can be looked upon either as the rate at which the furnace is losing momentum to the jet of emerging molecules or to the unbalanced gas pressure due to the removal of an element dS of the wall. The pressure range to which the effusion method is restricted can be extended by measuring the reaction and the gas flow simultaneously by a method due to Volmer. A light horizontal rod-like container is provided with two holes facing in opposite directions so that the reactions of the escaping gas form a couple tending to rotate the rod about a vertical axis. The container is supported by a vertical torsion wire (carrying a mirror) from a small silica microbalance. The rate of evaporation is deduced from the microbalance readings and the couple from the deflexion of a light beam reflected from the mirror. If A is the combined area of the holes, l their distance apart, θ the deflection, k the torsion constant of the suspension and p the pressure, the couple is $k\theta = Apl/2 \cdot \frac{1}{2}$. The factor $\frac{1}{2}$ comes from the consideration that the pressure p on the element of wall dS is due to the reversal of the momentum carried by the $n\bar{c}/4$ dS molecules which are escaping without having their momentum reversed. The rate of loss of mass g is

$$\frac{mn\bar{c}\mathrm{d}S}{4} = \sqrt{\frac{M}{2\pi RT}} \cdot Ap.$$

Eliminating Ap gives

$$M = \frac{\pi g^2 l^2 RT}{8k^2\theta^2}$$

Table 3.3

Metaldehyde	T°C	316.0	317.9	318.6	319.1
	M	176	181	181	178
Rhombic Sulphur	T°C	80.15	80.2	80.2	80.2
	M	247	256	258	258
Monoclinic sulphur	T°C	97.5	97.5	97.5	97.5
	M	278	257	278	250

The formula weight of metaldehyde $(CH_3CHO)_4$ *is 176.*
$M \simeq 256 = 8 \times 32$ *is the value found in solutions of sulphur.*

an expression which enables the molecular weight of the gas to be calculated from the measured values of g and θ. It is to be noted that one is not now restricted to the pressure range compatible with the effusion formula. If the wall is so thick that the holes are short capillaries they can be replaced by an effective area A and it is the unknown Ap which is eliminated to obtain the expression for M. The values shown in Table 3.3 were obtained experimentally by Volmer (1931) for the molecular weights of at the time chemically interesting substances.

The advance of technology has encouraged the application of the effusion method of measuring vapour pressures to less volatile substances at higher temperatures. For the convenience of the experimenter measurements are sometimes made in circumstances which only formally fulfill the conditions necessary for effusion. It is true that the mean free path is much greater than the size of the hole in the furnace wall, but this is achieved by making λ so large that there are few molecular collisions and one has to rely on the furnace wall behaving like a layer of gas many mean free paths thick (Knudsen's law discussed in section 3.13). This is not always true as shown by the experiments of Ward, Mulford and Kahn (1967), who used a radioactive method of determining the angular distribution of the atoms leaving the furnace. As seen in Fig. 3.9 there is a tendency for the aperture to form a pinhole camera image of the specimen in the furnace and the atoms emitted from the hole have an angular distribution which is demonstrably not a Knudsen cos θ distribution.

3.11 Rate of physical and chemical change in simple systems

The possibility of calculating the rate at which the molecules in a gas are striking a surface or colliding with each other enables some very

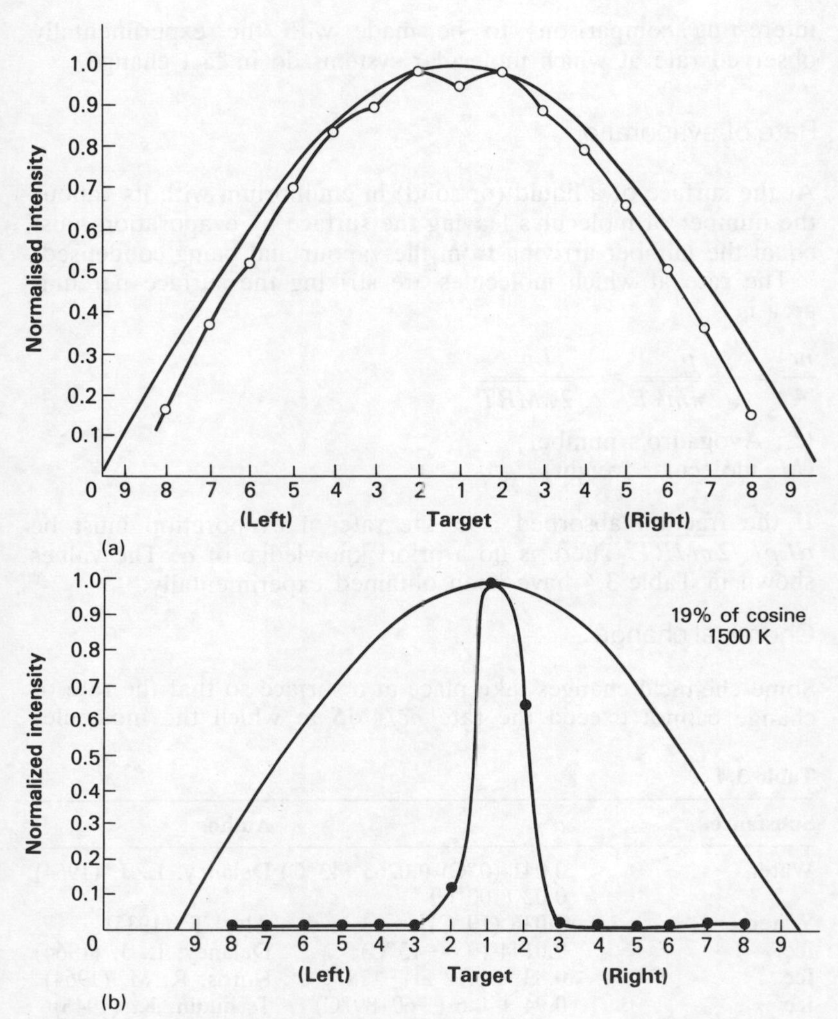

Fig. 3.9 Angular distribution of radioactive material during effusion.
(a) Gold from a stackpole graphite cell.
(b) Plutonium from a MgO cup.
(After Ward (1967)).

interesting comparisons to be made with the experimentally observed rate at which molecular systems do in fact change.

Rate of evaporation

At the surface of a liquid (or solid) in equilibrium with its vapour the number of molecules leaving the surface by evaporation must equal the number arriving from the vapour and being condensed.

The rate at which molecules are striking the surface per unit area is

$$\frac{n\bar{c}}{4} = \frac{p}{\sqrt{2\pi mkT}} = \frac{Lp}{\sqrt{2\pi MRT}}$$

(L; Avogadro's number)
(M; molecular weight)

If the fraction absorbed is α the rate of evaporation must be $\alpha Lp/\sqrt{2\pi MRT}$. There is no a priori knowledge of α. The values shown in Table 3.4 have been obtained experimentally.

Chemical change

Some chemical changes take place at a surface so that the rate of change cannot exceed the rate $n\bar{c}/4\ \mathrm{d}S$ at which the molecules

Table 3.4

Substance	α	Author
Water	0.041 (0 °C) 0.0265 (43 °C) 0.02 (100 °C)	Delaney, L. J. (1964)
Water	0.036 (10 °C)	Alty, T. (1935)
Ice	0.0144 (0 – −13 °C)	Delaney, L. J. (1964)
Ice	0.83 + 0.12 at −128 °C	Koros, R. M. (1964)
Ice	0.94 + 0.6 (−60–89 °C)	Tschudin, K. (1946)
Mercury (liquid)	1.0	Volmer, M. (1921)
Mercury (solid)	0.85 (−85 °C)	Volmer, M. (1921)
Sulphur	0.74	Bradley, R. S. (1956)
Benzophenone	0.28–0.69	Birks, J. (1949)
Di-n-butylphthalate	1.0	Birks, J. (1949)
Camphor (synthetic)	0.17	Birks, J. (1949)
Tridecylmethane	0.98 (25 °C)	Bradley, R. S. (1951)
Potassium chloride	0.7 (407–470 °C). 110 face 0.56	Bradley, R. S. (1953)

For a discussion see Hobbs, P. V. (1974) and Rutner, E., Goldfinger, P. and Hirth, J. P. (1964).

arrive at the surface. Some changes approach this limit. An example is the decomposition of ozone at a silver surface. In order to account for the rate at which ozone well diluted with oxygen disappears in the presence of a silver foil Strutt (1912) had to assume that every molecule colliding with the silver surface reacted with it and is decomposed. For homogeneous gaseous reactions (e.g. $H_2 + I_2 \rightleftharpoons 2HI$) which proceed by molecular collisions in the bulk of the gas, the position is quite different. The rate of collision between the molecules is enoromously greater than the rate at which they interact. This can be accounted for by the assumption that chemical change takes place through an intermediate complex which requires a high energy E for its formation so that only a small fraction $f(E/kT)\ e^{-E/kT}$, for which the energy of the colliding molecules exceeds E, are effective. The rate of reaction is then given by

$$\text{Rate} = S\nu f(E/kT)\ e^{-E/kT}$$

in which ν is the collision rate and S is a geometrical factor (steric hindrance) which is not expected to be very small (~1 per cent). This expression for the rate may be tested by measuring the dependence of the rate of reaction upon temperature. The exponential factor dominates the expression for the rate to such an extent that a plot of log (rate) against $1/T$ should give a straight line. This deduction is well borne out by experiment as shown in Fig. 3.10. The assumption of an intermediate complex also provides a simple explanation of homogeneous catalysis. A homogeneous reaction is one taking place in the bulk of the gas through molecular collisions and not on the surface of the vessel containing the reacting gases. In homogeneous catalysis the reaction is catalysed by the admixture of a small quantity of another gas uniformly mixed with the reactants. The classical example of homogeneous catalysis is the oxidation of SO_2 by the oxygen of the air containing a small amount of nitric oxide. The lead chamber process for making sulphuric acid which was based on this homogeneous catalysis has been almost entirely superseded by the surface catalysis of the same reaction at a V_2O_5 surface. The catalyst can provide an alternative path through different intermediate states. There is no obvious reason why the energy required for the formation of these intermediate complexes should not be less than the activation energy of the uncatalysed change. Owing to the overwhelming influence of the exponential terms the relatively small number of collisions with the gaseous catalyst is easily compensated for by a reduction in the energy of activation and the reaction may proceed much more rapidly in the presence of quite small ($\sim 10^{-3}$) amounts of catalyst.

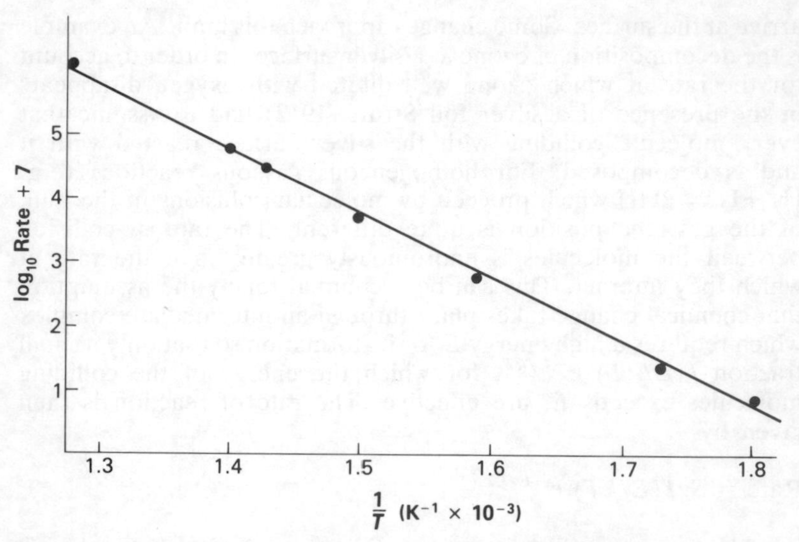

Fig. 3.10 Decomposition of hydrogen iodide. Plotted from the data of Bodenstein (1893).

Although the assumption of an intermediate state with a high activation energy E is generally thought to be correct the simple expression $S\nu e^{-E/kT}$ for the rate of a chemical change is only approximately true, and can be shown to lead to internal contradictions. It has been derived with too little regard to the laws of quantum mechanics and insufficient attention to the particular molecular structure of the complex. A similar derivation taking more account of these complications suggests that the exponential factor $e^{-E/kT}$ should be replaced by the often not numerically very different $e^{-G/RT}$ in which the activation energy E is replaced by the thermodynamically defined (section 5.6) Gibbs free energy G per mole. The consequent introduction of a factor e^{S} into the expression for the reaction rate is consonant with the general molecular explanation of entropy (S) discussed in Chapter 6.

This general explanation of homogeneous gaseous catalysis can be illustrated by the very simple reaction

$$\text{p-H}_2 \rightarrow \text{o-H}_2$$

even j   odd j

in which para-hydrogen is converted into ortho-hydrogen. The two forms of molecular hydrogen differ in the relative orientation of the proton spins. Nearly pure p–H_2 can be obtained by the desorption of H_2 from activated charcoal cooled in liquid hydrogen. It is quite stable and at s.t.p. it changes into the equilibrium mixture of 1/4 $p-H_2:3/4\ o-H_2$ so slowly that is doubtful if the direct para to ortho change due to direct molecular collisions has ever been observed. The reason for this is that a proton spin in inverted by a changing magnetic field and the magnetic field of a hydrogen molecule is minute. A direct calculation shows that the half life for a transition to the equilibrium mixture is about 10^8 seconds (3 years), during which time each molecule will have made about $10^8\ \bar{c}/\lambda = 10^{18}$ collisions. This is often described by saying the collision efficiency is 10^{-18}. At about 1000 °K the conversion to the equilibrium mixture proceeds at a conveniently measurable rate by the homogeneous gaseous reaction.

$$\underset{\downarrow}{H} + \underset{\downarrow\ \uparrow}{H\,H} = \underset{\downarrow\ \downarrow}{H\,H} + \underset{\uparrow}{H}$$

The small concentration of H atoms (10^{-2} per cent) is supplied by the thermal dissociation of molecular hydrogen

$$H + H \rightleftharpoons H_2$$

The concentration of H atoms can be calculated from the formulae of section 8.10. It is an equilibrium amount and so can be calculated without any reference to the rate at which it is formed. The experimental results are in good agreement with the simple theory and can be used to calculate the data in Table 3.5. After allowing for the increase in the concentration of H atoms with temperature the activation energy is found to be about 23 000 Joules per mole. Although the magnetic field of the hydrogen molecule is small there are plenty of molecules (O_2, NO, NO_2) with unbalanced magnetic moments which can act as homogeneous catalysts. In the presence of O_2 the direct $p \rightarrow o$ conversion proceeds according to the equation $u_t = u_0 \exp - (k[O_2]t$ in which u is the per cent $p-H_2$ − per cent $p-H_2$ at equilibrium and k

Table 3.5

Temp K	Collision efficiency × 10^3	Steric factor
873	2.87	0.06
923	2.57	0.05
975	4.00	0.06
1023	4.63	0.05

(Taken from Farkas, 1935, p. 64)

is the reaction constant. The collision efficiency varies slowly with temperature from 2.38×10^{-13} at 77.3 K to 7.61×10^{-13} at 623 K.

3.12 Vacuum pumps

Gaede and Langmuir were among the first to recognise that the representation of the properties of a gas in terms of the molecular mean free path had an important bearing on the design of vacuum pumps. At atmospheric pressure the steady orderly flow of a gas is determined as it is in liquids by equating the accelerating forces due to the pressure gradients to the retarding forces due to viscosity. The pressure exerted by a gas is due to molecular bombardment and for one elementary volume of gas to exert a force on a neighbouring elementary volume molecules from the one must cross the boundary between them and be stopped in the other. One cannot therefore apply the laws of mechanics in this way to elementary volumes whose sides are less than the mean free path. As soon as the pressure is so low that λ is greater than the size of the vessel the molecules will collide mainly with the walls and the terms pressure and pressure gradient cease to have any mechanical meaning in the body of the gas. While the term pressure is still used for convenience it is merely a convenient way of indicating the number of molecules per unit volume. When the pressure has fallen so that only molecular collisions with the walls need be considered the gas flow takes place by random wandering of the molecules. The flow of gas through a pipe by random wandering is usually referred to as molecular flow to distinguish it from the viscous and turbulent flow which occurs at higher pressures. While the transition from turbulent to viscous flow is fairly sharp, Roberts (1957), the transition from viscous to molecular flow is obviously spread out over a considerable pressure range because of the exponential spread $e^{-l/\lambda}$ of the lengths of the free paths. An ideal vacuum pump, for use when the pressure is low enough for molecular flow, consists of an area ΔS in the wall of the vessel which captures all the molecules which fall upon it. Gas will then be removed from the vessel at the rate of $n\bar{c}\Delta S/4$ mol/s, which is the rate they strike the area. Pumps of this type are very effective but of limited application. Early ones consisted of a side tube containing prepared charcoal immersed in liquid nitrogen. They are still used to maintain a high vacuum in sealed off electronic devices. Just before sealing a 'getter' is 'fired' which evaporates a layer of magnesium or barium on to the glass wall of the containing vessel. Molecules of the residual gas given off from the hot electrodes during operation react chemically with the layer and are so prevented from accumulating in the sealed off vessel. Spongy titanium, zirconium, or uranium, which form stable hydrides can also be used. The conventional speed of such a pump is easily calculated. The speed of a pump is conventionally expressed by comparison with a displacement pump which removes a fixed volume of gas each stroke. The volume of gas V removed each second is called the speed of the pump. Equating the Vn molecules per second with the $n\bar{c}\,dS/4$ mol/s flowing through ΔS shows that the speed of the molecular pump $V/s = \bar{c}\Delta S/4/s$. Since $\bar{c}$ for air is about 4.48×10^2 m/s at 0 °C the speed is about 112 m^3/per m^2 of exposed area. This is frequently quoted as 11.2 l/cm^2. For lighter molecules and higher temperatures the speeds will be correspondingly greater.

A very simple continuously acting pump has been devised after a great deal of experimental work by Gaede (1915) and Langmuir (1916) (the patent litigation connected with this invention is a classic). The usual form is shown in Fig. 3.11. It is simple and effective. It consists of an outer steel tube A fitted with a heater H. The upper part is fitted with a water jacket W. The whole of the apparatus is evacuated through a side tube S by means of a forepump or backing pump capable of reducing the pressure to 10 Pa (0.075 mmHg) which is the threshold of the molecular flow region. A shallow layer of the working fluid fills the bottom of the pump as shown by the broken line. The vapour rises through the central tube, is deflected by the hood or bell B, and returns down the annular space. It is condensed by the water cooled walls of the barrel of the pump and returns to the boiler (to be recycled) through the narrow gap between the outer wall and the central tube or chimney whose diameter is enlarged at the base to form a loosely fitting insert as shown. The working fluid is either mercury or a special oil which can be boiled for months without appreciable decomposition.

The pressure inside the boiler is a few centimetres of the working fluid and the vapour emerges from the jet at supersonic velocities so that there are relatively very few molecules going upwards into the chamber being evacuated. Molecules which hit the water cooled wall are condensed and so removed from the vapour stream. The result can, to a fair approximation, be described as a stream or jet of heavy molecules moving downwards parallel to the axis of the pump. The jet of molecules moving more or less parallel to the axis of the pump carries the gas with it to the base of the pump where all the working fluid is condensed and the remaining gas finds its way out through the side tube into the forepump. Only a small fraction of the uncondensed gas can flow back against the stream of heavy molecules which acts as a very effective barrier. Molecules diffusing

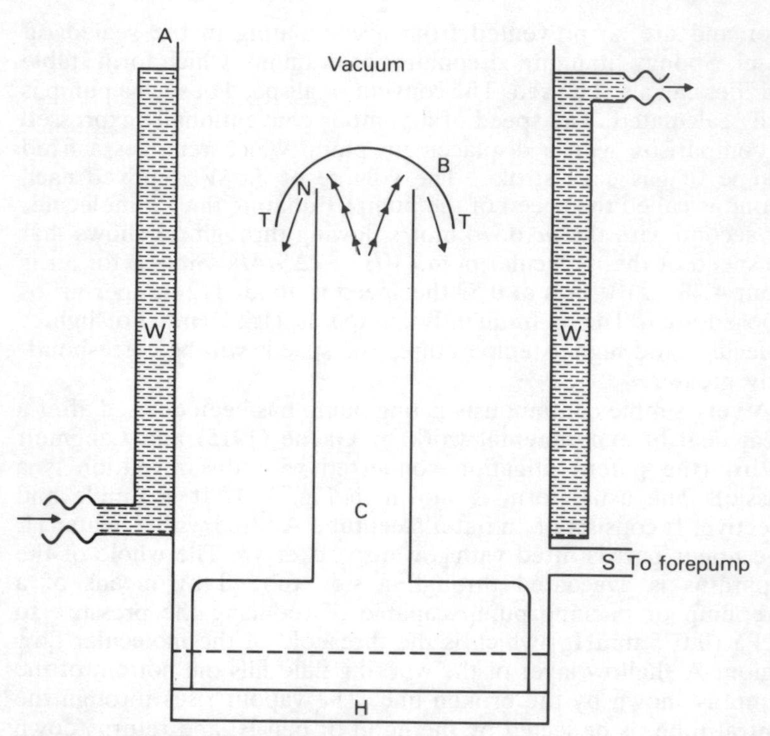

Fig. 3.11 Single stage condensation pump.
A = barrel, B = bell or hood, C = chimney, H = heater, N = nozzle, S = sidetube, T = throat, W = water jacket.

against this stream collide with fast, heavy molecules moving in the opposite direction and are reincorporated into the downward moving jet.

The pressure ratio (in contrast to the volumetric speed) of the pump, i.e. the lowest pressure reached by the forepump/lowest pressure in the evacuated vessel, is determined by the probability W that a molecule can traverse the jet stream without a collision and is approximately 1/W. If the pump and so the stream of molecules were indefinitely long this probability would approach zero even allowing for the attenuation of the jet by molecular collisions which direct molecules to the walls where they condense. An upper limit to the pressure ratio for a jet of length l is easily calculated. If the density n_0 of mercury atoms in the jet remained constant the collision cross section per unit area of jet would be $\Sigma = n_0 l \pi \sigma^2 = 1/\sqrt{2}\lambda_0$. The probability W would be $e^{-\sqrt{2}\alpha\Sigma}$ and the pressure ratio $e^{\sqrt{2}\alpha\Sigma}$ in which α (numerical value ≈ 0.65) is a factor which takes into account the difference in diameter between the molecules of the pumped gas and mercury atoms and makes some allowance for the increase in mean free path (e.g. Table 3.1) due to the supersonic velocity of the atoms in the jet. For a jet attenuated by the loss of mercury atoms by condensation on the water cooled walls Σ will be less. However, the drop in the collision cross section Σ can be compensated for by increasing the vapour flow. Thus on the simple theory of the mercury vapour condensation pump which has been outlined the pressure ratio of the pump will always increase rapidly with the vapour flow. This is contrary to the experimental facts and shows that some important factor has not been considered. The experimental behaviour has been described by Alexander (1946) who found that a fivefold variation in the flow of the mercury vapour produced no substantial change in the Mach number ($\approx$2). The pumping action does however depend upon the vapour flow and sets in rather suddenly (Fig. 3.13) at a mercury vapour flow determined by the pressure maintained by the forepump. The detailed behaviour of a pump in this threshold region was first described by Alexander who used a probe to measure the isobars of the pumped gas in the jet. He found that the anomalous behaviour for fixed vapour flow set in when the isobars, instead of ending up on the water cooled wall, began to run parallel to it as shown in Fig. 3.12. Alexander's conclusion, borne out by later work, was that the observed drop in pumping speed with gas flow was due to gas creeping back at the edge of the jet along the water cooled wall from the forepump to the evacuated space. A qualitative explantion of this leakage can be given in molecular terms. In the simple account the supersonic jet was assumed to be established in the neighbourhood of the throat and the subsequent influence of the water cooled wall was neglected. The wall is in effect a layer of liquid mercury and provides a continuous stream of mercury vapour at a saturated vapour pressure. The mean free path in this vapour is about 10 mm so that such a layer which is without pumpping action provides a path along which the gas can diffuse from the forepump to the vacuum vessel thus bypassing the pumping action of the central part of the jet. The existence of a layer near the wall in which the concentration of mercury vapour is determined by the temperature of the wall and not by the concentration of the vapour in the jet can be demonstrated by making the mercury vapour visible by means of a Tesla discharge. Working in this way Nöller (1955) found that if the walls of a glass pump were cooled with solid CO_2 there was no appreciable mercury vapour near the wall when the gas pressure was 0.13 Pa. Further investigation showed that the large increase in the pressure ratio which occurs when the flow of mercury vapour is increased is due to the suppression of the leakage

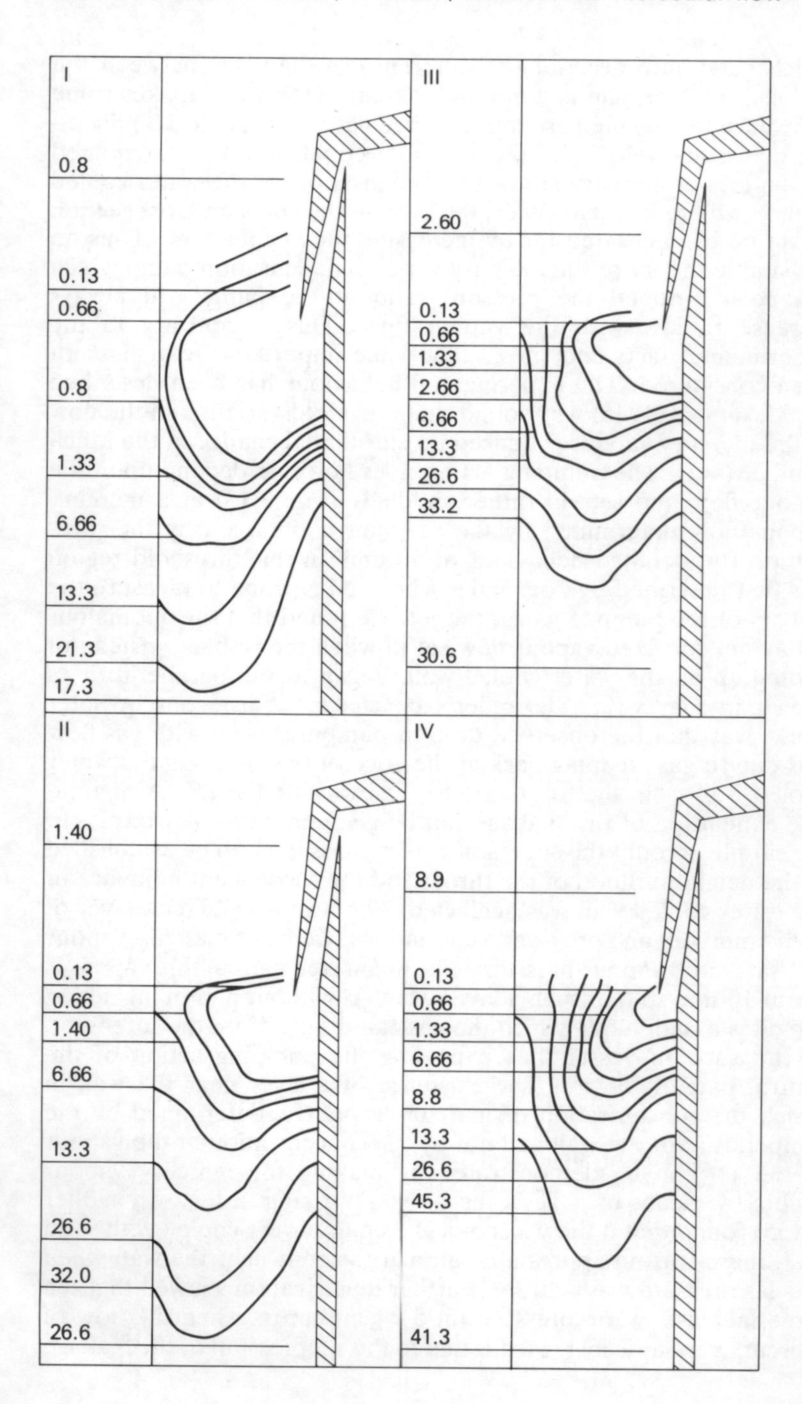
I
0.8
0.13
0.66
0.8
1.33
6.66
13.3
21.3
17.3
II
1.40
0.13
0.66
1.40
6.66
13.3
26.6
32.0
26.6
III
2.60
0.13
0.66
1.33
2.66
6.66
13.3
26.6
33.2
30.6
IV
8.9
0.13
0.66
1.33
6.66
8.8
13.3
26.6
45.3
41.3

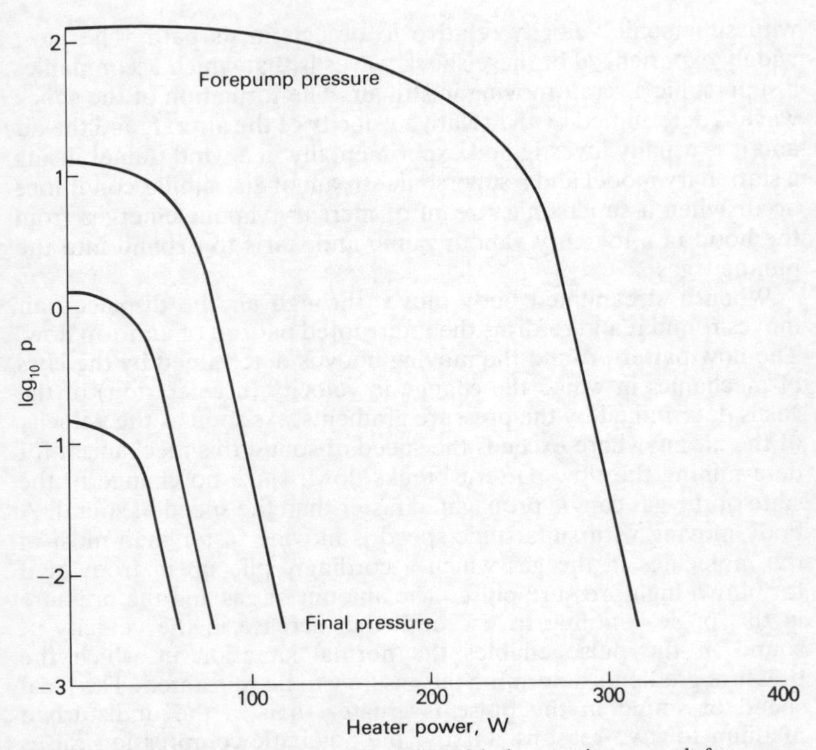

Fig. 3.13 Variation in pressure ratio with heater input and forepump pressure. Data from Reichardt (1955).

along the cold wall by the formation of a stationary shock wave. The sudden increase in density of the mercury vapour on going through the shock front where the shock wave meets the wall prevents the gas from diffusing back into the vacuum chamber. The formation of this supersonic plug has been demonstrated by Kutscher (1955) who used a d.c. discharge (1500 V, 5–10 mA) to make the mercury vapour and the pumped gas visible in the region of the jet. The formation of stationary shock fronts is characteristic of gas flowing

Fig. 3.12 Isobars of air in a mercury vapour pump.
The pumping speeds in l/s for the four diagrams are 135, 130, 83, 34, respectively. As the throat area was 147 cm^2 the volumetric speed under these conditions is considerably smaller than the theoretical 1600 l/s. The gap in the nozzle is 0.6 mm and the pressure of the mercury vapour = $\frac{3}{4}$ boiler pressure ~ 20 mm. It will be observed that the pumping speed drops when the isobar representing the minimum pressure comes away from the wall.

with supersonic velocity relative to objects in its path. The most widely experienced of these shock waves is that which accompanies a supersonic aircraft moving in still air. The formation of the shock wave is determined by the relative velocity of the aircraft and the air and it is usually investigated experimentally in a wind tunnel, using a stationary model and a supersonic stream of air: similar conditions occur when a supersonic stream of mercury vapour emerges from the hood in a mercury vapour pump and starts to expand into the pumped gas.

When a streamlined body moves through air the displaced air moves round it and reforms the interrupted pattern of uniform flow. The flow pattern round the moving body is determined by the laws of mechanics in which the change in velocity (acceleration) of the gas is determined by the pressure gradients. As soon as the velocity of the air anywhere exceeds the speed of sound this mechanism for determining the flow pattern breaks down since no change in the state of the gas can be propagated faster than the speed of sound. A body moving with supersonic speed is moving faster than most of the molecules in the gas which accordingly pile up in front of it forming a high pressure pulse. The amount of gas and the pressure in this pulse continue to rise until the increase in the velocity of sound in the pulse enables the normal situation in which the pressure gradients determine the gas flow to be reinstated. The local speed of sound in the pulse is greater than in the undisturbed medium for two reasons. Firstly, the adiabatic compression raises the temperature in the pulse ($v \propto \sqrt{T}$) and secondly the velocity of sound is not independent of the amplitude (as it is in the very low amplitudes used in speech) but rises rapidly with the amplitude of the sound wave. These two effects reinforce each other and a steady state is reached in which the body travelling with velocity V is preceded by a high temperature high pressure pulse propagated with velocity V which is the local velocity of sound in the conditions prevailing in the pulse. It is the dispersion of this high pressure travelling pulse into a conical sound wave whose vertex coincides with the moving body which constitutes the shock wave. The most easily observed analogy is the V-shaped bow wave created in smooth water by a ship as soon as it starts to move faster than the velocity of surface waves i.e. with a supersonic velocity. The theory of the formation of shock waves in three dimensions requires difficult mathematics for its exposition but a reasonable understanding of the subject can be obtained from an idealised one-dimensional treatment which is very much simpler. A stream of gas flowing with a supersonic velocity down a tube of uniform cross section is unstable; it can change from a low density high velocity stream into a stream transporting the same quantity of material at a higher density and a subsonic velocity. The change of density occurs suddenly in a transition layer a few mean free paths thick. The transition layer between the two states of flow is known as a shock front. The change of density ρ depends upon the Mach number M of the supersonic flow and for a perfect gas ($\rho_1 \rightarrow \rho_2$) is given by

$$\frac{\rho_2}{\rho_1} = \frac{M^2(\gamma + 1)}{M^2(\gamma - 1) + 2} \approx 2.3 \text{ for } M = 2 \text{ and } \gamma = 1.67$$

There is an increase of entropy when the supersonic stream becomes a subsonic one, some of the kinetic energy being converted into heat energy. This is in agreement (see Chapter 7) with the observation that the converse change from a subsonic to a supersonic flow is not observed to take place. The interested reader is referred to articles by Becker (1922), Penney (1950) and Freedman (1952).

Experience shows that the ultimate vacuum reached is determined more by the speed of the pump than by the limiting vacuum it can in theory attain. This is because it is difficult to avoid a continuous emission of gas occluded in the walls of the vessel into the vacuum. In this respect Gaede–Langmuir pumps are very favourably placed. The maximum speed given by the formula already deduced is 11.2*A* l/s in which A is the area of the annular space between the edge of the bell and the water-cooled wall of the barrel which can easily be 10 cm^2 for a small pump with a 5 cm internal diameter. The actual speed reached by such pumps is about 30 l/s showing that $\frac{2}{3}$ of the molecules diffusing into the annular space round the jet bounce back owing to unfavourable collisions with the molecules in the jet. This is not surprising since there must be a region in which the molecules moving radially have not reached the water-cooled walls and been removed from the jet by condensation. For large pumps (1000 l/s or more) careful design of the jets can decrease the fraction of molecules scattered back before they are incorporated into the jet to about 3/5. In small laboratory pumps the rate of evacuation is usually limited by other considerations to well below the maximum available at the throat of the pump.

It is obvious that while condensation pumps will efficiently remove gas from the space being evacuated they do so at the expense of filling the whole space with the vapour of the working fluid which escapes from the jet. If the working fluid is mercury whose vapour pressure at room temperatures is 0.13 Pa (10^{-3} mmHg) this is unacceptable and the mercury vapour has to be removed by means of a liquid nitrogen trap. If special organic oils are used for the working fluid whose room temperature vapour pressure can be as low as 1.3×10^{-7} Pa (10^{-9} mmHg) a simple water-cooled baffle above the throat of the pump usually produces acceptable vacuum conditions.

3.13 Molecular flow

The pump must be connected to the apparatus being evacuated by means of a connecting tube into which a stopcock or valve is usually incorporated. This connecting piece impedes the flow of the molecules from the apparatus into the throat of the pump and the effective pumping speed may well be determined by the maximum possible flow through the connecting tube. The design of suitable connections is an important part of vacuum engineering. To this end we must consider what happens when a molecule strikes the wall of a vessel. As a preliminary we examine the condition that the wall makes no difference to the state of the gas and makes no detectable difference to the motion of the molecules in the gas. This will be the case if the molecules leaving an elementary area $\mathrm{d}S$ of the wall have on the average the same velocity and energy distribution as the molecules coming through an elementary area $\mathrm{d}S$ in the body of gas. The number of molecules striking such an area has already been calculated; it is $n\bar{c}\,\mathrm{d}S/4$ per second. This then is the number of molecules which must leave the area $\mathrm{d}S$ of the real wall. This is an obvious conclusion; if this condition were not fulfilled gas molecules would steadily leave the gas and the walls would be acting like the 'getter' described in (section 3.12).

The angular distribution of the molecules leaving the wall must next be considered. The number of molecules passing through an area $\mathrm{d}S$ is independent of its orientation. This does not however mean that they arrive at $\mathrm{d}S$ at an equal rate from all directions. The rate of arrival of molecules at $\mathrm{d}S$ from an element $\mathrm{d}\tau$ of the gas is easily obtained. Consider the gas in the annular element of volume between r and $r + \mathrm{d}r$ and θ and $\theta + \mathrm{d}\theta$, for which $\mathrm{d}\tau = 2\pi r^2 \sin\theta \mathrm{d}\theta \mathrm{d}r$ (Fig.3.5b). The element of area $\mathrm{d}S$ subtends a solid angle $\mathrm{d}S \cos\theta/r^2$ at all points in this volume so that the rate at which molecules arrive at $\mathrm{d}S$ which had their last collision in $\mathrm{d}\tau$ is

$$\frac{n\bar{c}}{4\pi\lambda} \cdot \frac{\mathrm{d}S \cos\theta}{r^2} \cdot \mathrm{e}^{-r/\lambda} \cdot 2\pi r^2 \sin\theta \mathrm{d}\theta \mathrm{d}r.$$

Carrying out the integration with respect to r yields

$$\frac{n\bar{c}\,\mathrm{d}S \cos\theta}{4\pi} \cdot 2\pi \sin\theta\,\mathrm{d}\theta = \frac{n\bar{c}\,\mathrm{d}S \cos\theta}{4\pi}\,\mathrm{d}\omega$$

in which $d\omega$ is the solid angle included between cones of semi-vertical angle θ and $\theta + d\theta$. Thus, to imitate the imaginary area dS, the real area at the wall will have to emit molecules into each element of solid angle $d\omega$ at a rate given by $(n\bar{c}\, dS \cos\theta/4\pi)d\omega$. This represents a marked preference for emitting molecules normal to the surface and zero emission in a direction parallel to the surface. Finally, the molecules will have to be emitted again on the average with the same velocity distribution as that of the incident molecules; this will at the same time look after the momentum balance and ensure that the wall neither heats, cools nor accelerates the gas. The $\cos\theta$ distribution is usually referred to as 'Knudsen's cosine law'; it is familiar in optics as Lambert's law. It is the law which ensures that the brightness of an incandescent surface does not depend upon the angle from which it is viewed. It is not always accurately obeyed. It can be achieved in two quite separate ways best described in terms of an optical analogy. One way is through specular reflection; the other is diffuse scattering such as is shown by the magnesium oxide screens used in photometry. Specular reflection can only take place from sufficiently flat surfaces; the permissible degree of roughness depending on the angle of incidence. Reflection is a cooperative phenomenon in which the reflected beam builds up by the reinforcement of the separate elementary beams scattered in all directions from each element of the surface. A rough but on the average plane surface fails to reflect because the phase relationship between the scattered waves is disturbed by the hills and dales on the surface. It is clear from Fig.3.14 that the permissible degree of roughness is given by the relationship $h \sin\phi \ll \lambda$ between h the height of the hills, ϕ the glancing angle and λ the wavelength of the radiation. Since the wavelength of molecules given by de Broglie's formula $\lambda = h/mv$ is at most 0.1 nm one sees that specular reflection of molecules at ordinary surfaces is unlikely. It should be possible from the surface of very flat freshly cleaved crystal surfaces. This is found to be the case by direct experiments using molecular beams, and cleaved NaCl and LiF surfaces. Indeed the same surface can give quite good specular reflection for helium and hydrogen molecules and exact agreement with Knudsen's cosine law for atoms of Li, K and Cs. Thus, in view of the weight of the experimental evidence and the theoretical considerations, one has no hesitation in adopting Knudsen's cosine law as the basis of ordinary calculations.

If a volume of gas at low pressure is connected by a long tube of cross section S to a fast vacuum pump the process by which the vessel is evacuated is an example of random wandering in one dimension. Molecules which enter the tube have a negligible chance (strictly zero if the tube is slightly curved) of proceeding without

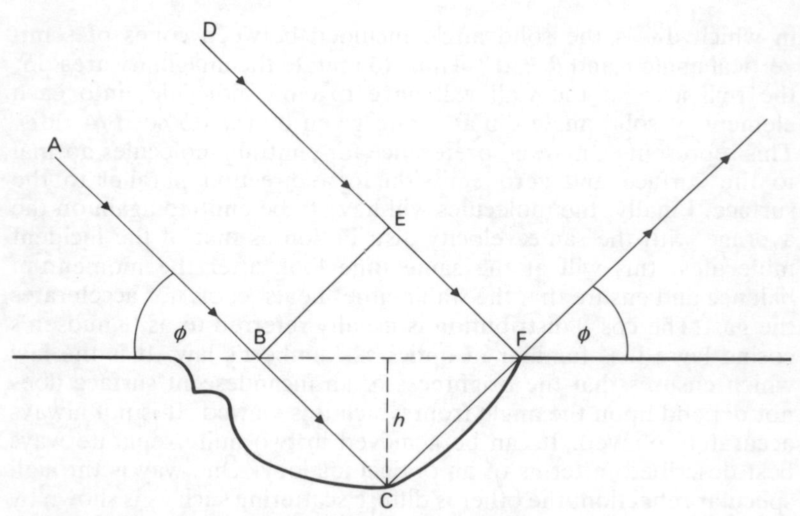

Fig. 3.14 Specular reflection at a rough surface.
The ray AB is scattered at C instead of B because the surface is uneven. The path BCF is longer than the path EF by BC + CF − EF = $2h \sin \phi$, because C is in a dale h deep. For specular reflection the paths must be the same and $2h \sin \phi \ll \lambda$.

collision to the pumped end of the tube. They will proceed by a zigzag path up and down the tube, sometimes emerging at the pump end and sometimes returning to the vessel from which they came. If a molecule which enters the tube has a probability P of emerging at the other the rate of evacuation of the vessel will be $P \cdot n\bar{c}S/4$ mol/s. If the tube is long the molecules will each make many collisions with the wall of the tube and one can define a mean free path λ_a as the mean distance moved along the axis of the tube between collisions. For long tubes no great error will be made in assuming that all the steps are of the same length λ_a and with this assumption P is readily calculated in terms of λ_a. Let the length of the tube l be an integral number of mean free paths so that $l = (a + b)\ \lambda_a$, a and b being integers. The progress of the molecule down the tube can be imagined to be determined in the following way: after each free path λ_a a true coin is spun to determine whether the molecule proceeds a further λ_a down the tube towards the pump, or returns a distance λ_a towards the vessel being evacuated. For a molecule at a point a λ_a from the entrance the probability of leaving the tube at the pump end is a/(a + b). This result is obtained by comparing the molecular problem with the probability of winning the well-known game of chance worked out in the example in section 6.2. Applying this result to the molecular problem gives at once that the rate of flow down the tube is

$$\alpha\frac{n\bar{c}S}{4}\cdot\frac{1}{\mathrm{a}+\mathrm{b}}=\alpha\frac{n\bar{c}S}{4}\cdot\frac{\lambda_{\mathrm{a}}}{l}$$

in which α is a factor to take into account the complicated situation at the entrance of the tube where the assumption of a constant λ_{a} is clearly inapplicable. For a cylindrical tube of radius r, $\lambda_{\mathrm{a}} = r$ and the flow down such a tube is $\alpha n\bar{c}\pi r^3/4l$. The value of α was first found indirectly by Knudsen (1900). The gas streaming through the tube has not got a Maxwell distribution. Knudsen approximated to the rate of molecular flow through the tube by applying a conventional kinetic theory calculation to a gas with a Maxwellian distribution distorted by the molecular flow. If one approximates to the unknown distribution by giving each molecule a drift velocity v one has, for each section of the tube, that the flux of molecules is given by $F = nvS = \pi nvr^2$ for a cylindrical tube. The density of molecules n is not constant but in the steady state the net flux is the same throughout the length of the tube. Each molecule striking the wall has an extra momentum mv so that the momentum per second transferred to a section of the cylindrical wall between l and $l + \mathrm{d}\,l$ is

$$\frac{n\bar{c}}{4}\cdot 2\pi r\mathrm{d}l\cdot mv=\frac{mF\bar{c}}{2r}\,\mathrm{d}l.$$

The whole momentum transferred to the wall of the tube is therefore $mF\bar{c}l/2r$ units per second. For a long tube the momentum of the emerging gas can be neglected and the momentum transferred to the walls is equal to the force pS which the gas would exert on the end of the tube if it were blocked and there was no flow. Equating the two expressions gives for a cylindrical tube that

$$pS=\pi r^2p=\frac{mF\bar{c}l}{2r}$$

$$\text{and } F=\frac{2\pi r^3}{m\bar{c}l}\cdot\tfrac{1}{3}nm\overline{c^2}=\frac{\pi^2n\bar{c}}{4}\cdot\frac{r^3}{l}.$$

The calculated flow is not very sensitive to the precise form of the assumed deviation from Maxwell's distribution; on the reasonable assumption preferred by Knudsen that the extra momentum of each molecule is on the average proportional to its velocity c, the constant $\pi^2/4$ is changed to $2\pi/3$ corresponding to a value for α of $\frac{8}{3}$. A less arbitrary derivation is given by Smoluchowski (1910).

The resulting formula

$$F = \frac{2\pi r^3}{3} \cdot \frac{r^3}{l} = \frac{\pi}{12} n\bar{c} \frac{d^3}{l}$$

agrees well with experiment and is used in design work. It is not applicable to short tubes and a correcting factor

$$\frac{1}{1 + \frac{8}{3}\frac{r}{l}}$$

due to Clausing is usually applied.

Worked example

Show that Clausing's correction gives the correct theoretical flow through a circular hole. For a very short tube r/l is very large, the correction becomes $3l/8r$ and the number of molecules flowing through the tube is

$$\frac{3l}{8r} \cdot \frac{\pi n\bar{c}d^3}{12l} = \frac{\pi r^2 n\bar{c}}{4}$$

which is the theoretical value $n\bar{c}S/4$.

Other methods of finding α give the same result. The most direct is due to Clausing (1932) who solved the integral equation which expresses the obviously true statement that the number of molecules arriving at any element of surface in the tube is the sum of those arriving directly from the source and those arriving after scattering from the walls. In this way Clausing obtained the numerical values for the probability of transmission P through short tubes shown in Table 3.6. These results agree well with the values found experimentally for smooth surfaces shown in Table 3.7. While this agreement is satisfactory it should be noted that all the methods of calculating the molecular flow down a tube depend upon assumptions which are known not to be strictly true: Knudsen's on

Table 3.6

l/r	1	2	3	4	5	6	7
P	0.672	0.514	0.420	0.357	0.311	0.275	0.248
l/r	8	10	12	14	16	18	20
P	0.225	0.191	0.166	0.147	0.132	0.119	0.109

(Clausing's results recalculated by J. C. Helmer, 1967.)

Table 3.7 Flow through circular cylinders

Material	l/r	P(experimental)	P(Clausing)
Flame polished glass	1.03	0.653	0.667
	2.04	0.536	0.509
	3.02	0.440	0.419
	5.03	0.314	0.314
Polished brass	0.97	0.684	0.679
	1.02	0.661	0.668
	1.93	0.496	0.522
	1.94	0.530	0.520
	1.95	0.531	0.519
	2.42	0.406	0.469
	2.90	0.439	0.428
	2.91	0.399	0.427
	4.79	0.287	0.324
	4.83	0.283	0.322
	4.84	0.308	0.321

(Taken from D. H. Davies, L. D. Leveson and N. Millerton, 1964.)

an assumed velocity distribution, Kennard's and Smoluchowski's on an assumed constant density gradient, and Clausing's on an assumed constant density across any diameter. The degree to which these assumptions are fulfilled can be judged from Helmers' calculation of the density $n(x)$ and the density gradient $\mathrm{d}n/\mathrm{d}x$ in a tube for which $l/r = 10$ given in Table 3.8. Te dependence of the molecular flow down a tube upon the distribution of the molecules near the ends is illustrated in the following example based upon a first approximation to an alternative treatment attributed to Sandler 1967.

Worked example

Show that if P_1 and P_2 are the probabilities of a molecule being transmitted down similar tubes of lengths l_1 and l_2, the probability P_{12} of transmission down both tubes in series is given by

$$\frac{1}{P_{12}} = \frac{1}{P_1} + \frac{1}{P_2} - 1$$

Use this result to show that if the number of molecules striking a surface is $n\bar{c}/4$ the probability P_1 for a tube of length l and radius r is

$$\frac{1}{1 + \dfrac{l}{2r}}.$$

[Worked example continues page 107]

Table 3.8 Value of n_x and $\frac{dn}{dx}$ in a tube for which $\frac{l}{r} = 10$

$\frac{x}{r}$	1	2	3	4	5	6	7	8	9	10
$n(x)$	0.830	0.746	0.665	0.564	0.503	0.422	0.341	0.258	0.173	0.085
$-r\frac{dn}{dx}$	0.086	0.084	0.081	0.081	0.081	0.081	0.081	0.083	0.085	0.088

If N molecules enter the first tube, P_1N emerge and P_1P_2N are transmitted down the combined tube. At the junction, $P_1N(1-P_2)$ molecules are reflected; of these $P_1^2N(1-P_2)$ return to the source and $P_1N(1-P_1)(1-P_2)$ are again reflected and arrive again at the mouth of the second tube to be partially reflected and partially transmitted. The total number transmitted allowing for these reflections at the junctions is thus

$$N(P_1P_2 + P_1P_2(1-P_1)(1-P_2) + P_1P_2(1-P_1)^2(1-P_2)^2 + \dots)$$

$$= \frac{NP_1P_2}{1-(1-P_1)(1-P_2)} = NP_{12}$$

and

$$\frac{1}{P_{12}} = \frac{1}{P_1} + \frac{1}{P_2} - 1 \text{ as stated.}$$

Consider a tube of radius r and length $l + \Delta l$, so that l_2 corresponds to a short section Δl. The value of P for this section can easily be calculated if one knows that number of molecules falling upon it. On the assumption stated $n\bar{c}/4 \times \frac{1}{2} \times 2\pi r\Delta l$ molecules per second will fall on this section as it will receive none from the empty space beyond the end of the tube; of these one half will be scattered forward and one half scattered back into the tube. The fraction of molecules reflected is thus

$$\frac{n\bar{c}\pi r\Delta l}{8} \bigg/ \frac{n\bar{c}}{4}\pi r^2 = \frac{\Delta l}{2r} = (1 - P_{\Delta l}).$$

Thus, $P_{\Delta l} = 1 - \dfrac{\Delta l}{2r}$ and $\dfrac{1}{P_{l+\Delta l}} = \dfrac{1}{P_l} + \dfrac{1}{1-\dfrac{\Delta l}{2r}} - 1$

which can be written as

$$\frac{d}{dl}\left(\frac{1}{P_l}\right) = \frac{1}{2r}$$

which on integration gives

$$P_l = \frac{1}{1 + \dfrac{l}{2r}}$$

since $P_l = 1$ when $l = 0$.
For a long tube $P_l = 2r/l$ which corresponds to a value of α the end correction of 2 instead of 8/3.

The slow diffusion of molecules down a tube seriously affects the pumping speed which can be achieved in practice. For example, the pumping speed through a metre of 50 mm diameter tubing cannot exceed 2.3 l/s. For this reason the taps used on the high vacuum side of a pumping system often consist of a flat sliding plate placed across the tube which leaves the pumping tube unobstructed when it is withdrawn into a side arm.

Worked example

What is the greatest rate at which a vessel containing low pressure air can be evacuated through a tube 10 mm diameter and 500 mm long?

The maximum flow through the tube is $\pi n\bar{c}d^3/12l = N$ molecules per second. The corresponding rate of pumping v is the volume occupied by these N molecules at the initial pressure at which there are n molecules per m^3, so that $v = N/n = \pi n\bar{c}d^3/12nl = \pi\bar{c}d^3/12l$ m^3/s. For the tube in question $d = 10^{-2}$ m, $l = 0.5$ m and for air at 0°C $\bar{c} = 448$ m/s and the maximum rate at which the vessel can be evacuated is $4.48\pi \times 10^{-4}/6 = 2.35 \times 10^{-4}$ m^3/s = 0.235 l/s.

The marked difference between molecular flow at low pressures which is proportional to the cube of the radius and viscous flow which is proportional to the fourth power of the radius should be noted.

Problems

1 A standard method of measuring the atmospheric pollution due to SO_2 is to expose a PbO_2 surface for several weeks and then analyse for the $PbSO_4$ formed. If $PbSO_4$ is formed at the rate of 1 mg per month per cm^2 of exposed surface, calculate the minimum amount of SO_2 (parts per million) in the polluted atmosphere.

2 A light wheel carrying four radial mica vanes is pivoted in an evacuated glass bulb. The vanes are blackened on one side and made reflecting on the other and are arranged so that the blackened surfaces face in the same sense round the periphery. Exposed to sunlight the wheel spins merrily so that the blackened surfaces move away from the sun. The vanes rotate because the blackened side is hotter than the reflecting side: molecules leaving the hot side have a greater value of $\bar{c}$ and so a greater momentum and there is a net torque on the vane. To be effective this must be transferred directly to the walls of the vessel. Make an estimate of the best gas pressure for the gas in the radiometer. Does it matter with what gas it is filled? Calculate approximately the pressure at which the 'radiometer effect' is neutralised by the pressure of the sunlight which is in the opposite sense. (The light pressure can be calculated on the assumption that each light quantum has a momentum $h\nu/c$ so that $p = \frac{1}{3}$ of the energy density of the radiation for isotropic radiation).

3 Calculate an upper limit to the absolute rate of oxidation of a B_4C control rod in air at 513 °C from the information that the heat of activation is about 1.59×10^5 J/mole and the particle size about 1 μm (Dominey J. *Chem. Soc. A.* 1968, **3**, 712).

4 Use the data of Fig. 3.10 to calculate the activation energy for the decomposition of hydrogen iodide.

5 If a porous pot fitted with a manometer is covered with an inverted beaker full of hydrogen the pressure in the pot is observed to rise. What explanation can you give of this observation? Discuss the specification of the porous material which will give a maximum effect and make a rough estimate of the way the pressure will change with time.

6 Suggest ways of measuring the speed of a vacuum pump (~1000 l/s) for air at 0.1 Pa.

7 Design an apparatus to measure how the molecular flow through short, wide tubes depends on the pressure drop and explain how you would try and overcome the obvious difficulties. (You may compare your design with that of D. H. Davies, L. L. Leveson and N. Millerton, *J. App. Phys.*, (1964) **35**, 529).

8 If the total effective cross section for collision of the molecules in a gas is Σ per unit volume show that the probability that a molecule going a distance l without collision is $e^{-\Sigma l}$.

9 Using the expression of section 3.7 for the density of gas diffusing from an origin show that
(a) the total amount of gas in the system remains constant
(b) the density of gas at a given place rises and then falls and calculate the time t at which the density is a maximum at a given radial distance r.

10 The internal surface of a two litre spherical glass bulb which is to be sealed after evacuation is covered with an absorbed layer of water molecules about 20 molecules thick, spaced a molecular diameter apart. This is slowly driven off when the walls are raised to 400 °C. If this quantity of gas has to be pumped through the constriction at which the vessel has ultimately to be sealed off at a pressure which never exceeds 0.01 Pa, make an approximate calculation of the minimum time required to pump the vessel down to a pressure at which the mean free path is 1000 times the diameter of the bulb. For the purposes of calculation take the constriction to be a capillary 2 mm diameter and 10 mm long. (The m.f.p. at s.t.p. for water vapour is 55 nm).

11 Calculate the minimum velocity which will enable molecules of hydrogen and helium to escape from the earth. If one interprets the composition of the upper atmosphere to mean that hydrogen can escape and helium is very unlikely to, what temperature does this suggest for the top of the atmosphere?

12 Integrate the barometric equation $dp/dh = -g\rho$ for an isothermal atmosphere allowing for the variation of g with h to obtain the modified barometric formula

$$p_h = p_0 \, e^{\left(\frac{-mgah}{kT(a+h)}\right)}$$

for the pressure p_h at a height h above a spherical earth of radius a in which g is the surface value of the acceleration due to gravity.
Calculate the value of h at which the helium pressure is 100 times the nitrogen pressure. Why do you think the result of your calculation is not in agreement with the observed facts? (Vide Kaye and Laby 1973).

Further reading

Becker, R, (1922), *Zeit. fur Physik*, **VIII**, 321.

Chapman, S. and **Cowling, T.G.** (1970) *Mathematical Theory of Non-uniform Gases*, CUP, London.

Condon, E.U. and **Odishaw, H.** (1967) *Handbook of Physics*, Sect. 5.6, McGraw-Hill, N.Y.

Dushman, S, (1962) *Scientific Foundations of Vacuum Technique*, Wiley N.Y.

Farkas, A. (1935) *Orthohydrogen, Parahydrogen and Heavy Hydrogen*, C.U.P. London.

Freedman, E.H. and **Greene, E.F.** (1957) *American Institute of Physics Handbook 2*, 231, McGraw-Hill, N.Y. and London.

Hirschfelder, J.O. Curtiss, C.F., and **Bird, R.B.** (1964) *Molecular Theory of Gases and Liquids*, Wiley, N.Y.

Jeans, J.H. (1925) *Dynamical Theory of Gases*, CUP, Cambridge.

Kennard, E.H. (1938) *Kinetic Theory of Gases*, McGraw-Hill, N.Y.

Laidler, K.J. (1950) *Reaction Kinetics*, McGraw-Hill, N.Y.

Penney, W.E. and **Pike, H.M.M.** (1950) *Shock Waves and the Propagation of finite Pulses in Fluids*, *Reports on Progress in Physics*, **XIII**, Physical Society, London.

Ramsey, N.F. (1956) *Molecular Beams*, Clarendon Press, Oxford.

Yarwood, J. (1955) *High Vacuum Technique*, Chapman Hall, London.

References

Alexander, P. (1946) *J. Sci. Inst.* **23**, 11.

Alty T. and **Mackay, C.A.** (1935) *Proc. Roy. Soc.* **149A**, 104.

Birks, J. and **Bradley, R.S.** (1949) *Proc. Roy. Soc.* **198A**, 226.

Bradley, R.S. (1951) *Proc. Roy. Soc.* **205A**, 555.

Bodenstein, M. and **Meyer, V.** (1893) *Chem. Ber.* **26**, 1146.

Clausing, P., (1932) *Ann. der Phys.*, **12**, 961.

Davies, D.H., Leveson, L.L. and **Millerton N.** (1964) *J. App. Phys.* **35**, 529.

Delaney L.J., Houston R.W. and **Eagleston L.C.** (1964) *Chem. Eng. Sci.* **19**, 105.

Helmer, J.C. *Vacuum Science and Technology* (1967) **4**, 360.

Hobbs, P.V. (1974) *Ice Physics*, Clarendon Press, Oxford.

Houston, R.A. (1929) *An Introduction to Mathematical Physics*, p. 101, Longman, Green and Co., London and N.Y.

Kaye, G.W.C. and **Laby, T.H.** (1973) *Tables of Physical and Chemical Constants*, Longman, London and N.Y.

Koros R.M. Beckers J.M., and **Bauderts M.** (1964) *Condensation and Evaporation of solids*, p. 681, Gordon Breach, N.Y.

Kutcher, H. (1955) *Zeit, für Angewandte Phys.* **7**, 229, 234.

Miller, R.C. and **Kusch, P.** (1955) *Phys. Rev.* **99**, 1318.

Milne, A.E. (1929) *Phil Mag.* **(7)**, **VII**, 273.

Neumann, K. (1932) *Zeit. für Phys. Chem.*, A171, 416.

Neumann K. and **Volker, E.**, (1932) *Zeit für Phys. Chem.* **A161**, 33.

Nöller, H.G. (1955) *Zeit. für Angewandte Phys.* **7**, 218, 225.

Reichardt, M. (1955) *Zeit für Angewandte Phys.* **7**, 300.

Richardson, G. (1928) *Phil Mag.* **(7) VI**, 1019.

Roberts, R.C. (1957) *American Institute of Physics Handbook*, 2.220, McGraw-Hill, N.Y.

Rutner E., **Goldfinger P.** and **Hirth J.P.** (1964). *Condensation and Evaporation of solids*, Gordon and Breach, N.Y.

Smoluchowski, von, (1910) *Ann. der Physik*, **33**, 1559.

Tschudin, K. (1946) *Helv. Phys. Acta.* **19**, 91.

Volmer, M and **Estermann, I.** (1921) *Zeit für Physik* **7**, 4.

Volmer, M. (1931) *Zeit. für Phys. Chem.* (*Festschrift*) **863**

Ward, J.W., **Mulford R.N.R.**, and **Kahn, N.** (1967) *J. Chem. Phys.* **47**, 1710.

CHAPTER 4

Transport of heat energy and momentum

4.1 Transport

If a gas is not at a uniform temperature it reaches a state of uniform temperature, through molecular collisions. This process would be described in terms of the physics of continuous matter as taking place by a combination of heat conduction and convection. The velocity of the convection currents is limited by the viscous drag exerted at the interfaces where there is a velocity gradient. If there is a temperature gradient, but no convection, heat energy flows through the gas by conduction. This heat energy is not an entity in itself but is always part of the energy of the molecules of the gas. If a molecule has a kinetic energy $= \frac{1}{2}mc^2$ it is not possible to unscramble this kinetic energy into a part which belongs to the gas and a part which belongs to the heat energy flowing through the gas. Nevertheless the net result of all the molecular collisions is that energy is transferred from the region of high temperature to the region of low temperature. The transfer takes place by the continual exchange of molecules between neighbouring layers of the gas which are at different temperatures but are only a few mean free paths apart. The free path of each molecule is terminated by a collision in which two molecules share their energy. Although there is no net transfer of matter, there is a net transfer of energy since molecules from the hot layer have on the average more energy to share than those from the cold layer. The process is analogous to the way goods are distributed by a road transport network, never remaining long in one lorry and redistributed at each depot. In gases the process is usually referred to as 'molecular transport'. The main transport phenomena in gases are heat conduction and viscosity. Clearly a long mean free path and a high molecular velocity favour rapid transport and the heat conductivity k and the viscosity η can be calculated in terms of λ and $\bar{c}$.

4.2 Viscosity

Consider a gas subject to a steady shearing by being placed between a fixed plate and a parallel moving plate so that there is a velocity gradient $\mathrm{d}v/\mathrm{d}z = a$ in the gas as shown in Fig. 4.1. The force F per unit area on either plate is $\eta\, \mathrm{d}v/\mathrm{d}z$ from the definition of η; the sense of the force is in the direction of v on the fixed plate and in the opposite direction on the moving plate. The fixed plate is thus gaining momentum from the gas and the moving plate losing momentum to the gas at the rate of ηa per unit area. The viscous

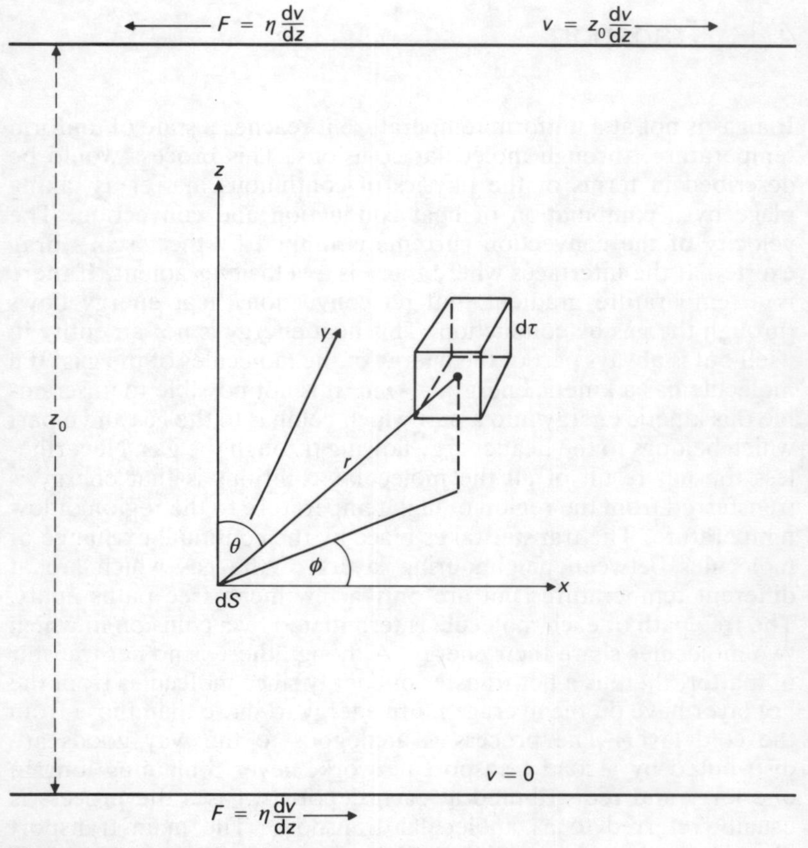

Fig. 4.1 Viscous drag between parallel plates.
The velocity of the gas is $v_0 + az$ in which v_0 is the velocity at the origin of the system of coordinates. Only $\mathrm{d}v/\mathrm{d}z = a$ enters into the calculation of the viscous drag and the momentum transfer.

process can thus be described as a flow of momentum across the gas and is a typical transport process. The momentum being transferred is always part of the momentum mc of the molecules between the plates. These molecules no longer have a Maxwell distribution function. Because $v \ll \bar{c}$ the distortion of the distribution function is very small. The state of the gas between the plates can also be described as one in which the establishment of the Maxwell distribution function by molecular collision is continually prevented by the influx (by exchange with neighbouring layers) of molecules with a distorted distribution function. Consider an elementary area $\mathrm{d}S$ parallel to the plates at the origin of a system of polar coordinates set up in the gas. Relative to the gas in the layer containing $\mathrm{d}S$, the $n\bar{c}\mathrm{d}S/4$ molecules which cross $\mathrm{d}S$ from the side of the moving plate all come from the region in which the gas has a net positive component of momentum in the direction of v while the $n\bar{c}\mathrm{d}S/4$ molecules crossing $\mathrm{d}S$ from the side facing the fixed plate all come from a region of net negative momentum. There is thus a net flow or transport of momentum through the area $\mathrm{d}S$. By equating the net flow of momentum to the known value $\eta\, \mathrm{d}v/\mathrm{d}z$ one can obtain an expression for η in terms of the molecular constants of the gas. The calculation may be carried out in the same way as was used in section 3.4 to calculate the flux of molecules. The drift velocity v is supposed to be negligibly small compared with $\bar{c}$ the mean molecular velocity. The deviation from Maxwell's distribution can thus be neglected except in the calculation of the drift momentum mv. The rate at which molecules are setting out from an element of volume $\mathrm{d}\tau$ towards $\mathrm{d}S$ is as before

$$\frac{n\bar{c}\mathrm{d}S \cos\theta}{4\pi\lambda r^2}\,\mathrm{d}\tau$$

molecules per second. Each of these molecules carries an extra momentum $mv = m(\mathrm{d}v/\mathrm{d}z)z = mar\cos\theta$. Since only $\mathrm{e}^{-r/\lambda}$ of these molecules arrive at $\mathrm{d}S$ the flux of momentum through the area from the element $\mathrm{d}\tau$ is

$$\frac{n\bar{c}\mathrm{d}S \cos\theta}{4\pi\lambda r^2}\cdot \mathrm{e}^{-r/\lambda}\cdot mar\cos\theta\,\mathrm{d}\tau$$

and the total momentum passing through $\mathrm{d}S$ from the whole of the gas is obtained by summing the contributions from all the elements of the gas; that is by integrating with respect to the three variables r, θ, ϕ. Since these variables occur independently the summation offers no mathematical difficulties and the total sum is the product of the simple integrals with respect to each variable. Thus the total momentum passing through dS from the whole of the gas is

$$\frac{mn\bar{c}a}{4\pi\lambda}\,\mathrm{d}S \underbrace{\int_0^\infty r\mathrm{e}^{-r/\lambda}\,\mathrm{d}r}_{\lambda^2} \times \underbrace{\int_0^\pi \cos^2\theta \sin\theta\,\mathrm{d}\theta}_{\frac{2}{3}} \times \underbrace{\int_0^{2\pi} \mathrm{d}\phi}_{2\pi}$$

$$= \frac{mn\bar{c}\lambda a}{3}\,\mathrm{d}S = \eta\frac{\mathrm{d}v}{\mathrm{d}z}\,\mathrm{d}S$$

Thus, the coefficient of viscosity is given by

$$\eta = \frac{mn\bar{c}\lambda}{3}.$$

The gain in momentum of an elementary layer of gas between z and $z + \mathrm{d}z$ is easily calculated since each molecule arriving at the layer has a path length in the layer $\mathrm{d}r = (\mathrm{d}z/\cos\theta)$ and, therefore, a probability of being stopped and giving up its momentum to the layer of $(\mathrm{d}z/\lambda\cos\theta)$. Thus, the total momentum gained by the layer per second per unit area is

$$\frac{mna\bar{c}}{4\pi\lambda^2}\,\mathrm{d}z \int_0^\infty r\mathrm{e}^{-r/\lambda}\,\mathrm{d}r \times \underbrace{\int_0^\pi \cos\theta \sin\theta\,\mathrm{d}\theta}_{0} \times \int_0^{2\pi} \mathrm{d}\phi$$

and is identically zero. This shows that the stratified gas with a linear velocity gradient $\mathrm{d}v/\mathrm{d}z = a$ is stable and once this state is set up it will remain unchanged if the moving plate continues to move with a velocity v. This conclusion is not valid near the plates since the range of integration for r is then limited on the side nearer to the plate. Hence, the drift velocity of the gas near to the stationary plates does not decrease to zero, but varies in some way as shown in Fig. 4.2. This is quite a different behaviour from that often postulated in the hydrodynamics of a continuous medium in which it is assumed that there is no slip at the boundary. The increased velocity near the boundary increases the momentum brought up to the surface and so the tangential force acting on it. This increase is necessary in order to give a consistent result since it is obvious that if there is a constant linear velocity gradient right up to the boundary there will only be a drag of $F/2$ instead of F. It is sometimes assumed that the slip at the boundary exactly produces the missing factor 2, This is not quite true and a correction must be applied.

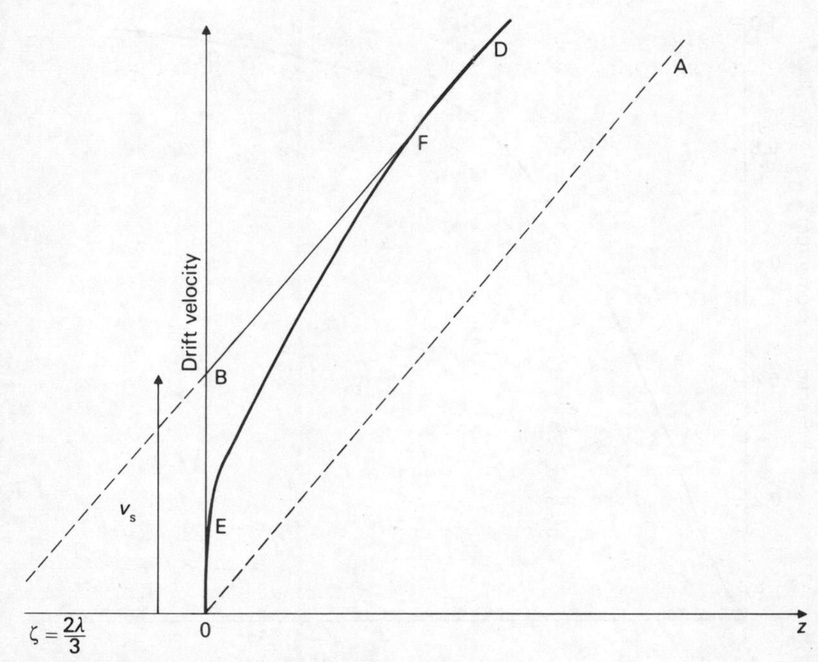

Fig. 4.2 Momentum transfer and drift velocity.
(a) OA = simple linear gradient.
OB = slip velocity, EFD = true velocity,
BD assumed velocity.

Integration shows that the fraction of the flux of momentum which comes from the layers not more than q free paths away is given by an expression whose behaviour is shown in Fig. 4.3. It is evident that the disturbance due to the wall will not extend beyond a few mean free paths and the correction for slip is negligible except at low pressures. A more detailed account is given in section 4.5. While an elementary layer is in mechanical equilibrium in a linear velocity gradient, this cannot be true of its energy balance since momentum and energy are different functions of the velocity. If u, v, w are the molecular velocities in a gas at rest, the velocities in a gas with a drift velocity V_x along the x axis will be $u + V_x$, v, w and its kinetic energy will be $\frac{1}{2}m(u + V_x)^2 + v^2 + w^2$. Averaged over a large number of molecules the excess energy due to the horizontal motion of the layer is $\frac{1}{2}mV_x^2$ since u is sometimes positive and sometimes negative. In the case of viscous flow which we are considering $V_x = za$ and the rate of gain of energy per unit area by a layer of gas between z and $z + \mathrm{d}z$ by transport from the neighbouring

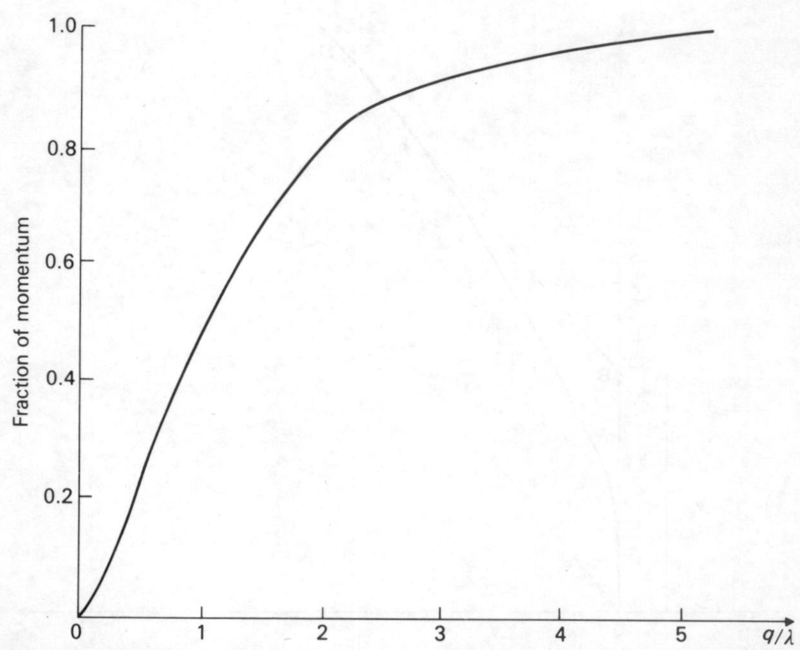

Fig. 4.3 Fraction of the total momentum flux which is received from layers less than q mean free paths from dS.

layers is

$$\frac{1}{2}\frac{\mathrm{d}z\, mna^2\bar{c}}{4\pi\lambda^2}\int_0^\infty r^2 \mathrm{e}^{-r/\lambda}\,\mathrm{d}r \times \int_0^\pi \cos^2\theta \sin\theta\,\mathrm{d}\theta \times \int_0^{2\pi} \mathrm{d}\phi$$
$$= \frac{mn\bar{c}\lambda a^2 \mathrm{d}z}{3}$$

This is independent of the position of the layer and the gas between the plates is gaining energy at the rate of $(mn\bar{c}\lambda/3)a^2 z_0$ J/s/m^2. The rate at which the external force is doing work on the gas is

$$Fv = \eta a \times z_0 a = \frac{mn\bar{c}\lambda}{3} \cdot a^2 z_0$$

Thus, the whole of the work done on the gas by the external forces (which being viscous forces have no potential) is converted into the internal energy of the gas, whose temperature rises steadily.

The attenuation of sound affords another example of the conversion of mechanical energy into heat energy by the interchange of molecules between layers of different velocities. In a plane sound wave a common displacement $\xi = \xi_0 \cos 2\pi(vt - x/\lambda)$ is superimposed on common displacement $\xi = \xi_0 \cos 2\pi (\nu t - x/\lambda)$ is superimposed on the chaotic motion of the molecules of the gas. Since all the displacements are equal and parallel it is tempting to say that there is no velocity gradient perpendicular to the motion and therefore that there is no viscous degradation of mechanical energy into heat. This conclusion is not correct. Owing to the dependence of ξ upon x each undisturbed cube of the gas of side a is distorted into a rectangular block of sides a, a and $a(1 + \mathrm{d}\xi/\mathrm{d}x)$ and is periodically changing its shape and volume. Since a change of shape always involves a shear there will always be a degradation of mechanical energy in a sound wave. The calculation of the work done against the viscous forces requires a wider mathematical knowledge than might have been expected and was carried out by Kirchoff and Stokes. The interested reader is referred to Rayleigh's *Sound* (1926). An approximate molecular treatment leads to a similar conclusion. The expression $\eta \times$ (velocity gradient)^2 for the rate of dissipation of energy per unit volume in viscous flow does not depend upon the direction of the velocity gradient with respect to the direction of flow and applies equally well to a sound wave in which the velocity gradient $\mathrm{d}^2\xi/\mathrm{d}x\mathrm{d}t$ is parallel to the flow of the gas in the wave. One therefore expects a dissipation of mechanical energy $\eta(\mathrm{d}u/\mathrm{d}x)^2$ per unit volume in both cases. Lord Rayleigh's more profound analysis gives a value of the attenuation coefficient α in the relationship $\xi = \xi_0\, \mathrm{e}^{-\alpha x} \cos 2\pi\, (\nu t - x/\lambda)$ which corresponds to $\frac{4}{3}\eta(\mathrm{d}u/\mathrm{d}x)^2$ for the rate of dissipation of energy in a sound wave. There is a comparable loss due to heat conduction between the peaks and the troughs of the waves and the full expression for the attenuation coefficient is

$$\frac{4\pi\nu^2}{2c^3\rho}\left[\frac{4\eta}{3} + (\gamma - 1)\frac{k}{C_p}\right]$$

in which k is the heat conductivity. The attenuation is rather small for accurate measurement but for air it is said to be about half as much again as the classical theoretical value (see Kaye and Laby (1975) and Am. Inst Phys. Handbook (1957). At low frequencies some of this discrepancy can be accounted for by the neglect of attenuation by diffusion in which molecules from the dense parts of the wave diffuse into the regions of lower density.

4.3 Heat conductivity

The transport of internal energy by diffusion in a gas not in a steady state can be looked upon as the process of heat conduction. If a gas with no mass motion is maintained in a stratified state between two parallel plates with a linear temperature gradient in the gas, the exchange of molecules between the layers will cause heat energy to flow through the gas which will behave like a simple conductor of heat. To achieve this condition it is important to avoid convection. This can be done by having the plates horizontal and arranging that the top plate has the higher temperature. Convection is also reduced by working at about 1/10 atm pressure, below which convection in the volumes of gas usually dealt with in a laboratory is very small. The reason for this is that the forces promoting convection are proportional to the density while the opposing viscous forces are more or less independent of the gas pressure. (For a full discussion see Rogers (1976) and for a simplified account McKensie (1976)).

The heat conductivity can be calculated approximately in terms of the mean free path as follows. With the same choice of axes as was used to discuss viscosity let the temperature T of each layer be given by $T_z = T_0 + \mathrm{d}T/\mathrm{d}z \cdot z$. If c_v is the heat capacity per unit mass at constant volume the heat capacity of one molecule is mc_v and the average excess energy due to the temperature gradient is $z\,\mathrm{d}T/\mathrm{d}z\; mc_v$ per molecule. Then as before the flux of energy across the plane element $\mathrm{d}S$ is

$$mc_v \frac{\mathrm{d}T}{\mathrm{d}z} \cdot \frac{n\bar{c}}{4\pi\lambda} \cdot \mathrm{d}S \int_0^\infty r e^{-r/\lambda}\,\mathrm{d}r \times \int_0^\pi \cos^2\theta \sin\theta\,\mathrm{d}\theta$$

$$\times \int_0^{2\pi} \mathrm{d}\phi = \frac{mc_v n\bar{c}\lambda}{3} \frac{\mathrm{d}T}{\mathrm{d}z}\,\mathrm{d}S = k \frac{\mathrm{d}T}{\mathrm{d}z}\,\mathrm{d}S$$

and k the heat conductivity is given by

$$k = \frac{mc_v n\lambda\bar{c}}{3}.$$

In this calculation $\bar{c}$ has been taken as that appropriate to the mean temperature of the gas. This is a reasonable approximation because as we have seen all the transport comes from molecules in a layer not more than a few λ away. Thus, at room temperatures a 1 per cent variation in $\bar{c}$ corresponding to a 2 per cent variation in temperature = 5.4 °C occurring in such a thin layer would imply at atmospheric pressure a temperature gradient of 5×10^6 K/m, quite outside the range of experimental conditions contemplated. Similarly n the number of molecules per unit volume has been assumed constant. The distortion of the Maxwell distribution by the temperature gradient has also been neglected. (Strictly speaking a collection of molecules not having a Maxwell distribution do not have a temperature). Further to avoid difficulties in the integration it has been assumed that each molecule has the same share of the excess energy due to the temperature gradient. To take these errors properly into account means effectively abandoning the mean free path approximation and using more advanced methods.

4.4 Variation with temperature and pressure

These expressions for the viscosity and heat conductivity of a gas first given by Maxwell (in 1861 and 12 years before van der Waals published his equation) had a considerable impact on the experimental physics of the time. They directed attention to the existence in gases of phenomena which while well known in liquids had hardly been considered in connection with gases. Not only did they predict correctly the order of magnitude of the coefficients of viscosity and heat conductivity for gases but also indicated that these quantities would vary with temperature in a way which was quite different from that which occurred in liquids. The dependence of η and k upon $\bar{c}$ the mean molecular velocity implies that both viscosity and heat conductivity should be proportional to the square root of the absolute temperature. This predicted increase in the viscosity of a gas with rising temperature is in marked contrast with the behaviour of liquids in which the viscosity falls rapidly with increasing absolute temperature T approximately in agreement with an exponential relationship

$$\eta = A\mathrm{e}^{-B/T} \quad \text{or} \quad A\mathrm{e}^{B/T-T_0}.$$

The difference in behaviour with respect to temperature is because viscosity in gases and liquids have quite different molecular explanations, and are only formally related through the definition of the coefficient of viscosity. The viscosity of a gas is a transport phenomena which does not depend upon attractive molecular forces for its existence. The viscous force between the moving layers of a gas is due to the exchange of molecules of different momenta. In contrast to this the forces between the moving layers of a liquid are the actual forces between the molecules; indeed in very viscous liquids like pitch or glass the phenomenon of viscosity is better described in terms of a relaxation time in which the stresses in a deformed solid gradually disappear as the molecules take up new positions. The strong variation with temperature is for the most part to be explained in terms of the expansion of the liquid and the rapid fall-off of the attractive force between the molecules with distance.

Another somewhat surprising prediction is that both viscosity and heat conductivity should be independent of the pressure of the gas, since the factor $n\lambda$ is insensitive to pressure. In the case of hard spheres

$$\lambda = \frac{1}{\sqrt{2}\pi n\sigma^2}$$

so that

$$n\lambda = \frac{1}{\sqrt{2}\pi\sigma^2}$$

which depends on the size of the molecules and not on the number per unit volume. The preliminary confirmation of these almost paradoxical predictions was looked upon as a triumph for the elementary kinetic theory. This paradox is resolved if one thinks in terms of transport; less molecules transport the same amount because they travel further at the lower pressure.

4.5 Wall effects

The theoretical value of the coefficients of viscosity (η) and heat conductivity (k) have been calculated by considering the transport of momentum and heat energy through the main body of the gas. To obtain η and k from the experimental results some consideration must be given to the conditions near the surfaces between which the transport is taking place.

Viscosity and the correction for slip

As mentioned in section 4.2, the drift velocity a few mean free paths from the wall cannot obey the simple linear law that it does in the main body of the gas. The variation could be calculated on the assumption that each layer must lose as much momentum as it gains. It would be a mathematical exercise of little physical significance unless it took into account the deviation from Maxwell's distribution function near the wall. Fortunately, a simple approximation is adequate for the purpose of interpreting the experimental results. One makes the assumption (a common one in this type of problem) that the velocity gradient a is everywhere the same, and that the wall is displaced by a distance ζ. This is equivalent to saying that the drift velocity near the wall can be approximated to by the simple expression

$$v = v_s + az$$

in which v_s is the slip at the real wall and v is the drift velocity at a distance z from the real wall. The value of v_s is adjusted to give the correct force on the wall. The rate at which the molecules of the gas are bringing up tangential momentum to each unit area of the walls is $\frac{1}{2}\eta a$ due to the velocity gradient and mv_s for each of the $n\bar{c}/4$ molecules which are striking the wall per second, due to the slip. Equating this to the known drag which can be taken as ηa for the purpose of calculating a correction gives at once that

$$\frac{mn\bar{c}v_s}{4} + \frac{1}{2}\eta a = \frac{mn\bar{c}v_s}{4} + \frac{1}{6}nm\bar{c}\lambda a = \eta a = \frac{1}{3}nm\bar{c}\lambda a$$

so that

$$v_s = \frac{2\lambda a}{3} = \frac{2\eta a}{nm\bar{c}} \quad \text{and} \quad \zeta = \frac{2\lambda}{3} = \frac{2\eta}{nm\bar{c}}$$

On this assumption the formal calculation of the viscous drag between two plates d apart has to be made between the displaced plates and the expression $F = \eta v/d$ for the force per unit area has to be replaced by

$$F = \frac{\eta v}{d + 2\zeta} = \frac{\eta v}{d + \frac{4\lambda}{3}} = \frac{\eta v}{d + \frac{4\eta}{nm\bar{c}}}$$

if the real plates are moving with a relative tangential velocity v. As seems first to have been explicity stated by Millikan the last expression can be tested experimentally without knowing the value of λ. This is another way of saying that the value of λ used in the correction must be obtained from the viscosity data and not from some other source.

Worked example

A rotating cylinder viscometer consists of an outer rotating cylinder (internal radius b) and an inner coaxial cylinder (radius a) suspended on a torsion wire. Find the torque per unit length of the inner cylinder due to the viscous drag of the gas between cylinders. Each element of gas distance r from the axis will flow in a circular path with an angular velocity ω so that the tangential velocity $v = \omega r$. The velocity gradient in the direction perpendicular to the flow is

$$\frac{dv}{dr} = \frac{d(\omega r)}{dr} = \omega + r\frac{d\omega}{dr}.$$

If ω is independent of r the gas is rotating as a whole like a solid body, no viscous forces are called into play and $dv/dr = \omega$. The effective velocity gradient which shears the gas and determines the viscous forces is

$$\frac{dv}{dr} - \omega = r\frac{d\omega}{dr}.$$

Consider an annular ring of gas of unit length. Since this is in steady motion there is no net torque acting upon it. To achieve this the torques acting on the inner and outer surfaces which are of opposite sign must have the same numerical value. Accordingly the torques $\pm T$ acting on the inner and outer surfaces of the annular ring, given by tangential force × arm must not depend on the radius r at which they are applied, so that

$$T = \underbrace{\eta}_{\text{viscosity}} \times \underbrace{2\pi r}_{\text{area}} \times \underbrace{r\frac{d\omega}{dr}}_{\text{effective gradient}} \times \underbrace{r}_{\text{arm}} = 2\pi\eta r^3\frac{d\omega}{dr} = \text{constant}$$

The differential equation $d\omega/dr = T/2\pi\eta r^3$ integrates at once to give

$$T = \frac{4\pi\eta\Omega a^2 b^2}{b^2 - a^2},$$

in which T is the torque on the outer surface of the inner cylinder and the constant of integration has been chosen to that $\omega = \Omega$ when $r = b$ and $\omega = 0$ when $r = a$. If there is slip ζ at the walls, $b \to b + \zeta$ and $a \to a - \zeta$ and to the first order in ζ

$$T = \frac{4\pi\eta\Omega a^2 b^2}{b^2 - a^2}\left\{1 - \frac{2\zeta(a^3 + b^3)}{ab(b^2 - a^2)}\right\}$$

Deviations from the simple law were first looked for by Maxwell using an oscillating disc viscometer. He worked over a limited pressure range, found, owing to an arithmetical slip, that no correction was necessary to account for his results, and used this observation to support the (erroneous) view that the repulsive force between molecules varied inversely as the fifth power of the distance between them. Fifty years later Millikan (1923) working at lower pressures was able to detect and measure accurately the correction for slip using a rotating cylinder viscometer. He found that the apparent decrease in η at low gas pressures was well accounted for by the approximate theory. Deviations which had been previously observed were usually expressed in terms of a fraction f introduced by Maxwell. He divided the molecules into two classes; a fraction f which gave up all their drift momentum to the wall and the remaining fraction $(1 - f)$ which lost no momentum parallel to the wall and were specularly reflected. On this assumption the slip velocity is easily shown to be given by

$$v_s = \frac{2a}{3}\left(\frac{2}{f} - 1\right)\lambda$$

The experimental results expressed in this notation are given in Table 4.1. One can sum up these results by saying that for metal surfaces there is no specular reflection; all the 'slip' can be accounted for by the distortion of the linear velocity gradient in the gas. For other rather ill-defined surfaces there is evidence that up to 10 per cent of the molecules retain their tangential momentum after they have collided with the wall.

Table 4.1

Gas	**Wall**	***f***	**Method and year**
Air	Machined brass	1.00	Rotating cylinder, 1923
	Mild steel	1.00	Rotating cylinder, 1957
	Aluminium	1.00	Rotating cylinder, 1957
	Mercury	1.00	Falling drop.
	Oil	0.90	Falling drop.
	Old shellac	1.00	Rotating cylinder, 1923
	Fresh shellac	0.79	Rotating cylinder, 1923
	Glass	0.89, 0.77	Capillary tube, 1875, 1909, 1946
	Oil on aluminium	0.90	Rotating cylinder, 1951
CO_2	Machined brass	1.00	Rotating cylinder, 1923
	Old shellac	1.00	Rotating cylinder, 1923
	Oil	0.92	Falling drop.
He	Oil	0.87	Falling drop.
N_2	Mild steel	1.00	Molecular beam, (1957).
	Glass	0.97	Molecular beam, (1957).
	Aluminium	1.00	Molecular beam, (1957).

(Taken in part from Methods of Experimental Physics, (Ed. Estermann, I., 1959).

Heat conduction and the correction for accommodation

In an experiment to measure the heat flow between parallel plates the linear temperature gradient as in the analogous case of viscous drag cannot be maintained to within a few mean free paths of the hot and cold plates. The situation can be dealt with as before by supposing the linear temperature gradient a to be maintained between imaginary plates displaced by ζ so that the heat flow Q per unit area between plates d apart maintained at temperatures T_1 and T_2 is given by

$$Q = \frac{k(T_2 - T_1)}{d + 2\zeta}$$

In calculating ζ two factors have to be taken into account. Firstly, the nonlinear temperature gradient near the plates necessary to maintain thermal equilibrium. This is a function of λ only and is the origin of a temperature jump at the plates. Secondly, the degree to which the molecules arriving at a plate reach thermal equilibrium with it before they leave again to join the main body of the gas. This aspect of the problem is represented by an accommodation coefficient defined by

$$\alpha = \frac{T_{\text{gas}} - T_{\text{reflected}}}{T_{\text{gas}} - T_{\text{plate}}}$$

in which $T_{\text{gas}} - T_{\text{plate}}$ is the temperature jump between the gas and the plate and $T_{\text{reflected}}$ is the mean temperature of the molecules diffusely reflected from the plate back into the gas. If the molecules come into thermal equilibrium with the plate before returning to the gas, $T_{\text{reflected}} = T_{\text{plate}}$ and the accommodation coefficient is unity. The accommodation coefficient α is the analogue of Maxwell's f so that the temperature jump is given by

$$\Delta T = \mathrm{A}\left(\frac{2 - \alpha}{\alpha}\right) a\lambda$$

in which A is a constant.

The calculated value of A depends upon how the approximations to allow for the deviations of the gas near the surface plates from a Maxwellian distribution are made. A detailed discussion is given in Kennard (1938) who gives

$$\mathrm{A} = \frac{2k}{\eta(c_p + c_v)} \approx 1$$

To determine α it is not necessary to find ζ from an elaborate series of measurements of the variation of the heat conductivity with pressure and then use the theoretical relationship between ζ and α to calculate the accommodation coefficient. It can be determined directly by measuring the heat lost from a hot wire in a gas whose pressure is so low that the molecules leaving the wire reach the walls of the container without making a molecular collision, thus eliminating the correction for the temperature jump between the gas and the wire.

Worked example

What is the energy lost per second by conduction per metre in helium at 0.1 Pa by a wire 200 μm diameter and accommodation coefficient 0.8 maintained at 1000 °C to a coaxial cylinder 10 mm, diameter and unit accommodation, coefficient maintained at 25 °C. How does this loss compare with the energy lost by radiation? (neglect end effects and take the emissivity to be $\frac{1}{4}$).

From the tables $\lambda = 1.73 \times 10^{-7}$ at s.t.p. Taking atmospheric pressure $\approx 10^5$ Pa gives a mean free path at 0.1 Pa equal to $1.73 \times 10^{-7} \times 10^{+6} = 1.73 \times 10^{-1}$ m: this is much larger than the radius of the cylinder and it is legitimate to neglect molecular collisions and assume that all molecules reaching the wire have a Maxwellian distribution corresponding to 25 °C. Molecules strike the wire at a rate corresponding to $n\bar{c}/4$ per unit area. From the tables at s.t.p. $n = 2.68 \times 10^{25}$ and $\bar{c} = 1.202 \times 10^3$ for He. Thus

$$\underbrace{2.68 \times 10^{25} \times 1.2 \times 10^3 \times 2.5 \times 10^{-1}}_{\frac{n\bar{c}}{4}\text{ at s.t.p.}} \times \underbrace{\sqrt{\frac{273}{298}}}_{\substack{\text{temperature}\\ \text{correction}\\ \text{to } 25\,^{\circ}\text{C}}} \times \underbrace{\frac{1.0 \times 10^{-1}}{1.01 \times 10^5}}_{\substack{\text{pressure}\\ \text{correction}\\ \text{to } 0.1\text{ Pa}}}$$

$$= 7.62 \times 10^{21}$$

helium molecules strike unit area of the wire each second. The area of the wire is $\pi \times 200 \times 10^{-6} = 6.28 \times 10^{-4}$ m^2 and the rate at which molecules are striking the wire is $7.62 \times 10^{21} \times 6.28 \times 10^{-4} = 4.78 \times 10^{18}$ helium molecules per second. These molecules leave the wire having increased their temperature by $\alpha\,(T_{\text{wire}} - T_{\text{gas}}) = 0.8 \times 975 = 780$ K. Each molecule therefore carries away $3k \times 780/2$ J more than it brought up and energy must be supplied to the wire at the rate of $1.5 \times 1.38 \times 10^{-23} \times 7.8 \times 10^2 \times 4.78 \times 10^{18}$ J/s $= 0.08$ W if it is not to cool.

The rate of emission of radiant energy for a black body is σT^4 W/m^2 so that the heat energy radiated by the filament is approximately

$$\underbrace{5.669 \times 10^{-8}}_{\substack{\text{Stefan's}\\ \text{constant}}} \times \underbrace{2.5 \times 10^{-1}}_{\substack{\text{coefficient}\\ \text{of emissivity}}} (\underbrace{1273^4}_{\substack{\text{(wire}\\ \text{temperature)}}} - \underbrace{298^4}_{\substack{\text{(ambient}\\ \text{temperature)}}}) \times \underbrace{6.28 \times 10^{-4}}_{\text{area}}$$

$$= 23.3 \text{ W}.$$

In spite of the large number of measurements which have been made in this way there is still a considerable degree of difficulty in interpreting them. This is because a measurement of the heat loss having been made, one still has to decide on the true area and nature of the surface to which it refers. Thus, the measured accommodation coefficient of hydrogen on clean nickel is 0.25; after exposure to oxygen so that it is covered by an invisible film of oxide 3 nm thick the accommodation coefficient has risen to 0.48. Similarly, the accommodation coefficient of hydrogen on black platinum is 0.74 and on bright platinum is 0.32; it is not at all obvious how much of this change to attribute to changes in the effective area and how much to intrinsic changes of the surface. This uncertainty makes it difficult to compare experimental and theoretical values of the accommodation coefficient. Some typical values of α are shown in Table 4.2. It will be seen from the table that there is a revived interest in the numerical values of the accommodation coefficient. This is partly because of its practical importance in the design of space capsules, as the rate they heat up on entering the atmosphere depends upon α, and partly because of its intrinsic theoretical interest. Many attempts have been made to calculate the accommodation coefficient; it is essentially a calculation of the transfer of energy from a single molecule to the quantised vibrations of a crystal lattice. The early (1914) work of Baule treated the problem as a single collision between a moving and a stationary molecule.

Table 4.2 Accommodation coefficients

Gas	**Surface**	α	T/**K**	**Author**	**Date**
H_2	Nickel (clean)	0.25	liquid air	D. R. Hughes, and	1928
H_2	Nickel (oxidised)	0.48	liquid air	R. C. Bevan,	1928
He	Nickel (flashed)	0.085	room temp.	J. K. Roberts,	1932
He	Nickel (unflashed)	0.20	room temp.	J. K. Roberts,	1932
He	Tungsten (flashed)	0.057	room temp.	J. K. Roberts,	1932
He	Tungsten (flashed)	0.025	79	J. K. Roberts,	1932
He	Aluminium	0.073	305	L. B. Thomas,	1961–66
He	Lithium	0.024	90	L. B. Thomas,	1961–66
Ne	Molybdenum	0.055	298	L. B. Thomas,	1961–66
Kr	Molybdenum	0.510	298	L. B. Thomas,	1961–66
Xe	Molybdenum	0.956	298	L. B. Thomas,	1961–66

(Taken in part from Methods of Experimental Physics (Ed. Estermann, I, 1959) and in part from F. O. Goodman and H. Y. Wachman, 1967.)

The simple relationship

$$\alpha = \frac{m_1^2 + m_2^2}{(m_1 + m_2)^2}$$

which he obtained between the accommodation coefficient and the masses of the molecules gives the right order of magnitude in spite of the complete neglect of the lattice forces. The agreement which can be obtained between a semi-empirical theory and the experimental results is shown in Figs. 4.4 and 4.5.

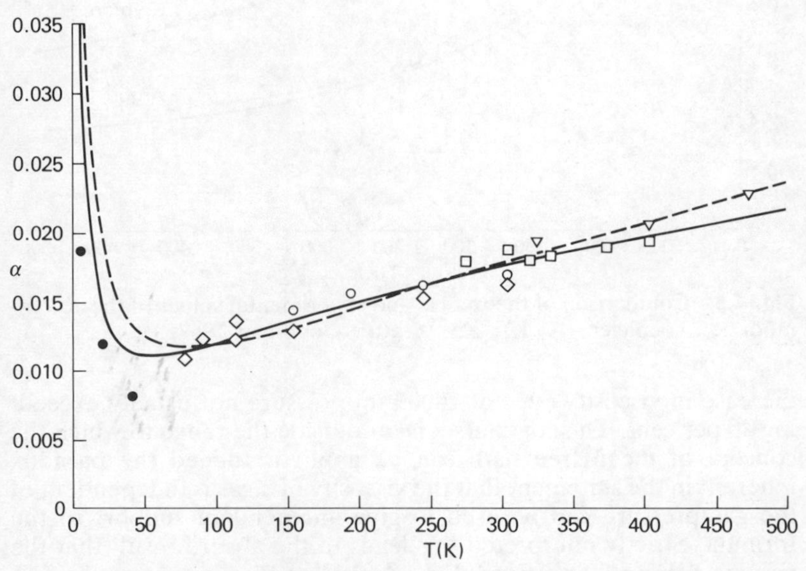

Fig. 4.4 Comparison of theoretical and experimental values of the accommodation coefficient He–W after Goodman 1967.

4.6 Comparison with experiment

Although the dependence of η and k upon temperature and pressure is roughly in agreement with Maxwell's predictions, subsequent careful experiment has shown that the simple formulae for the viscosity and heat conductivity of nearly perfect gases are not in all respects accurately obeyed. The absence of a pressure coefficient at constant temperature is in good agreement with experiment, the

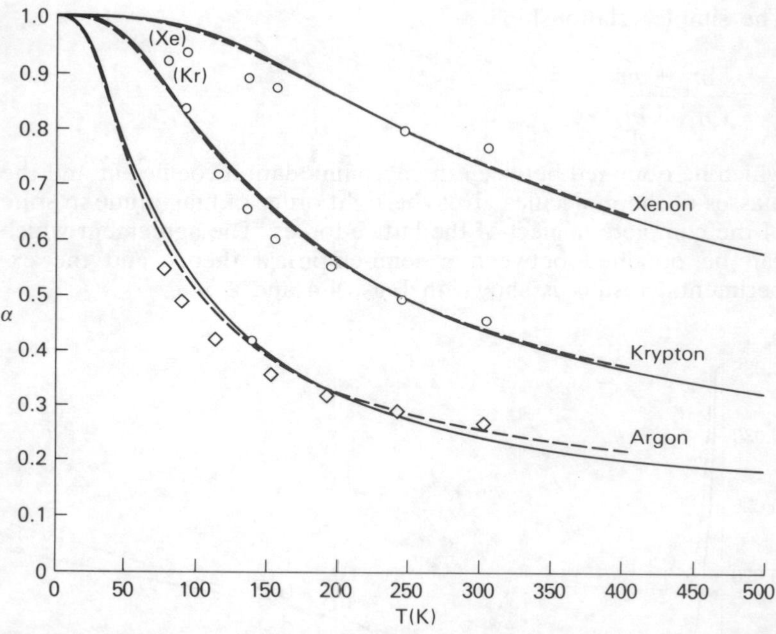

Fig. 4.5 Comparison of theoretical and experimental values of the accommodation coefficient A, Kr, Xe–W after Goodman 1967.

increase in viscosity even at 1000 atm pressure not usually exceeding 40 per cent. This, of course, is far outside the range to which the concept of mean free path can be applied. Indeed the paradox inherent in the statement that the viscosity of a gas is independent of the gas pressure can be used to give independent support to the formula. Strictly interpreted this leads to the absurd result that the viscous damping of a pendulum would not disappear in a perfect vacuum. The absence of a pressure coefficient depends upon $n\lambda$ being independent of pressure. One would expect this relationship to be more accurately obeyed the lower the pressure as the lower the pressure the less ambiguous the meaning of the term free path. However, the constancy of $n\lambda$ will fail at low pressures as the free path cannot exceed the linear dimensions of the vessel in which the gas is confined. When the pressure is as low as this there will be few collisions between molecules of the gas and all transport of energy or momentum will take place directly from one plate to the other. In these circumstances both η and k will clearly be proportional to n the number of molecules available for the transport and so to the pressure. This indeed is found to be the case and is made the basis of the Pirani–Hale gauge for measuring low pressures. The fact that the pressure at which the change from one law to another takes place varies in the correct way with the size of the vessel in which the measurements are being made is excellent supporting evidence for the view that heat conductivity and viscosity in gases are indeed transport phenomena. Although the temperature coefficient of viscosity and heat conductivity predicted by the simple formulae has the right sign and is of the right order of magnitude the simple variation with $\sqrt{T}$ is not in agreement with experiment. For example, the variation of the heat conductivity of nitrogen with temperature is well represented by the semi-empirical formula

$$k = 2.28\sqrt{T} - 13.8 \text{ mW/mK}$$

as is shown in Table 4.3.

Table 4.3 Thermal conductivity of nitrogen

T/K	93.2	173.2	223.3	273.2	293.2	373.2	473.2
(Observed × 10^3)	8.36	16.3	20.3	23.8	25.5	30.5	35.5
(From formula)	8.2	16.2	20.3	23.8	25.2	30.3	35.8

The constant term in the formula has no theoretical significance and merely represents the best fit to a shallow S-shaped curve. The fit for oxygen is about as good and for the rare gases not so good. A direct comparison between the experimental and calculated values of η and k is not very helpful because the numerical value of λ is uncertain: it is difficult to measure and ill-defined theoretically except for the hard sphere model. Also the simple formulae are known to be in error. In deriving the expression for the viscosity it was assumed that a molecule gives up the whole of the extra momentum which it is transporting in a single collision and immediately acquires the momentum appropriate to its new position. This is not so, the extra momentum being shared more or less equally between the colliding molecules. This effect is cumulative and increases the rate of transport of momentum by about 25 per cent. This phenomenon is usually referred to as the 'persistence of velocity' and can be expressed in terms of a fictitious mean free path $\lambda_{\text{transport}}$ to be used in calculating the viscosity from the simple formula. Similar considerations apply to the calculation of the heat conductivity and approximately the same value of $\lambda_{\text{transport}}$ should reconcile the experimental and theoretical values of the heat conductivity and the viscosity. This conclusion can be tested by comparing the experimental value of the dimensionless quantity $k/\eta c_v = \varepsilon$ the simple theoretical value of which is unity.

Table 4.4 Variation of ε with temperature

Gas	Temp. K						
	50.2	**194.7**	**273.3**	**373.2**	**491.2**	**579.1**	**Mean**
He	2.43	2.42	2.44	2.44	2.43	2.42	2.43 ± 0.01
A	2.50	2.52	2.52	2.51	2.50	2.48	2.50 ± 0.01
Xe	—	2.58	2.61	2.58	2.54	—	2.58 ± 0.01

(Taken from Handbuch der Physik, 1958.)

Reference to Table 4.4 shows that this conclusion is not well substantiated. However, the equally important prediction that ε is independent of the temperature at which the measurements are made is in good agreement with the experimental data. The constancy of the ratio in spite of the wide variation in the individual values of k and η confirms the theoretical view that these quantities have the same fundamental molecular explanation and represent two ways of measuring the transport of the properties of one layer to another by means of molecular interchange. This view is reinforced by the observation that the discrepancy in the numerical value of the ratio (2.4 instead of 1) is greatest for the rare gases which have a simple molecular structure and least (1.55 instead of 1) for the complicated molecules like carbon dioxide and ethylene (ethene) and varies systematically with the complexity of the molecule. The reason for this discrepancy is well understood to be that the heat conductivity has not been correctly calculated. The heat conductivity was calculated on the assumption that each element of the gas had the same value of n, λ, and $\bar{c}$ and that the molecules in a hot layer all had the same extra energy $c_v\Delta T$. This assumption is approximately true for complicated molecules with a large number f' of degrees of freedom of internal vibration and rotation. The energy $mc_v\Delta T$ is made up of two parts, $f'(k\Delta T/2)$ of internal energy and $3k\Delta T/2$ of kinetic energy of translation. If $f' \gg 3$, $c_v\Delta T$ will have an appreciable part of its value not depending directly on the velocity of translation c and the conditions assumed in the simple derivation are approximately fulfilled. For monatomic gases, ($f' = 0$) they are not fulfilled at all and it is not surprising that $k/\eta c_v$ shows the greatest discrepancy from simple theory for such gases. For monatomic gases it is evident that the fast molecules transport a higher proportion of the heat energy than the slow molecules. Not only have they more energy to transport but they transport it faster because of the high value of c and further since their mean free path λ_c is greater than the mean free path λ of all the molecules. Correspondingly, the slow molecules transport very little energy; they have little to transport and do not get very far with it since their mean free path λ_c is less than the mean free path λ. Thus, in general one can expect that monatomic gases will have a greater value for ε than polyatomic gases. Beyond this point elementary discussion is not very effective. At most a rough estimate of the errors inherent in a simple mean free path treatment can be made. The fundamental difficulty is that a gas with a temperature gradient has to satisfy two conditions; the pressure must be the same everywhere and there must be no net flux of molecules along the temperature gradient. These conditions are not compatible with a Maxwellian distribution in which the temperature varies along the gradient. Some account may be taken of these difficulties in the following way. If the pressure is everywhere constant it would require $mc_p\Delta T$ J/mol to convert one layer of gas into an adjacent one so that it would be better to put $k = \frac{1}{3}nm\bar{c}\lambda c_p$ thus increasing the calculated heat conductivity by a factor $\gamma = 5/3$ for a monatomic gas and 7/5 for a diatomic gas. Furthermore the assumption that all the molecules in a layer had the same extra energy $mc_v\Delta T$ was made for mathematical convenience. Acceptable for molecules with many internal degrees of freedom, it is clearly wrong for monatomic gases. The correction on this account will increase the heat conductivity by a factor which should not be very different from the 4/3 found in the simplified example of section 3.4. The correction due to the increase in the mean free path with velocity is ~4 per cent according to Jeans, and can be ignored in a rough computation. The result of this discussion is to suggest that the properly calculated value of ε might be as great as 2.2 for monatomic gases and less for diatomic gases. This is quite close to the experimental value. This agreement, although encouraging, is not significant. It merely confirms the view that the rather large discrepancy between the experimental and mean free path values of ε is compatible with heat conductivity in gases being entirely due to the transport of energy by the exchange of molecules between neighbouring layers. The solution to this problem was first found by Maxwell. Using a different method he confined himself to molecules repelling each other according to an inverse fifth power law for which the flux of energy can be calculated without knowing the exact form of the distribution function. He obtained a multiplying factor of $2.5 = \varepsilon$. The calculations have been extended by Enskog and by Chapman using a Lennard–Jones potential. Their value of ε does not differ much from Maxwell's value. For hard spheres $\varepsilon = 2.2$. The value of ε for a polyatomic gas can be estimated by linear interpolation. The resultant formula

$$\frac{k}{\eta c_v} = \frac{2f' + 15}{2f' + 6} = \frac{9\gamma - 5}{4}$$

suggested by Eucken (1913) agrees with the experimental results Table 4.5 better than might be expected.

Table 4.5 Variation of ε with molecular structure

Gas	ε	$\frac{1}{4}(9\gamma - 5)$
H_2	2.02	1.92
N_2	1.96	1.91
C_2H_2	1.54	1.54
CO_2	1.64	1.68
N_2O	1.73	1.68
CH_4	1.77	1.695
NH_3	1.455	1.715
SO_2	1.50	1.64

(All data at 0°C and taken from Wright Patterson Air Force Base OHIO 1960 data as recalculated by S. Chapman and T. G. Cowling, 1970.)

Worked example

Obtain Eucken's interpolated formula for ε.

The heat conductivity of a gas is observed to be higher than the simple theoretical value $k = \frac{1}{3}nm\bar{c}\lambda c_v$. Instead of interpreting this in terms of a fictitious mean free path $\lambda_{transport}$ Eucken interpreted it in terms of a fictitious heat capacity c_v' to be used only in calculating heat conductivities. He then calculated c_v' in a way which expressed the view that in heat conduction kinetic energy of translation is transported more rapidly than the internal energy of the molecules. Each degree of freedom contributes $kT/2$ to the energy of a molecule so that for a gas whose molecules have f degrees of freedom the energy u is $fkT/2$ and $\mathrm{d}u/\mathrm{d}T = mc_v = fk/2$. Eucken split the f degrees of freedom into 3 degrees of translation and $f - 3$ of internal energy. In calculating c_v' he weighted the translational degrees of freedom by Maxwell's factor 5/2 to get $mc_v' = ((3 \times 5/2) + (f - 3))k/2$.

Accordingly

$$\varepsilon = c_v'/c_v = \frac{9 + 2f}{2f}$$

f can be replaced by γ through the relationship $\gamma = (f + 2)/f$ to give $\varepsilon = (9\gamma - 5)/4$. For a monatomic gas $\gamma = 5/3$ so that $\varepsilon = 5/2$ as is to be expected.

These difficulties inherent in calculations in the presence of a temperature gradient do not apply to the calculation of viscosity since all the molecules of a particular layer may reasonably be supposed to have the same excess momentum mv. Usually $v \ll \bar{c}$

and λ and $\bar{c}$ are not changed much from their Maxwellian values and the assumptions of the simple calculation are more nearly fulfilled than they are in calculating the heat conductivity. The persistence of velocity and other minor effects can be taken into account by adopting a fictitious mean free path and the modified expression $\eta = \frac{1}{2}mn\bar{c}\lambda$ gives a moderately satisfactory agreement between theory and experiment. Nevertheless for many gases the viscosity increases more rapidly with temperature than corresponds to the simple $T^{1/2}$ law, an empirical law $\eta \propto T^{\mathrm{n}}$ in which $\mathrm{n} > \frac{1}{2}$ reproduces the experimental results fairly well. (Some data are quoted in Problem 4.18). This relationship can in part be interpreted as being due to an increase in the mean free path with increasing temperature. Such an effect which cannot be explained in terms of the hard sphere model is to be expected as soon as the molecules are represented as centres of force. An increase in the mean free path with temperature is equivalent to a corresponding decrease in the collision cross section $\pi\sigma^2$. That this decrease actually occurs for point molecules of mass m and relative velocity V whose mutual potential energy is μ/r^{n} is easily demonstrated for the small fraction of the molecules which make nearly head-on collisions. As found in section 3.1 the distance of closest approach of such a pair of molecules is $(4\mu/\mathrm{m}V^2)^{1/\mathrm{n}} = r_{\mathrm{min}}$. The distance of closest approach depends upon V and an approximation to its mean value can be obtained by replacing V^2 by its mean value

$$\bar{V}^2 = 2\overline{c^2} \text{ to give } \bar{r}_{\mathrm{min}} = \left(\frac{2\mu}{m\overline{c^2}}\right)^{1/\mathrm{n}} = \left(\frac{2\mu}{3kT}\right)^{1/\mathrm{n}}.$$

If the colliding molecules had been hard spheres the initial and final motions would have been the same as that of the point molecules and the distance of closest approach would have been independent of T. One can therefore say that for head-on collisions the point molecules can be replaced by hard spheres of diameter $\sigma = (2\mu/3kT)^{1/\mathrm{n}}$, a rather strong dependence of cross section (and so free path) upon temperature. If one tries to extend this calculation to all types of collision one runs into difficulties. While the molecular orbits can be calculated, the hard spheres of diameter equal to the distance of closest approach are no longer deflected through the same angle as the point molecules they are intended to represent. Many detailed calculations have been carried out to take this into account; a very ingenious dimensional argument due to Lord Rayleigh enables some of the main results to be retrieved without going through the difficult mathematical analysis. Let the molecules be represented by centres of force repelling according to an inverse power law μ/r^s. Then since the viscosity is to be independent of the density it can depend only on the mass of the molecule, the mean velocity $\bar{c}$ and the force constant μ, the dimensions of which are m, l/t and ml^{s+1}/t respectively.

Combining the solution of the usual dimensional equation

$$\eta = \frac{ml}{t} = (m)^a(\bar{c})^b(\mu)^c = m^a\left(\frac{l}{t}\right)^b\left(\frac{ml^{s+1}}{t}\right)^c$$

with the experimental result that $\eta = AT^n$ gives at once, since the temperature $T \propto (\bar{c})^2$, $2n = (s+3)/(s-1)$ and $n = \frac{1}{2} + 2/(s-1)$. The same result can be obtained even more simply by assuming that the drift momentum is only transferred in nearly head-on collisions.

On this assumption

$$\lambda \propto \frac{1}{\sigma^2 \text{min}} \propto \frac{1}{r^2 \text{min}} \propto T^{2/n}$$

and η which is proportional to λ and $\bar{c}$ depends upon T as $T^{1/2+2/n} = T^{1/2+2/s-1}$ if s is the exponent in the law of force μ/r^s. For hard spheres $s = \infty$ and $n = \frac{1}{2}$ as found from simple theory; in principle a careful measurement of n can be used to find the corresponding value of s. Some typical results are shown in Table 4.6.

Table 4.6 Law of force from viscosity data.

Gas	He	Ne	A	Xe	H_2	N_2	O_2	CO	CO_2	CH_4
n	0.66	0.66	0.94	0.95	0.67	0.75	0.76	0.75	0.94	0.97
s	13	13	6	6	13	9	9	9	6	7

The values of n in the relationship $\eta \propto T^n$ are determined from measurements made over the temperature range 0–600 °C. The empirical law is not well enough obeyed to justify calculating s to better than the nearest integer.

Calculations by Chapman, Enskog and others enable the transport phenomena in a gas to be calculated over a wide range of temperature in terms of the two parameters ε and σ of a Lennard–Jones potential. Viscosity and diffusion coefficients can be calculated so as to differ by not more than 5 per cent from the experimental value. The more difficult calculation of the heat conductivity is accurate to about 10 per cent. These accurate calculations have had two main results. Firstly, they confirmed that viscosity, heat conduction and diffusion in gases at low pressure, in spite of the moderate numerical agreement with the predictions of elementary theory, really are transport phenomena to be accounted for in terms of random molecular wandering. Secondly, they revealed the existence of a minor phenomenon known as 'thermal diffusion' which depends explicitly on the law of force between the particles. This has turned out to be of practical importance in the separation of isotopes. If a linear temperature gradient is maintained in a mixture of two gases a steady state will be reached with a constant heat flow across the gas. No satisfactory analaysis of this steady state can easily be made. This is because there are are two equilibrium conditions which depend quite differently on the molecular velocities c_1 and c_2 of the two species of molecules in the gas, with masses m_1 and m_2. The ratio n_1/n_2 of molecules per unit volume is determined primarily by a diffusion process. To a first approximation this is governed by the diffusion equations in which the properties of the molecules enter through the expression $\lambda\bar{c}$. Mechanical equilibrium in the gas mixture is determined by equality of pressure throughout the gas so that

$$\tfrac{1}{3}n_1m_1\overline{c_1^2} + \tfrac{1}{3}n_2m_2\overline{c_2^2} = \text{total gas pressure} = \text{constant}.$$

When the temperature is uniform this is a neutral condition; because $m_1\overline{c_1^2} = m_2\overline{c_1^2}$ it is satisfied by any state of the gas for which $n_1 + n_2$ is the same throughout the gas. The constant ratio n_1/n_2 corresponding to a uniform composition is reached and maintained by diffusion. Since pressure varies as the temperature T and the diffusion coefficient as $\sqrt{T}$ it is unlikely that the same equilibrium conditions will obtain in a temperature gradient. They do not, and the now incompatible equilibrium conditions are reconciled by allowing the composition n_1/n_2 of the gas to vary along the temperature gradient. The new equilibrium depends upon so many small interelated factors that it cannot properly be discussed in terms of the mean free path model. The weakness of a simple mean free path model in dealing with mixtures is shown by considering the experimental values of the heat conductivity and viscosity of a mixture of helium and argon. The viscosity goes through a maximum of 22.9 μ Pa s which is greater than the value for helium (22.0 μ Pa s) or argon (19.4 μ Pa s). Similarly, the heat conductivity goes through a minimum for a 1:3 mixture of helium and argon. Clearly such results cannot be accounted for on a simple mean free path model. However, if the two molecular species are isotopes so that there is only one field of force governing the three types of collisions, a plausible qualitative account can be constructed. For hard spheres the mean free path which enters the expression for the diffusion coefficient through the expression $\lambda\bar{c}$ is made up from two parts; the mean free path proper $1/\sqrt{2}\pi n\sigma^2$ determined by the molecular size and Maxwell's distribution and an additional part due to the persistence of velocity. For hard spheres the latter is due to the lack of symmetry produced by the motion of the centre of gravity of colliding molecules. As molecules from the hot parts of the gas have on the average a greater velocity than those from the colder parts the diffusion of heavy molecules against the temperature gradient is favoured and at equilibrium there are proportionately more light molecules in the high temperature region. This is in agreement with experiment for mixtures of ^{4}He and ^{3}He and other rare gases fairly well represented by a hard sphere model. Changes in the diffusion coefficient produced by giving the repulsive forces a finite range as in the inverse power model favour a concentration change of opposite sign; as a result thermal diffusion disappears for an inverse fifth power law. Since this was the only law of force seriously considered by Maxwell in his pioneer researches it is usually supposed that this is the reason he overlooked the phenomenon which was not discovered theoretically and independently until fifty years later by Chapman and Enskog. That an inverse fifth power law is critical for thermal diffusion can be made plausible by a dimensional argument. Since viscosity η and diffusion coefficient D depend in simple theory on the product $\lambda\bar{c}$ they should vary with temperature in the same way. One can therefore write $D = \mathrm{A}T^{1/2+2/\mathrm{s}-1}$ which for $\mathrm{s} = 5$ becomes $D = \mathrm{A}T$. In other words for a fifth power law of force the equilibrium conditions in a mixture remain compatible when the temperature is changed and there is no need for a compensating concentration gradient to maintain the equilibrium in a temperature gradient. It is rather surprising that thermal diffusion was not discovered experimentally since analogous effects in liquids and metals are well known. The variation in the concentration of copper sulphate in an aqueous solution in which a temperature gradient is maintained was discovered by Soret in 1887. The analagous effect in metals in which an e.m.f. is developed in a wire in which there is a temperature gradient was predicted theoretically and verified experimentally by Thomson (Lord Kelvin) in 1854. The separation produced by thermal diffusion is always smaller than that which can be achieved by direct diffusion. Its practical importance depends upon the fact discovered by Dickel and Clausius (1938) that the separation can be much more easily increased by cascading than is the case with a conventional separation using diffusion. The separation column consists of a long (several metres) vertical tube with a hot wire running along the axis. The radial temperature gradient between the wire and the wall of the tube produces a small radial separation by thermal diffusion. The vertical convection currents continually draw off the separated material which is further separated as it moves slowly along the axis of the column; about 20 theoretical stages per metre can be obtained in a well designed column. The simplicity of the arrangement makes it attractive for laboratory work on a small scale. On an industrial scale the very large power consumption prevents it competing with a cascaded diffusion plant. The degree of agreement which can be obtained between exact calculations in which the coefficients σ and ε of a Lennard–Jones potential are chosen to agree with the experimental values of the heat conductivity and viscosity are shown in Fig. 4.6.

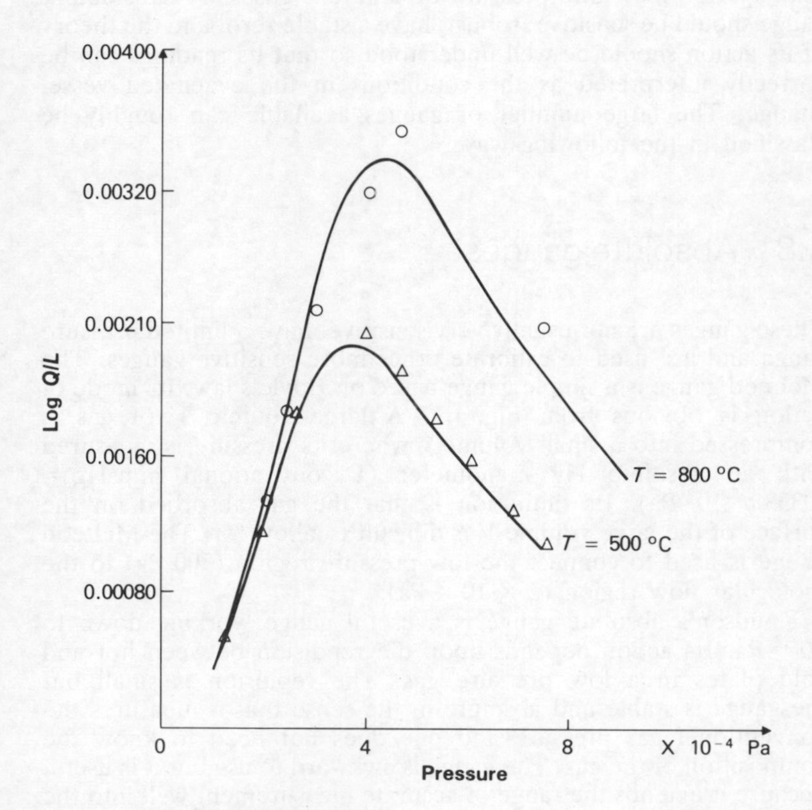

Fig. 4.6 Thermal diffusion.
Separation factor $\log_e$ Q per unit length as a function of pressure for the $^{14}NN^{14}$ – $^{15}NN^{14}$ separation. The lines – are calculated from theory. The column was 7.32 m long, 19.05 mm diameter and the central wire was 1.59 mm in diameter.
After Rutherford *et al* 1969.

4.7 Vacuum gauges

The kinetic theory provides the theoretical basis for many manometers designed to operate in the molecular flow region, $10^{-1} - 10^{-7}$ Pa. As already explained these manometers are really indicators of the number of molecules per unit volume. Owing to the usually ill-defined composition of the residual gases in a highly evacuated vessel whose origin is usually layers of gas absorbed on the walls and gas dissolved by metal parts during their manufacture, it does not matter that they are often not absolute and give different readings for the same pressure of different gases. A satisfactory gauge should be sensitive, robust, have a stable zero, and the theory of its action should be well understood so that its readings can be correctly interpreted as the conditions in the evacuated vessel change. The large number of gauges available can roughly be classified in the following way.

4.8 Absolute gauges

These gauges are not usually very sensitive, have a limited pressure range and are used to calibrate other more sensitive gauges. The McLeod gauge is a simple gauge based on Boyle's law. Its mode of action is obvious from Fig. 4.7. A large volume V of gas is compressed into a small volume v where its pressure is measured with an absolute Hg manometer (1 conventional mm Hg = 1.333×10^2 Pa). Its limitation is that the gas absorbed on the surface of the large volume V is difficult to allow for. The McLeod gauge is used to connect the low pressure region (500 Pa) to the molecular flow region (2×10^{-2} Pa).

Knudsen's absolute gauge is a useful gauge working down to 10^{-3} Pa. Its action depends upon the repulsion between hot and cold plates in a low pressure gas. The repulsion is small but the gauge is stable and absolute in the sense that it measures the conventional gas pressure and one does not need to know the composition of the gas. The gauge is awkward to use but it is useful because it extends the range of accurate measurement well into the region of molecular flow.

Worked example

Two large parallel plates at temperatures T_1 and T_2 are placed in a gas whose temperature is T_1 and whose pressure $p = nkT_1$ is so low that

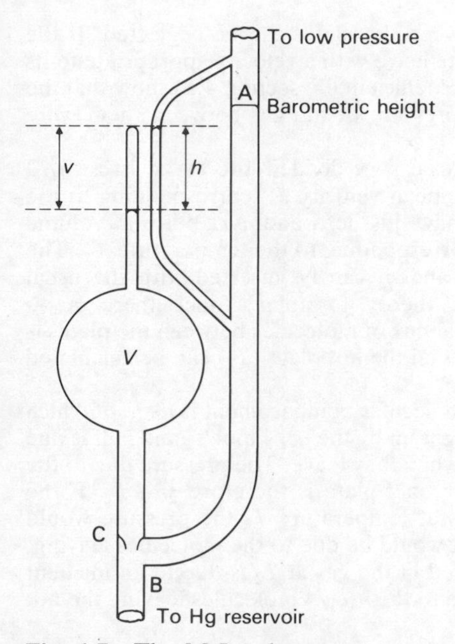

Fig. 4.7 The McLeod gauge.
The gauge is connected to the vessel to be evacuated through a liquid nitrogen trap which prevents mercury vapour entering the vessel and ensures that the gas in the gauge is not contaminated with water vapour. The pressure to be measured is sufficiently low for the top of the mercury column to stand at about the barometric height above the free surface of the mercury in the reservoir. The effective level of this surface can be changed mechanically so that the level of the mercury in the gauge is either at A or B. To operate the gauge the level is lowered to B so that the bulb is connected to the evacuated vessel through the side arm. The mercury level is then slowly raised to A. As the mercury level passes the side arm it reaches the re-entrant tube C and cuts off a definite volume V of gas in the gauge. When the mercury level reaches A in the side arm this gas has been compressed into the capillary tube at the top of the bulb and occupies the small volume v which can be read off on a scale placed behind the capillary tube. The pressure of this gas p is measured by h the depression below the level of the mercury surface in a parallel capillary tube connected to the evacuated vessel; the two capillaries are made from the same piece of tubing so that the meniscus correction due to surface tension is the same in both and can be neglected. If P is the pressure being measured and so the initial pressure of the gas in the volume V, then by Boyle's law $PV = pv$ and $P = g\rho hv/V$ Pa in which ρ is the density of the mercury. The volume v of the gas is obtained from the measured length and diameter of the cylinder which it occupies. In practice the capillary tube must be calibrated (in the usual way with a weighed mercury column) before it is sealed onto the main volume. For a detailed account of the McLeod gauge the reader is referred to *Vacuum Practice* by Dunoyer (1926).

molecular collision in the gas between the plates can be neglected. If the molecules after collision with a plate leave with a velocity appropriate to its temperature (unit accommodation coefficient c f. section 4.5) show that the plates repel each other as if the pressure of the gas between them were $p/2(1 + \sqrt{T_2/T_1})$.

The molecules between the plates can be divided into two classes: $n_1/2$ molecules per unit volume with a mean velocity $\bar{c}_1$, corresponding to the temperature T_1 of the plate they have just left, and $n_2/2$ per unit volume molecules with mean velocity $\bar{c}_2$ corresponding to the temperature T_2. The factor $\frac{1}{2}$ is introduced so that n_1 and n_2 can be inserted with the usual meaning into the simple kinetic theory formulae. Since there is no accumulation of molecules, the total flux of molecules between the plates is zero and $n_1\overline{c_1} = n_2\overline{c_2}$. The pressure on the hot plate (T_2) can be calculated as follows:

The pressure on a plate due to molecular bombardment is nkT, of which one half is due to momentum brought up by the $n\bar{c}/4$ mol/s, and half is due to the momentum they take away when they leave. The pressure due to the $n_1\bar{c}_1/4$ molecules arriving from the cold plate is therefore $\frac{1}{2}n_1kT_1$. If the plate had been immersed in a gas at temperature T_2 the pressure would have been n_2kT_2 of which one half would be due to the molecules leaving. This pressure on the plate immersed in the gas at T_2 is due to an incident flux $n_2\bar{c}_2/4$ so that the pressure due to the $n_1c_1/4$ molecules leaving the hot plate is

$$\frac{n_2kT_2}{2} \times \frac{n_1\bar{c}_1}{n_2\bar{c}_2} = \frac{n_2kT_2}{2}.$$

The pressure on the hot plate is therefore

$$\frac{n_1kT_1}{2} + \frac{n_2kT_2}{2}.$$

The flux of molecules into the space between the plates from the surrounding gas is $nc_1/4$. The outward flux is

$$\frac{1}{2}\left(\frac{n_1\bar{c}_1}{4}\right) + \left(\frac{n_2\bar{c}_2}{4}\right).$$

In the steady state the fluxes must be equal which gives at once that $n_1 = n$ and

$$n_2 = n \cdot \frac{\bar{c}_1}{\bar{c}_2} = n\sqrt{\frac{T_1}{T_2}}.$$

Substitution for n_1 and n_2 gives at once that p_{T_2} the pressure on the hot plate is given by

$$p_{T_2} = \frac{nkT_1}{2} + \frac{nkT_2}{2}\sqrt{\frac{T_1}{T_2}} = \frac{p}{2}\left(1 + \sqrt{\frac{T_2}{T_1}}\right)$$

4.9 Viscosity gauges

These gauges measure the viscous damping of an oscillating system. The simplest form consists of a fine (10 μm) silica fibre sealed to a firm base in the evacuated vessel, which can be set vibrating by tapping the base. At s.t.p. the vibrations would be completely damped, but in a high vacuum the damping is quite small and the fibre will vibrate for an hour or more. The relationship between the observable decrement and the pressure (Section 4.10) is obtained empirically. Owing to their simplicity these gauges are useful for measuring the change in pressure with time in highly evacuated sealed vessels such as X-ray tubes. Silica fibres are used because silica has a very low internal viscosity and the damping is almost entirely due to the gas.

Worked example

A fused silica fibre, Youngs modulus E and density ρ, of diameter a = 10 μm and length l = 20 mm is anchored at one end to the wall of a highly evacuated sealed glass vessel. It can be set into vibration by tapping the base of the vessel so that the transverse motion y a distance x from the base is given by

$$A(x^3 - 3lx^2)\cos \omega t\, e^{-bt}$$

The angular frequency is approximately given by

$$\omega = \frac{1.194^2\pi^2 a}{16l^2}\sqrt{\frac{E}{\rho}}$$

If the damping is almost entirely due (as it is for fused silica) to the collision of the fibre with the molecules of the residual gas in the evacuated vessel, calculate the approximate relationship between b and the gas pressure. Estimate the range of pressures for which this Langmuir gauge is useful.

In the low pressure range in which the gauge is useful the damping constant b is so small that it is legitimate to treat the motion as undamped in calculating the rate of dissipation of energy. If a small body of mass M and area S oscillates with a velocity amplitude $v \ll \bar{c}$ the number of molecules striking the surface will on the average not be very different from the $n\bar{c}S/4$ which would strike a stationary surface of the same area S. If these molecules are momentarily absorbed onto the surface they will leave it with an additional velocity equal to the velocity of the surface and the rate of transfer of energy from the oscillating body to the gas is approximately $n\bar{c}S/4 \times \frac{1}{2}m\bar{v}^2 = -\mathrm{d}E/\mathrm{d}t$ in which $\overline{v^2}$ is the mean square velocity of the surface. The kinetic energy of the body is $\frac{1}{2}Mv^2$ and if it is executing S.H.M.

under its internal elastic forces its total mechanical energy $E = M\overline{v^2}$ and

$$-\frac{1}{E}\frac{\mathrm{d}E}{\mathrm{d}t} = \frac{mn\bar{c}S}{8M}$$

Each element of the vibrating fibre can be treated in this way so that the rate of loss of energy per unit length can be calculated since the motion is known. If as in this example all parts of the fibre vibrate with s.h.m. each element of the fibre makes the same contribution to $-(1/E)\,(\mathrm{d}E/\mathrm{d}t)$ and it is not necessary to calculate the actual energy of the fibre so that one can write

$$-\frac{1}{E}\frac{\mathrm{d}E}{\mathrm{d}t} \approx 2\mathrm{b} = \frac{mn\bar{c}4\pi al}{8\rho a^2\pi l} = \frac{mn\bar{c}}{2\rho a}$$

and $1/b = \tau$ the time for the amplitude of vibration to fall to 1/e th of its initial value is $4\rho a/mn\bar{c}$. For nitrogen at 0.1 Pa, $\mathrm{n} = 2.69 \times 10^{25}/1.01 \times 10^6 = 2.66 \times 10^{19}$, $\mathrm{m} = 28 \times 1.66 \times 10^{-27}$, $\bar{\mathrm{c}} = 4.55 \times 10^2$ and $\rho = 2.65 \times 10^3$ so that

$$b = \frac{28 \times 1.66 \times 10^{-27} \times 2.66 \times 10^{19} \times 4.55 \times 10^2}{4 \times 2.65 \times 10^3 \times 10^{-5}}$$

$$= 5.31 \times 10^{-3}$$

and $\tau = 1.88 \times 10^2$ s.

The useful range of the gauge is between 1 Pa and 0.01 Pa, below which pressure it is no longer legitimate to neglect the internal mechanical losses, which depend upon how the fibre is secured to its base as well as the internal viscosity of the silica.

4.10 Thermal conductivity gauges

These measure the thermal conductivity of the gas and depend upon the change in the conductivity $k = \frac{1}{3}mc_v n\bar{c}\lambda$ with pressure which occurs as soon as λ is commensurate with the size of the vessel. Conductivity gauges are nearly all of the Pirani–Hale type in which the active element is a fine tungsten filament, well supported and mounted in a bulb connected to the vacuum vessel through a wide tube (Fig. 4.8). The filament is connected so as to form one element of a Wheatstone bridge. In a good vacuum there is little loss of heat by conduction through the gas and the filament is maintained at a relatively high temperature by the current through the bridge. On

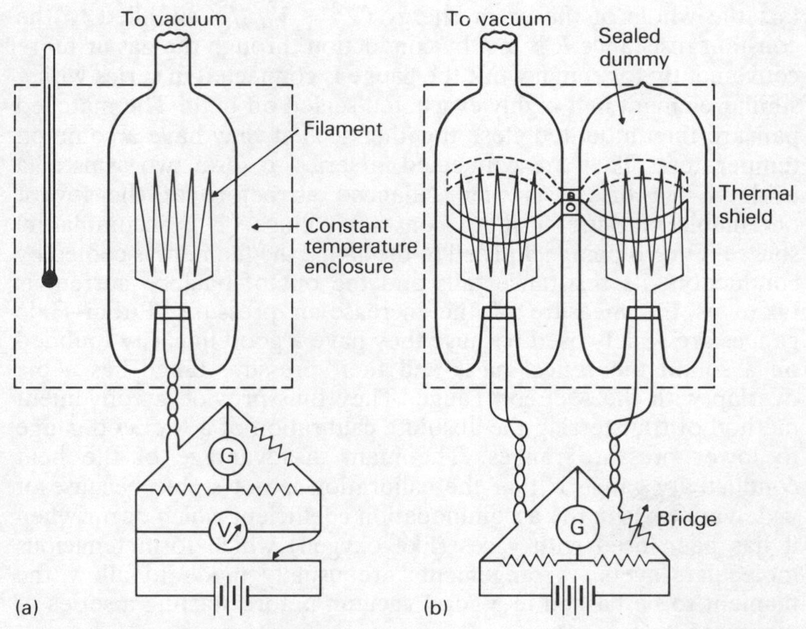

Fig. 4.8 The Pirani–Hale gauge.
The theoretically most advantageous way of using the gauge is shown in Fig. 4.8(a) in which the filament is kept at a constant temperature as the pressure of the gas in the gauge varies. This procedure reduces uncertainties due to changes in the accommodation coefficient of the wire during a series of measurements. The three fixed resistors of the bridge are well ventilated and made from an alloy with a low temperature coefficient of resistance. A more convenient way of using the gauge is shown in Fig. 4.8(b) in which a dummy sealed-off gauge is used to compensate for fluctuations in the ambient temperature. Both versions have to be calibrated against a McLeod gauge and the calibration depends upon the gas.

gas being admitted, the temperature of the filament falls and its resistance decreases thus disturbing the balance of the bridge. The gauge may be used in two ways. **Either** the active, element is surrounded by a fixed temperature jacket and the bridge is balanced by varying the voltage across (and so the current through) the bridge. If V_0 is the voltage at which the bridge balances at a very low pressure and V the voltage at which it balances at a pressure p then

$$\frac{V^2 - {V_0}^2}{{V_0}^2} = Ap$$

in the pressure range $10 - 10^{-5}$ Pa. This relationship expresses the fact the whole of the extra energy $(V^2 - V_0^2)/R$ supplied to the constant resistance R is lost by conduction through the gas **or** more conveniently for rough work the gauge is connected in series with a similar element in a highly evacuated sealed off bulb. The matched pair are then mounted close together so that they have a common temperature. They are connected in series to form two arms of a bridge. The bridge is then balanced as before at the lowest obtainable pressure, using a constant voltage (2 V accumulator) source. On gas being admitted to the gauge the filament is cooled by conduction, its resistance falls and the out of balance current is taken as the measure of the increase in pressure. Pirani–Hale gauges are widely used because they have a good linearity founded on a sound theoretical basis and their pressure range has a big overlap with the McLeod gauge. They thus provide a convenient method of transferring the absolute calibration of a McLeod gauge to lower pressure ranges. The main disadvantage of the heat conductivity gauge is that the calibration is not stable because of wide variations in the accommodation coefficient which occur when it has been used with gases (like oxygen) which form tenacious molecular layers. Arrangements are usually made to allow the filament to be flashed in a good vacuum before starting a series of measurements.

4.11 Ionisation gauges

These are the best (and indeed almost the only gauges) gauges for measuring high vacua below 10^{-3} Pa. If a beam of fast electrons ($v \approx c/10$) passes through a low pressure gas the electrons occasionally collide with and ionise the neutral molecules in the evacuated space. It is to be expected that

$$\frac{\text{the number of positive ions formed per second}}{\text{the number of electrons passing through the gas per second}}$$

$$= \text{a constant characteristic of the gas and the electron velocity.}$$

If the positive ions are collected to form a positive ion current to a subsidiary electrode one will have that

$$\frac{\text{positive ion current}}{\text{electron current}} = Ap$$

The constant A is, of course, not known *a priori* and will vary from gas to gas, but remains constant for constant conditions. Since the linearity of the gauge at very low pressures is theoretically well established it only needs to be calibrated at one pressure. The variation of A with the gas is not of great importance in practice since at the low pressures at which the gauges are used ($\lambda \approx 1$ km) the experimenter rarely wants to know either the pressure p or the molecular density n. Presented with these quantities he would proceed to calculate the probability that an electron would cross the

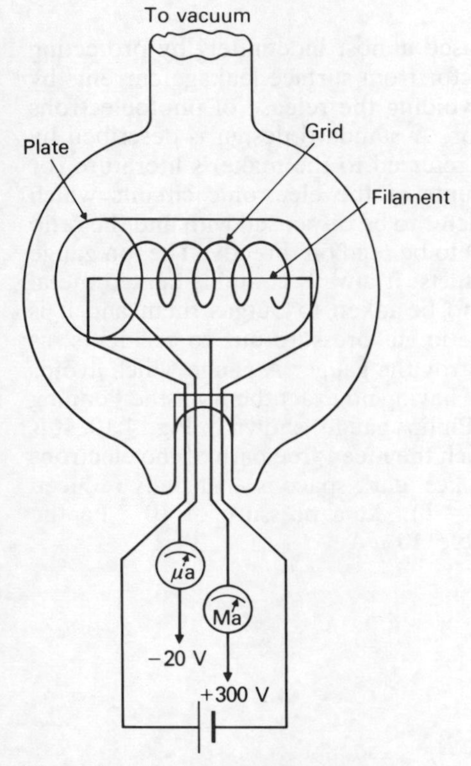

Fig. 4.9 The ionisation gauge.
The electrons from the hot filament flow through the grid, from which they are repelled, to the plate. Positive ions formed by collision with the gas whose pressure is being measured are collected by the negatively charged grid. The ratio of the positive grid current to the plate current is a measure of the amount of gas in the gauge. At very low pressures steps must be taken to ensure that the very small current measured in the grid circuit is due to the arrival of positive ions and not to surface leakage in the pinch or to the emission of photoelectrons from the grid.

vacuum vessel without an ionising collision: this is just the quantity which the ionisation gauge measures directly. A simple ionisation gauge consists of an old fashioned triode (circa 1920) consisting of a filament, widely spaced helical grid, and a coaxial cylindrical plate (Fig. 4.9). It is run with an abnormally high anode voltage (300–400 V) and an abnormally low (−20−−30 V) grid voltage which collects all the positive ions. For such an ion gauge

$$p \approx 2\tfrac{1}{2}\,\frac{\text{positive ion current}}{\text{electron current}}\ \text{Pa}.$$

The sensitivity may be increased almost indefinitely by protecting the positive ion current collector from surface leakage currents by means of a guard ring and avoiding the release of photoelectrons from the positive ion collector. A standard design is described by Alpert (1953). The reader is referred to the maker's literature (or his own ingenuity) for accounts of the electronic circuits which enable the sensitive galvanometer to be dispensed with and the ratio (ion current/electron current) to be read off directly. The ion gauge suffers from two practical defects: it always contains heated metal parts so that great care has to be taken to outgas them and it is unstable against a sudden rise in gas pressure due to accidents, as this may cause arcing and destroy the gauge. A gauge which avoids these defects at the expense of having no exact theory is the Penning gauge (often referred to as a Philips gauge) shown in Fig. 4.10. It is a simple discharge tube in which the mean free path of the electrons and so the length of the Crookes' dark space is artificially reduced by a magnetic field (4×10^{-2} T). At a pressure of 10^{-3} Pa the current is an easily measurable 10 μA.

Problems

1 A gas, pressure p, flows steadily and isothermally through a long uniform tube of length l. If the mass flow is everywhere proportional to $-p(\mathrm{d}p/\mathrm{d}l)$ show that $\mathrm{d}p^2/lp$ is constant and that the flow is proportional to $(p_1^2 - p_2^2)$ in which p_1 and p_2 are the initial and final pressures at the ends of the tube.

2 If following Maxwell one looks upon the viscosity of a very viscous fluid as a measure of the relaxation time τ with which the

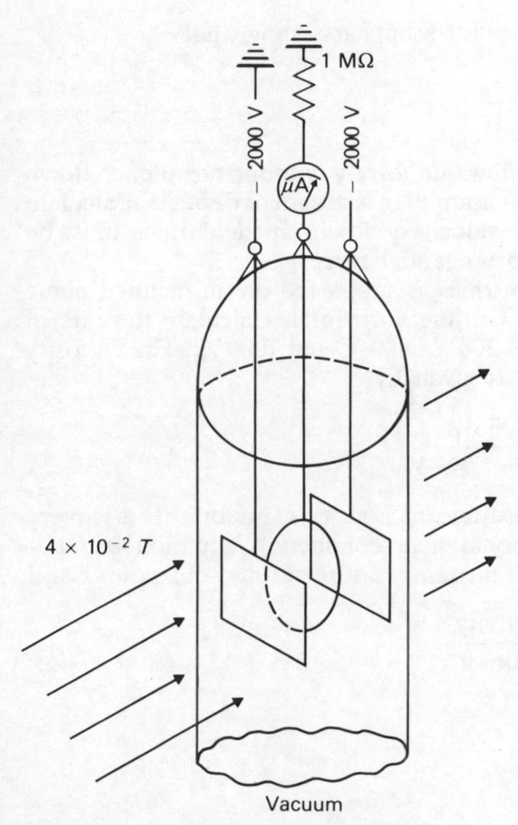

Fig. 4.10 The Penning gauge.
A gas discharge between parallel plates in a cylindrical tube was much used sixty years ago to indicate the pressure in an evacuated system. It ceases to be useful when the pressure is below 0.1 Pa as the electrons make too few ionising collisions to sustain the discharge. This lower limit can be extended to a useful $10^{-3} - 10^{-4}$ Pa by operating the discharge in a magnetic field. The electrons pursue helical paths round the lines of force and so make more ionising collisions in their passage from cathode to anode. This effect is enhanced if the electrons are forced to migrate across the lines of force by using a ring shaped anode as in the discharge tube illustrated in which the currect through the tube is an easily measured 10 μA at 10^{-3} Pa. The great merit of the Penning gauge is that it is robust, has simple electrodes which can easily be outgassed in an induction furnace and is not ruined by a sudden pressure rise as manometers using a hot filament usually are.

stress in a deformed solid disappears, show that

$$\tau = \frac{\text{viscosity}}{\text{shear modulus}} = \frac{\eta}{G}.$$

3 Liquid of viscosity η flows uniformly without turbulence down an inclined plane (inclination θ) in a stream of depth d. Calculate the rate of flow. (The velocity gradient at the surface must be zero since there is no tangential force).
A sheet of glass 5 mm thick is supported on an inclined plane making an angle 20° with the horizontal. Calculate the rate of flow down the plane at 300 °C, 400 °C and 1000 °C. The viscosity of soda glass at T/K is given by

$$\log_{10}\eta = -1.422 + \frac{3612}{T - 543.5}.$$

4 Verify by direct substitution Fourier's important discovery that the one dimensional heat conduction equation $\partial\theta/\partial t = a(\partial^2\theta/\partial x^2)$ in which θ is the temperature at time t at a point x and

$$a = \frac{\text{thermal conductivity}}{\text{heat capacity} \times \text{density}}$$

is solved by

$$\theta = \theta_0 \sin\frac{2\pi x}{\lambda}\,\mathrm{e}^{-(4\pi^2 at/\lambda^2)}$$

5 Calculate the numerical value of α the attenuation coefficient of sound from the expression in section 4.2 and use it to estimate
(a) distance from a band at which its musical balance will be lost.
(b) the practical range of sonar using 1 cm waves.

6 Knudsen assumed in calculating the molecular flow in a tube that the drift velocity $\propto c$. What would be the calculated viscosity of a gas if this assumption were made in the usual calculation of viscosity?

7 What correction must be made to Knudsen's absolute manometer if the plates have an accommodation coefficient α?

8 Estimate the values of n and s in Table 4.6 from the following data for CO_2

T°C	0	20	50	100	200	300	400	500	600
$\eta/(\mu$ Pas)	13.6	14.7	16.2	18.6	23.0	27.0	30.7	34.1	37.3

Further reading

Condon, E.K. and **Odishaw, H.** (1967) *Handbook of Physics*, Chap. 4, McGraw-Hill, N.Y.

Goodman, F.O. and **Wachman, H.Y.** (1976) *Dynamics of Surface Scattering*, Academic Press, N.Y.

Grew, K.E. and **Ibbs, T.L.** (1952) C.U.P. London.

Hobbs, P.V. (1974) *Ice Physics*, Clarendon Press, Oxford.

Jones, R.C. and **Furry, W.H.** (1946) *Rev. Mod. Phys.*, **18**, 153.

Kennard E.H. (1938) *Kinetic Theory of Gases*, Chap. 8, McGraw-Hill, N.Y.

Rutner, E. Goldfinger, P. and **Hirth J.P.** (1964) *Condensation and Evaporation of Solids*, Gordon and Breach N.Y.

References

Chapman, S. and **Cowling, T.G.** (1970) *Theory of nonuniform gases*, p 249 Cambridge University Press, London.

Clusius, K and **Dickel, G.** (1938) *Naturwissenschaften*, **26**, 546.

Clusius, K and **Dickel, G.** (1939) *Zeit für Phys. Chem.* **44**, 451.

Condon E.U. and **Odishaw, H.** (1967) *Handbook of Physics*, 5–61, McGraw-Hill, N.Y.

Estermann, I. (1959) *Methods of Experimental Physics*, Vol. I, p 161, Academic Press, N.Y.

Goodman, F.O. and **Wachman, H.Y.** (1967) *J. Chem. Phys.* **46,** 2376.

Handbuch der Physik XII, 401 (1958) XI (1) S. Flugge editor. Springer Verlag Berlin.

Jeans, J.H. (1925) *Dynamical Theory of Gases*, Cambridge University Press, London.

McKensie, P.P. and **Richter F.** (1976) *Scientific American*, November, **235**, 72.

Millikan R.A. (1923) *Phys Rev*, **XXI**, 224.

Rogers, R.H. (1976) *Reports on Progress in Physics*, 39, **3.**

Rutherford, W.H. Roos, W.J. and **Kaminsky K.J.** (1969) *J. Chem. Phys.* **50,** 5539.

CHAPTER 5

Temperature entropy and heat flow

5.1 Reversible changes

As explained in Section 1.9 all gases, real as well as perfect, obey an equation of state. For a perfect gas, $pv = rT$ which can be written as $pv/r = T = 2u/3r$. For a real gas these simple relationships are modified by the effect of the forces between the molecules and can be written in a similar form as $F(p, v) = T = f(u, v)$ in which u is the internal energy and T the temperature measured on a perfect gas scale. These variables can be made the coordinates in a three dimensional p, v , T space in which the equation of state defines a surface. Any state of the gas in equilibrium is represented by a point in this surface (Fig. 5.1) and the changes in the gas which we are going to consider can be represented by a curve lying in this surface. Such changes are called by definition 'reversible changes'. The restriction to points lying in the p, v, T surface is not a trivial one and implies that throughout the change the gas remains in a state of equilibrium, and that the coordinates of the point representing the instantaneous state of the gas are independent of whether the gas is expanding or being compressed. Any gas undergoing a rapid change cannot be represented by a single moving point on the p, v, T surface. This is quite obvious because the finite flow of the gas will call viscous forces into play and there will be corresponding pressure gradients and the 'pressure of the gas' can no longer be represented by a single quantity p. The slower the change the more accurately it can be represented by a point moving in the surface representing the equation of state. The closeness of the approximation in any particular change can be estimated by an experimental physicist.

It is important to grasp that for any gas the p, v, T surface is a definite one which can be constructed by a series of static measurements and that a line in it represents a change (say an expansion) which can be approximated to in the laboratory. The arguments we are going to use are essentially geometrical ones applying to the line in the surface, not to the real gas following an approximately similar

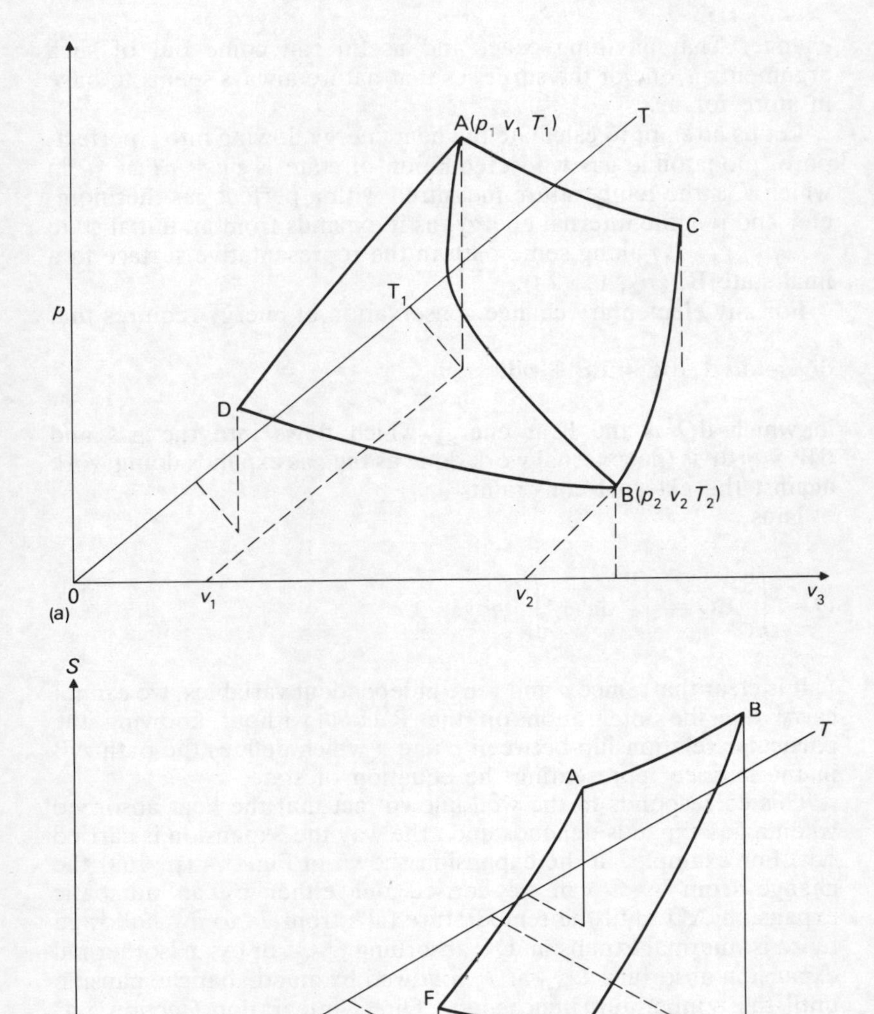

Fig. 5.1 Surfaces of state.

(a) AB is a reversible path in the surface ACBD representing the equation of state in a v, T, p space.

(b) The path AB of Fig. 5.1a is represented in a v, T, S space. S_B is greater than S_A showing that during the expansion heat has been added to the system.

change. That anything exact and useful can come out of such arguments is one of the surprises that nature always seems to have in store for us.

Let us attempt to calculate the heat energy flowing into a perfect, pure, monatomic gas whose equation of state is $pv = rT = \frac{2}{3}u$ in which T is the temperature measured with a perfect gas thermometer and u is the internal energy, as it expands from an initial state A, (p_1, v_1, T_1) along some path in the representative surface to a final state B, (p_2, v_2, T_2).

For any elementary change conservation of energy requires that

$$dQ = du + dW = du + pdv$$

in which dQ is the heat energy which flows into the gas and $dW = pdv$ is the external work done as the gas expands doing work against the external constraints

Thus,

$$Q = \int_A^B dQ = \int_A^B du + \int_A^B pdv.$$

It is clear that, since p and v are independent variables, we cannot carry out the integration on the R.H.S. without knowing the particular relationship between p and v which defines the path AB in the surface representing the equation of state.

This corresponds to the well-known fact that the heat absorbed when a gas expands depends upon the way the expansion is carried out. For example, in the expansion shown in Fig. 5.4 (p. 168) the change from A–C can be carried out either by an adiabatic expansion AD until the temperature falls from T_1 to T_2, followed by an isothermal expansion DC absorbing Q_{DC}, or by an isothermal expansion absorbing Q_{AB} at T_1 followed by an adiabatic expansion until the temperature reaches T_2. Direct integration (section 5.4) shows that for a perfect gas the heat absorbed along the two paths is quite different and $Q_{AB}/Q_{DC} = T_1/T_2$.

The heat energy Q which has flowed into a gas is thus a quantity which depends not only on the state of the gas but also on the way in which that state has been reached. This statement should cause no surprise since the flow of heat no longer represents a flow of 'caloric' into the gas. The existence of a measurable quantity Q which depends not only on the initial and final state of the gas but also upon which of the many reversible paths between them the gas has followed is a situation well known in mathematical physics. It is described by saying that dQ is an incomplete differential, a total or complete differential being one which can be integrated without additional knowledge.

In the present example, in which the working substance is a perfect monatomic gas, one can make use of the simple equation of state to convert dQ into a total differential.

If $dQ = du + p\,dv$

then $$\frac{dQ}{pv} = \frac{du}{pv} + \frac{p\,dv}{pv}$$

and substituting from $pv = rT = \frac{2}{3}u$ this becomes

$$\frac{1}{r}\int_A^B \frac{dQ}{T} = \frac{3}{2}\int_A^B \frac{du}{u} + \int_A^B \frac{dv}{v} = \frac{3}{2}\log\frac{u_B}{u_A} + \log\frac{v_B}{v_A}$$

The auxilliary function $1/pv$ is called an integrating factor. The quantity $\int_A^B dQ/T$ is a quantity whose change is independent of the path chosen in going from A to B so long as it remains in the surface and dQ/T is correspondingly said to be a total differential. It is usually written as

$$dS = \frac{dQ}{T} \text{ and } \int_A^B \frac{dQ}{T} = \int_A^B dS = S_B - S_A.$$

The quantity S_A is characteristic of the state A and does not depend upon how the state was reached. It is called the entropy of the state, a term invented by Clausius to mean heat energy which cannot be converted into work. By usage it has come to have the more precise meaning given here. Since to every point on the p, v, T surface a value of S can be assigned, one can construct a new surface in a v, T, S space.

The equation to this surface is

$$S = \frac{3r}{2}\log T + r\log v + K.$$

Except for the constant of integration K this is as good an equation of state as $pv = rt$ or $pv = \frac{2}{3}u$. It is important to realise that S is a property of state and is a definite quantity for the gas in a definite state, e.g. with a given value of pv and so through the equation of state of T and u. It does not depend upon how the gas got into this state. It need not have reached it along a reversible path passing through a series of equilibrium states and so wholly contained in the p, v, T surface. The gas can have changed from state A, (p_1, v_1, T_1) to state B, (p_2, v_2, T_2) quite quickly by some process easily realised in the laboratory; if it finally arrives at the point B it will have the entropy S_B appropriate to that state.

It is only if we want to *calculate* the value of $S_B - S_A$ that we must carry out the integration along a path lying in the surface defined by the equation of state. The actual path chosen in the surface will be eliminated during the course of the calculation.

5.2 Entropy

The existence of entropy has been demonstrated for a perfect gas and its value calculated except for an integration constant. A similar demonstration can be given for any gas for which the equation of state is known. A formal proof can be given and the following intuitive proof is helpful in understanding the meaning of an integrating factor.

Consider a stretch of mountainous country (Fig. 5.2a) in which each point of the surface is represented by cartesian coordinates x,

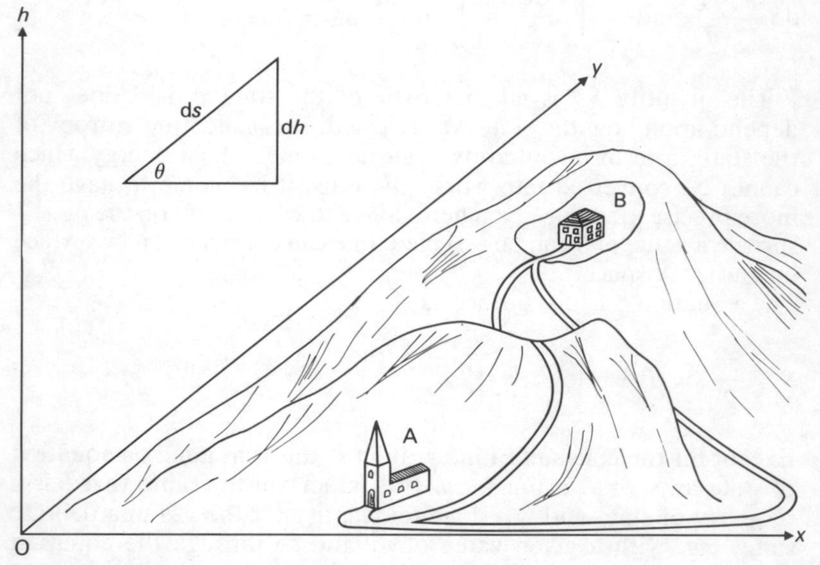

Fig. 5.2 Heights are measured from the base plane x, y. The difference in height between A and B is the same along the two paths, although they are different in length.

y, in a horizontal plane and h in a vertical plane. If a traveller at A, (x_1, y_1, h_1) wishing to go to B, (x_2, y_2, h_2) asked how far it was to B he would be told it depended on which way he took; the long easy way through the valleys or the short hard one through the pass.

If ds was an element of his path then the total path $s = \int_A^B ds$ and ds is an incomplete differential which cannot be evaluated unless the path is known. The difference in height between A and B is clearly $h_2 - h_1$. The element of height dh is a total differential and

$$\int_A^B dh = h_B - h_A$$

can be found without knowing the path. It will be the same for all paths starting at A and ending at B. Now it is clear (Fig. 5.2) that dh and ds are related by the relationship $dh = \sin\theta \, ds$, in which θ is the slope of the path. Thus $\sin\theta = f(x, y)$, which is a function of position in the horizontal plane, is an integrating factor which converts the incomplete differential ds into the total differential dh. It is clear that this integrating factor always exists and that its existence has nothing to do with the particular profile of the mountains. All that is necessary is that they are sufficiently regular that one can determine a definite slope, $\sin\theta$ (by rolling a ball say) at each point on the surface. The mountains do not have to be real mountains and can represent the surface defined by the equation of state of a real gas with three parameters of state p, v, T. Mathematicians will recognise that such an artificial space is an 'affine space'. Care has to be taken in abstract arguments not to attribute to it the metrical properties of ordinary space. This is because p, v, T have no common measure (metric) and the 'angle between two lines' and 'the distance between two points' has no unambiguous meaning. The bearing of this upon thermodynamical arguments is discussed for example by F. Weinbold (1976).

Thus, this very obvious discussion about paths in a mountainous district is sufficient to show that all differentials such as dQ which are functions of two variables only (p and v say) can always be converted into an exact differential dS by multiplying by an integrating factor $f(p, v)$.

Actually there are many integrating factors since a whole series of new ones can be obtained by logarithmic differentiation of any function of an integrating factor. There is therefore nothing remarkable in finding that the simple function $1/pv$ is one of the integrating factors of dQ when the working substance is a perfect gas.

What is remarkable is that pv is not a general function of p and v but that particular combination which occurs in the equation of state and so can be replaced by the single variable T, the temperature of the system.

It will now be shown that this result is not confined to the special case of a perfect gas and that the integrating factor which converts dQ into a perfect differential dS is always some function of the arbitrary temperature only. This is clearly not a mathematical theorem to be proved by ingenious manipulation of symbols: it is only true when the symbols p, v, T in the general relationship $F(p, v) = T$ are in some way restricted so that they represent, the pressure volume and temperature of a real gas. A little consideration suggests that the restricted variable is the temperature T which must conform to the fundamental property of temperature that bodies in equilibrium and in thermal contact must have the same arbitrary temperature. The argument must therefore involve two systems.

Since the ensuing argument is quite abstract it will be as well to outline the stages in which it is carried out before considering the details.

1 One starts off with two systems for each of which the equations of state and the arbitrary temperature θ is known. Each system is known to have a fifth variable of state s called the entropy. The problem of finding the entropy is the mathematical problem of finding the integrating factor of dQ. It can always be solved if the equations of state in terms of the variables of state p, v, u, T are known. The molecular explanation of T is not known except for a perfect gas and the molecular explanation of s is quite unknown. The investigation is not initially directed towards improving our understanding of the molecular explanation of either s or T; of course, it leads to this in the end.

2 The two systems are now combined into a single system with a common arbitrary temperature θ and the entropy s of the joint system calculated in terms of the entropies s_1, s_2 of the separate systems. This means finding the integrating factor of dQ, the heat added reversibly to the joint system.

It turns out that the restriction imposed by the condition that both systems have a common temperature implies that the integrating factor which would in general be an unrestricted function of the three independent variables s_1, s_2, θ is restricted to the special form $1/f(\theta)\, h(s_1, s_2)$ in which $f(\theta)$ is a function of the common temperature θ only. This mathematical result makes a completely new definition of temperature possible. The temperature T is defined as $T = f(\theta)$ and $1/T$ can be shown to be the universal integrating factor which converts dQ into the total differential dS.

That temperature can be given this rather abstract mathematical meaning is one of the great discoveries of the nineteenth century, and is the basis of the science of thermodynamics. Once discovered it is found to have far reaching consequences which have been verified experimentally.

Like many applications of mathematical reasoning to physics it involves giving a definite physical meaning to what were originally abstract mathematical symbols. This identification may modify or limit the axioms which govern the system and certain mathematically possible relationships may have to be excluded on physical grounds.

In these circumstances the distinction between a mathematical proof and an experimental demonstration of a theorem becomes blurred. If the axioms are 'bent' to fit the physics the mathematical 'proof' degenerates into a convenient framework in which to describe the experimental results. If one is a purist and sticks rigidly to the axioms the 'proof' often ceases to have any connection with the physical world. As Gibbs who took a purist view of the matter put it 'Here there can be no mistake in regard to the agreement of the hypothesis with the facts of nature for nothing is assumed in that respect. The only error into which one can fall is the want of agreement between the premises and the conclusions and this with care one may hope in the main to avoid'. The nice adjustment of these two points of view so that progress is possible, can usually be made in more than one way. Physicists should never forget that abstract mathematical demonstrations only redisplay what has been agreed to in the axioms.

The mathematical section which follows depends upon a particular way of making the adjustment and can be omitted on a first reading by those not very interested in abstract mathematical arguments. It is its conclusion and not its mathematical content which is important.

The required demonstration can be carried out in the following way due to Born and Carathéodory. The properties of partial differential coefficients upon which the proof depends are outlined in Chapter VII.

5.3 Entropy of a real gas

Consider a system (Fig. 5.3) of two real gases in separate cylinders (1 and 2) with separate pistons controlling the volumes, but having a

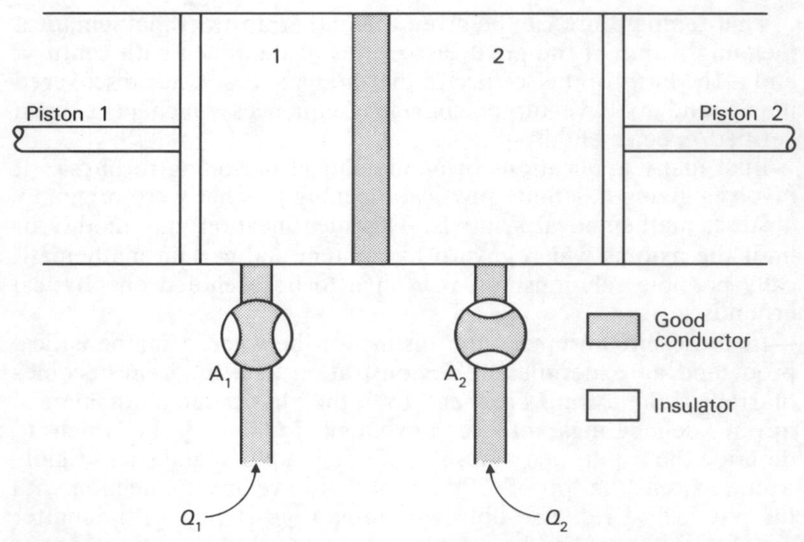

Fig. 5.3 The flow of heat energy into the separate cylinders is controlled by the insulating 'taps' A_1 and A_2. The gases in the cylinders are always at the same temperature so that no heat energy flows through the well-conducting cylinder head which separates them.

common cylinder head made from a good conductor of heat such as copper or gold.

The state of the system can be varied by expanding or compressing the gases by means of their respective pistons and by adding quantities of heat Q_1 and Q_2 reversibly by conduction through the cylinder walls in the usual way.

For reversible changes, which are the only changes we shall consider, the gases will have a common temperature θ, measured on some arbitrary scale, since they are in thermal contact through the well-conducting cylinder head. The variables of state of the first gas are p_1, v_1 its pressure and volume, u_1 its internal energy and θ_1 the temperature. As has just been shown, there is also a fifth variable of state the entropy s_1 which is some function of *any two* of the other independent variables. Choose them to be θ_1 and u_1 so that $s_1 = g_1(u_1, \theta_1)$ or $u_1 = f_1(s_1, \theta)$.

Similarly, the second gas will have variables of state p_2, v_2, u_2, θ_2, s_2 in which s_2 the entropy of the second gas is some function $g_2(u_2, \theta_2)$ of the independent variables u_2 and θ_2.

Of these ten variables only three and not four are independent because for reversible changes $\theta_1 = \theta_2 = \theta$ the common temperature of the two gases.

The gas in the two cylinders can thus be described as forming a single system with three independent variables.

It does not matter which three are chosen to represent the state of the composite system and we choose s_1, s_2 and θ so that the equation of state of the system is of the form $F(s_1, s_2, \theta) = 0$.

In a change in which heat dQ_1 is added to the first gas, and dQ_2 to the second gas, the law of conservation of energy will give for dQ the heat added to the system as a whole

$$dQ = dQ_1 + dQ_2 = du_1 + p_1 dv_1 + du_2 + p_2 dv_2.$$

Writing this in terms of the selected variables s_1, s_2, θ by writing

$$u_1 = f_1(s_1, \theta), \quad v_1 = F(s_1, \theta) \text{ so that}$$

$$du_1 = \left(\frac{\partial u_1}{\partial s_1}\right)_\theta ds_1 + \left(\frac{\partial u_1}{\partial \theta}\right)_{s_1} d\theta \text{ etc.}$$

one finds that

$$dQ = \left[\left(\frac{\partial u_1}{\partial s_1}\right)_\theta + p_1\left(\frac{\partial v_1}{\partial s_1}\right)_\theta\right] ds_1 + \left[\left(\frac{\partial u_2}{\partial s_2}\right)_\theta + p_2\left(\frac{\partial v_2}{\partial s_2}\right)_\theta\right] ds_2$$
$$+ \left[\left(\frac{\partial u_1}{\partial \theta}\right)_{s_1} + \left(\frac{\partial u_2}{\partial \theta}\right)_{s_2} + p_1\left(\frac{\partial v_1}{\partial \theta}\right)_{s_1} + p_2\left(\frac{\partial v_2}{\partial \theta}\right)_{s_2}\right] d\theta.$$

Since p_1 and p_2 are not independent variables but are themselves functions of the three independent variables

s_1, s_2, θ this equation is of the form

$$dQ = F(s_1, s_2, \theta)\, ds_1 + G(s_1, s_2, \theta)\, ds_2 + H(s_1, s_2, \theta)\, d\theta$$

in which F, G, H are functions which could actually be found for a simple equation of state and must exist whether we can easily find them or not.

The heat added dQ is therefore an incomplete differential of a function of three independent variables s_1, s_2 and θ.

Assume for the moment without discussion that dQ can be converted to a total differential ds by means of an integrating factor $1/t(s_1, s_2, \theta)$ in the same way that we know the incomplete differentials dQ_1, dQ_2 can be converted to perfect differentials ds_1, ds_2.

The three expressions for the entropy,

$$ds_1 = \frac{dQ_1}{t_1(s_1, \theta)}$$

$$ds_2 = \frac{dQ_2}{t_2(s_2, \theta)}$$

$$ds = \frac{dQ}{t(s_1, s_2, \theta)}$$

are related by the relationship

$$dQ = dQ_1 + dQ_2$$

so that

$$t(s_1, s_2, \theta)\, ds = t_1(s_1, \theta)\, ds_1 + t_2(s_2, \theta)\, ds_2$$

and

$$t(s_1, s_2, \theta) \frac{ds}{d\theta} = t_1(s_1, \theta_1) \frac{ds_1}{d\theta} + t_2(s_2, \theta) \frac{ds_2}{d\theta}$$

Since s_1, s_2 and θ are independent variables, θ can change while s_1 and s_2 are unchanged, so that if $ds_1 = ds_2 = 0$

$$t(s_1, s_2, \theta)\left(\frac{\partial s}{\partial \theta}\right)_{s_1 s_2} = 0$$

and $(\partial s/\partial \theta)_{s_1 s_2}$ must itself be zero.

This implies that the entropy s is explicitly independent of θ and must be a function $s(s_1, s_2)$ of s_1 and s_2 only.

Thus,

$$ds = \frac{t_1(s_1, \theta)}{t(s_1, s_2, \theta)}\, ds_1 + \frac{t_2(s_2, \theta)}{t(s_1, s_2, \theta)}\, ds_2$$

is independent of θ.

This can only happen if the three integrating factors, $t_1(s_1, \theta)$, $t_2(s_2, \theta)$, $t(s_1, s_2\theta)$ all contain θ as a common factor $f(\theta)$

and

$$t_1(s_1, \theta) = f(\theta)\, h_1(s_1)$$
$$t_2(s_2, \theta) = f(\theta)\, h_2(s_2)$$
$$t(s_1, s_2, \theta) = f(\theta)\, h(s_1, s_2)$$

Now, introduce new variable S_1, S_2, and T defined so that

$$T = \mathrm{C}f(\theta)$$

$$S_1 = \frac{1}{\mathrm{C}} \int h_1(s_1)\, \mathrm{d}s_1 \quad \mathrm{d}S_1 = \frac{1}{\mathrm{C}} h_1(s_1)\, \mathrm{d}s_1$$

$$S_2 = \frac{1}{\mathrm{C}} \int h_2(s_2)\, \mathrm{d}s_2 \quad \mathrm{d}S_2 = \frac{1}{\mathrm{C}} h_2(s_2)\, \mathrm{d}s_2$$

Since T is a function of θ only, S a function of s_1 only and S_2 a function of s_2 only, the new variables T, S_1, S_2 are functions of state and can be used to describe the state of the two systems 1 and 2. In this notation

$$T\, \mathrm{d}S_1 = \mathrm{C}f(\theta) \cdot \frac{1}{\mathrm{C}} h_1(s_1)\, \mathrm{d}s_1 = f(\theta)\, h_1(s_1)\, \mathrm{d}s_1 = \mathrm{d}Q_1$$

$$T\, \mathrm{d}S_2 = \mathrm{C}f(\theta) \cdot \frac{1}{\mathrm{C}} h_2(s_2)\, \mathrm{d}s_2 = f(\theta)\, h_2(s_2)\, \mathrm{d}s_2 = \mathrm{d}Q_2$$

so that $1/T$ is an integrating factor which converts the incomplete differentials $\mathrm{d}Q_1$, $\mathrm{d}Q_2$ into the total differentials $\mathrm{d}S_1$, $\mathrm{d}S_2$.

This is what we set out to prove, namely that the integrating factor, known to exist for any pure substance, can always be expressed as a function of temperature only as was the case for a perfect gas. One notes that no use has been made of the special properties of the gaseous state so that the demonstration applies to any pure substance which obeys an equation of state. The constant C can be adjusted so that T has an agreed value, e.g. 273.16, at any selected arbitrary temperature, say the triple point of water. The T so obtained is defined to be the absolute temperature of the system at the temperature θ measured on an arbitrary scale.

Since $T = \mathrm{C}f(\theta)$ is a function of the arbitrary temperature θ only, it will have all the equilibrium properties of temperature which have been discussed. When two bodies have the same θ they will have the same T and will be in equilibrium and so T defined purely as an integrating factor of $\mathrm{d}Q$ can be treated as a temperature in the ordinary sense. The two gases (1 and 2) which have been considered were in general two real gases; the changes in entropy are dQ_1/T and dQ_2/T in which T is the common absolute temperature. If gas 1 is a perfect gas it is already known that the change in entropy is $dQ_1/T_{\text{perfect gas}}$. A comparison of the two expressions shows that the absolute temperature T is identical with $T_{\text{perfect gas}}$ measured on the perfect gas scale.

Absolute temperature adjusted to agree with the perfect gas scale is designated by K in honour of William Thomson (Lord Kelvin). He based the scale upon Carnot's Theorem (section 5.4) and advocated (in vain) the presently adopted method of determining the size of the degree through a single arbitrary number, i.e. 273.16 K for the triple point of water. The present International Scale is a perfect gas scale still based upon dividing the interval between the boiling and melting point of water into 100 deg. The figure 273.16 °K for the triple point of water represents the best compromise between the two systems.

Establishing a perfect gas scale by extrapolating pv to zero pressure is one of the best ways of establishing the absolute scale of temperature, but it is not the only way. As will be shown in section 7.8, $T = Cf(\theta)$ can be found by measuring with an uncorrected gas thermometer the cooling which occurs when a real gas expands through a porous plug. Some experimentally established fixed points on the absolute scale are given in Table 5.1.

Table 5.1

Defining fixed point	Assigned value (K)	Estimated uncertainty (K)
Triple point of hydrogen	13.81	0.01
Boiling point of equilibrium hydrogen	20.28	0.01
Boiling point of neon	27.102	0.01
Triple point of oxygen	54.361	0.01
Boiling point of oxygen	90.188	0.01
Triple point of water	273.16	Exact by definition
Triple point of benzoic acid	395.52	
Boiling point of sulphur	717.824	
Freezing point of copper	1357.6	
Temperature of melting tungsten	3660	

(Taken from Metrologia, 1969.)

The entropy of the combined system must now be considered. By analogy with S_1 and S_2 define a quantity S so that

$$S = \frac{1}{\mathrm{C}} \int h(s_1, s_2)\, \mathrm{d}s$$

in which s_1, s_2, s have the meaning previously assigned to them. This is only possible if the function $h(s_1, s_2)$ can be expressed as a function of s only. The condition for this can be established in the following way.

For brevity write $h(s_1, s_2) = h, \quad s(s_1, s_2) = s.$ Then

$$\mathrm{d}s = \left(\frac{\partial s}{\partial s_1}\right)\mathrm{d}s_1 + \left(\frac{\partial s}{\partial s_2}\right)\mathrm{d}s_2$$

$$\mathrm{d}h = \left(\frac{\partial h}{\partial s_1}\right)\mathrm{d}s_1 + \left(\frac{\partial h}{\partial s_2}\right)\mathrm{d}s_2$$

If h can be expressed as a function of s only

$$\mathrm{d}h = \frac{\mathrm{d}h}{\mathrm{d}s}\ \mathrm{d}s = \frac{\mathrm{d}h}{\mathrm{d}s}\left(\left(\frac{\partial s}{\partial s_1}\right)_{s_2} \mathrm{d}s_1 + \left(\frac{\partial s}{\partial s_2}\right)_{s_1} \mathrm{d}s_2\right)$$

In the two expressions for $\mathrm{d}h$ the coefficients of $\mathrm{d}s_1$ and $\mathrm{d}s_2$ must be the same since s_1 and s_2 are independent variables so that

$$\frac{\mathrm{d}h}{\mathrm{d}s}\left(\frac{\partial s}{\partial s_1}\right)_{s_2} = \left(\frac{\partial h}{\partial s_1}\right)_{s_2}$$

$$\frac{\mathrm{d}h}{\mathrm{d}s}\left(\frac{\partial s}{\partial s_2}\right)_{s_1} = \left(\frac{\partial h}{\partial s_2}\right)_{s_1}$$

and the condition that h is a function of s only can be written as

$$\left(\frac{\partial h}{\partial s_1}\right)_{s_2}\left(\frac{\partial s}{\partial s_2}\right)_{s_1} = \left(\frac{\partial h}{\partial s_2}\right)_{s_1}\left(\frac{\partial s}{\partial s_1}\right)_{s_2}$$

It is not difficult to show (e.g. problem 5.9) that the conservation of energy expressed through the relationship $\mathrm{d}Q = \mathrm{d}Q_1 + \mathrm{d}Q_2$ ensures that the condition is fulfilled.

Accordingly there is no ambiguity in the definition of S and one can write as one has already done for $\mathrm{d}S_1$ and $\mathrm{d}S_2$ that

$$T\mathrm{d}S = \mathrm{C}f(\theta)\frac{1}{\mathrm{C}}\, h(s)\, \mathrm{d}s = f(\theta)\, h(s)\, \mathrm{d}s = \mathrm{d}Q$$

Thus $1/T$ is the integrating factor which converts the incomplete differential dQ into the total differential dS. The entropy of the combined system is therefore S and $dS = dS_1 + dS_2$ equivalent to $S = S_1 + S_2 + \text{constant}$. The relationship $S = S_1 + S_2$ obtained by choosing the constant to be zero is, as will be shown in the next chapter, the starting point for a further development of the subject.

Since the argument by which the absolute scale of temperature has been set up is of necessity involved and abstract it is as well to recapitulate the various steps in the argument.

1 The argument justifying the existence of a fifth variable of state (to be called the entropy) is first of all carried through for a perfect gas. Molecular theory is used to justify the relationship $pv = \frac{2}{3}u$. The integrating factor is found to be $1/pv = 1/T$ and can thus be written as a function of the perfect gas temperature only.

2 Mathematical considerations allow the argument to be extended to real gases. The argument depends upon the readily proved theorem that all incomplete differentials, depending upon two variables only, have integrating factors. However, in general this factor is a function of both variables.

3 By considering a composite system and introducing the restrictions imposed by the conservation of energy and the experimentally determined properties of temperature it is shown that for a class of pure substances the integrating factor can be written as a function of temperature only as it was for a perfect gas.

4 This enables a new definition of temperature to be made which does not depend on the properties of a perfect gas or on the assumptions of molecular theory.

5.4 Historical background

The discovery that simple substances have a fifth variable of state, the entropy S, related to the reversible heat intake dQ by the relationship $TdS = dQ$ in which T is a quantity endowed with all empirical attributes of temperature but defined without any reference to these is a landmark in the development of physics. From it follow a large number of important relationships, for historical reasons called ‘thermodynamic relationships’, between the variables of state.

The discovery was not made in the way which has been described, which is due to much later work by Carathéodory (1909) and Born (1921). At the beginning of the nineteenth century steam engines were replacing windmills and water wheels as prime movers in workshops and factories. The steam engine is still the most important source of mechanical power and transforms heat energy (section 1.7) obtained from the combustion of coal (or other fuel, including nuclear fuel) into mechanical work.

Early steam engines were obviously inefficient owing to leaking joints and unlagged pipes; even when these defects had been remedied by improved engineering practice, their thermal efficiency defined by

$$\frac{\text{mechanical work done}}{\text{heat energy supplied to the boiler}}$$

was lamentably low. Good railway locomotives in their heyday (~1920) had a thermal efficiency of 7–8 per cent. This low efficiency is in marked contrast with the comparable efficiency of a battery-operated electric motor which can easily reach 70–80 per cent. A rational explanation of the low efficiency of heat engines was first given by Sadi Carnot, a French military engineer and the son of a Napoleonic hero. He thought deeply about the nature of heat and was the first to have a theoretical understanding of the way a heat engine transforms heat energy into mechanical work. For this purpose he imagined an ideal cyclical engine in which the working substance was returned to its original state after completing each cycle.

The essential features of Carnot's cycle are that it is reversible, and that heat energy is only exchanged between the engine and its surroundings at two temperatures T_1 and $T_2 (T_1 > T_2)$. Heat Q_1 is taken in during an isothermal expansion at T_1 and heat Q_2 is given out during an isothermal compression at T_2.

In 1824 Carnot enunciated his now well-known theorem that the efficiency of all engines working in a Carnot cycle is a universal function of T_1 and T_2 and is independent of the working substance. Carnot's proof was incomplete. He showed by a general argument that the Carnot cycle was the most efficient cycle for a heat engine but not that all Carnot engines had the same efficiency whatever the working substance. This seemed so obvious to Carnot that he offered no formal proof. A modern account of Carnot's work is given by Wilson (1981). Later expositors of Carnot's work, notably Clapeyron, completed the proof using the caloric theory of heat. What Carnot himself thought is not known as he died aged 36 and the importance of his work was not recognised until after his death.

There is evidence in his notebooks that he favoured, and was planning experiments to prove, the mechanical theory of heat.

In Fig. 5.4 ABCD represents (not to scale) the pressure–volume relationship for a substance following Carnot's reversible cycle.

The features of the diagram which make it a Carnot cycle are that AB and CD are isothermals and AD and BC are adiabatics. In transversing the cycle in a clockwise direction heat Q_1 is taken in along AB and heat Q_2 along CD; Q_1 and Q_2 have opposite signs since AB represents an expansion and CD a compression. No heat is transferred to the working substance at any intermediate temperature.

Since the working substance had returned to its initial state A its internal energy is unchanged. Conservation of energy requires that $Q_1 + Q_2$ the heat energy which has flowed into the system during the cycle has all left the system as external work W, so that $Q_1 + Q_2 = W$. The thermal efficiency of the cycle is W/Q_1 and this must according to Carnot be a function of T_1 and T_2 only. Carnot's theorem thus requires that

$$\frac{Q_1 + Q_2}{Q_1} = F(T_1, T_2) = 1 + Q_2/Q_1$$

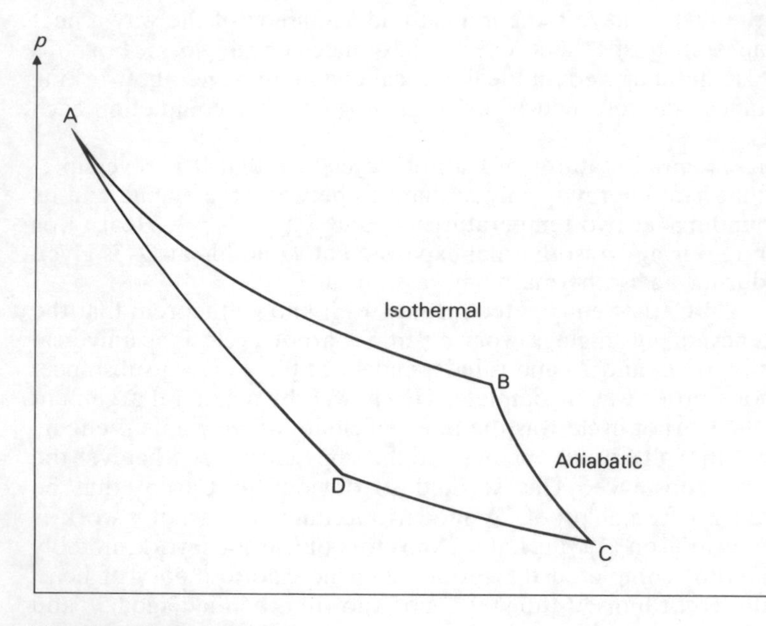

Fig. 5.4 Carnot's cycle.

whatever the working substance gas, liquid, or solid. The theorem is demonstrably true if the working substance is a perfect gas for which $pv = rT$

The equations of the four sections of the diagram are then

for AB, $pv = rT_1$
for BC, $pv^{\gamma} = k_1 = pv \cdot v^{\gamma-1} = rT\, v^{\gamma-1}$
for CD, $pv = rT_2$
for DA, $pv^{\gamma} = k_2 = pv \cdot v^{\gamma-1} = rT\, v^{\gamma-1}$

At B $T = T_1$ and $v = v_B$ so $k_1 = rT_1\, v_B{}^{\gamma-1} = rT_2\, v_C{}^{\gamma-1}$ and $(v_B/v_C)^{\gamma-1} = T_2/T_1$. Similarly $(v_D/v_A)^{\gamma-1} = T_1/T_2$ so that $v_B/v_A = v_C/v_D$.

For a perfect gas the internal energy depends only on the temperature so that all the heat energy flowing into the gas during an expansion at constant temperature flows out again as the work done $= \int p\mathrm{d}v$ against the external pressure.
Thus

$$Q_1 = \int_A^B p\mathrm{d}v = rT_1 \log \frac{v_B}{v_A}$$

and

$$Q_2 = \int_C^D p\mathrm{d}v = rT_2 \log \frac{v_D}{v_C} = -rT_2 \log \frac{v_C}{v_D}$$

But since $\dfrac{v_C}{v_D} = \dfrac{v_B}{v_A}$

$$\frac{Q_1}{T_1} = -\frac{Q_2}{T_2} \quad \text{and} \quad \frac{Q_1}{Q_2} = -\frac{T_1}{T_2} = F(T_1, T_2)$$

as Carnot's theorem requires.
This can be rewritten as

$$\frac{Q_1}{T_1} + \frac{Q_2}{T_2} = 0$$

and this is the most useful form of Carnot's theorem. An alternative formulation is to express the thermal efficiency η in terms of the defining temperatures T_1 and T_2. By definition

$$\eta = \frac{W}{Q_1} = \frac{Q_1 + Q_2}{Q_1} = 1 + \frac{Q_2}{Q_1} = 1 - \frac{T_2}{T_1}$$

$$= \frac{T_1 - T_2}{T_1}.$$

This way of expressing Carnot's theorem is useful in discussing the behaviour of engines working in other theoretical cycles like the Rankine cycle approximated to by steam engines.

The second part of Carnot's theorem, that the result is independent of the working substance, is also fulfilled since the molecular weight which would identify the perfect gas has not been used in the proof. No way of modifying this proof to include other working substances has ever been found. Although no simple proof of Carnot's theorem based upon the mechanical theory of heat could be devised, it became clear, particularly to the brothers James and William Thomson, that Carnot's theorem was not only true but could be applied to many branches of physics not obviously connected with simple heat engines.

On the basis of this very wide applicability of Carnot's theorem a new general law (the second law of thermodynamics) was put forward governing the conversion of heat energy into mechanical work. The second law is a general statement about heat engines and states that 'It is not possible to construct a self-acting machine working in a cycle which raises a weight, cools a reservoir and nothing else'. No attempt is made to explain why the construction is impossible. From this general law Carnot's theorem was easily deduced as a corollary. Once Carnot's theorem is accepted the existence of entropy and the role of temperature as a universal integrating factor is readily established in simple cases such as that of a pure substance.

To establish for a pure substance that S the entropy defined by $S = \int dQ/T_{\text{perfect gas}}$ is a function of state, it is sufficient to show that $\oint dS = \oint dQ/T$ round any possible reversible closed cycle is zero. From Carnot's theorem this is true for all Carnot cycles (Fig. 5.4) since $\oint dQ/T = Q_1/T_1 + Q_2/T_2 = 0$. This result can be extended to any reversible closed cycle taking in heat over a range of temperatures by approximating to it by a series of Carnot cycles as shown in Fig. 5.5. The arguments of the calculus are then used to show that in the limit the saw toothed cycle approximates sufficiently to the actual cycle to establish the theorem.

This approach to the subject through the theory of heat engines was undoubtedly encouraged by the circumstance that the bulk of

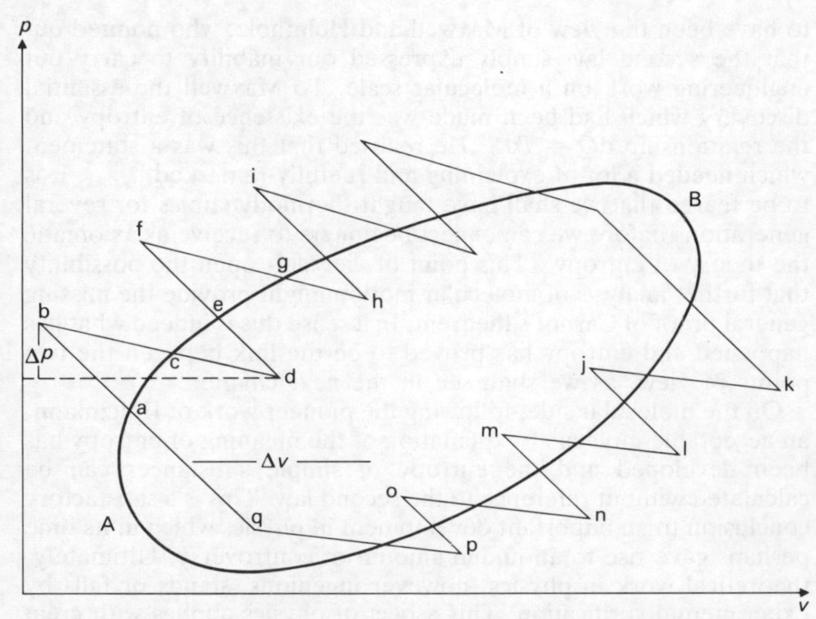

Fig. 5.5 AB represents the pv relationship of a reversible cycle. It is required to show that $\oint(\mathrm{d}Q/T)$ round the cycle is zero. That this is approximately so can be shown by covering the pv plane by a network of adiabatic and isothermal curves of which fedop, ihmn are typical adiabatics and fghjk, bcdml are typical isothermals. The cycle AB is then approximated to by the saw toothed cycle abcdefghi . . . . kjlmnopq . . . . in which ab, df, hi, etc. are adiabatics and bd, fh, etc. are isothermals. For this cycle $\oint(\mathrm{d}Q/T) = 0$ because each pair of sections of the same toothed cycle like defgh and mno are part of a Carnot cycle fghmnode. If Q_1 is taken in along fgh at T_1 and Q_2 along no at T_2 $(Q_1/T_1) + (Q_2/T_2) = 0$ by Carnot's theorem. Therefore each pair of isothermal legs of the saw toothed cycle of which fh and no are typical members contribute nothing to $\oint(\mathrm{d}Q/T)$ round the saw toothed cycle. The adiabatic legs also contribute nothing by definition. The value of $\oint(\mathrm{d}Q/T)$ round the reversible cycle AB therefore differs from zero by the contributions q/T from the small cycles of which abcdea is a typical member. These contributions are of the second order $(\Delta p\ \Delta v)$ and in the limit as the spacing Δv between the adiabatics is indefinitely reduced, can be neglected, thus giving an intuitive proof of the proposition that $\oint(\mathrm{d}Q/T) = 0$ round a closed reversible cycle.

students who had to be instructed in this matter were engineers interested in the design of heat engines as prime movers in the great industrial revolution which was taking place around them. Looking back one can see that this search for a new principle of nature, to get round a temporary difficulty may have been a mistake. This appears to have been the view of Maxwell and Helmholz, who pointed out that the second law simply expressed our inability to carry out engineering work on a molecular scale. To Maxwell the essential discovery which had been made was the existence of entropy and the relationship $dQ = TdS$. He realised that this was a statement which needed a lot of explaining and ruefully remarked '. . . . it is to be feared that we shall have taught thermodynamics for several generations before we can expect beginners to receive as axiomatic the theory of entropy'. This point of view left open the possibility that further analysis of molecular motion might provide the missing general proof of Carnot's theorem. In a sense this is indeed what has happened and entropy has proved to be the link between the two points of view, as we shall see in the next chapter.

On the molecular side, following the pioneer work of Boltzmann, an acceptable molecular explanation of the meaning of entropy has been developed and the entropy of simple substances can be calculated without reference to the second law. This is a satisfactory conclusion to an important development in physics which in its time perhaps gave rise to an undue amount of controversy. Ultimately, theoretical work in physics, however ingenious, stands or falls by experimental verification. This aspect of physics applies with great force to the second law which states boldly that certain easily imagined processes cannot be carried out. This has always been taken as a challenge by inventors. For a century answering correspondence on this subject was a well known chore for the holders of the more famous chairs of physics. Even so distinguished a scientist as the late J. B. Haldane at one time maintained that the second law could not be applied to contracting muscle because on contraction it did a lot of work but suffered a minimal rise in temperature. Some inventors made a direct attack on Carnot's theorem. A practicable heat engine was imagined using a working substance of postulated properties. There is no difficulty in choosing these so that the engine's efficiency is much greater than the $(T_1 - T_2/T_1)$ allowed by Carnot's theorem. Kipling (1910) has an ingenious version in his science fiction story 'With the Night Mail'. The answer to this attack is to get the inventor to find the mythical substance. It has never been found because the second law implies a relationship between vapour pressure and latent heat of evaporation which is violated by the imaginary working substance. Nevertheless, the second law as a great principle of physics is an embarrassment. If universally true it involves the ultimate stagnation of the universe: the so-called 'heat death'. This may be true but physicists may be pardoned for hoping it is not. Indeed, in some countries the possibility of a 'heat death' is officially excluded as leading to proofs of the existence of God and is relegated to a limbo of unacceptable bourgeois ideas.

It is therefore not surprising that attempts have been made to establish the relationship between entropy and temperature in some other way. It is common ground that the mechanical theory of heat has to be supplemented by some further assumption if it is to be usefully employed in solving simple problems. The surprising discovery which has been made is that all that is necessary is a simple mathematical restriction on the variables of state. One has to assume that the integrating factor of dQ, which always exists and for a perfect gas is $1/T$, retains this simple form in the general case when there are forces between the molecules. What has to be justified is the relationship $dQ = TdS$ for reversible changes. It is from this premise that all the important results known as 'thermodynamic relationships' are deduced. As a mathematical student interested in the theory of integrating factors Carathéodory was critical of the exposition of the theory of entropy in the lectures he attended. It seemed to him that the lecturers unversed in the niceties of his favourite subject made unwarranted assumptions, which if true could be used to establish that entropy was a function of state without relying on Carnot's theorem. For this reason Carathéodory asked what are the minimum assumptions required for establishing the existence of entropy. His answer will now be given. The mathematical demonstration that $1/T$ is a universal integrating factor of dQ given in section 5.3 contained a reservation. It was assumed that an incomplete differential dQ depending upon three independent variables (s_1, s_2, θ) could be converted into a total differential (ds) by an integrating factor $t(s_1, s_2, \theta)$ of the three independent variables. This is not necessarily so. There is always an integrating factor if dQ depends on two variables only, but for more than two variables other conditions have to be fulfilled. According to Carathéodory these conditions are always fulfilled when the variables are the pressure, volume, temperature, etc., of a real system. The discovery of an integrating factor depends upon being able to identify the adiabatics, i.e. to solve the equation $dQ = 0$ (cf. Piaggio, 1936). For two variables the adiabatic path, in the notation usually used in this branch of mathematics, is determined by the differential equation

$$X dx + Y dy = 0$$

and for three variables

$$X dx + Y dy + Z dz = 0$$

in which X, Y, Z are functions of the independent variables x, y, z. Equations of this type are often referred to as Pfaffians. (J. F. Pfaff

1765–1825; a friend of Gauss and after him the most distinguished German mathematician of his time).

Worked example

Express the adiabatic relationship $dQ = 0$ for a perfect gas and for the three variable system of Fig. 5.3 in a Pfaffian form.
For a perfect gas $PV = RT$ and

$$dQ = 0 = dU + PdV = C_v\,dT + \left(\frac{RT}{V}\right)dV$$

which is in the required form with $x = T$, $y = V$, $X = C_v$ = constant for a perfect gas and $Y = RT/V$, a function of independent variables T and V. Assume for simplicity that the arrangement with two pistons of Fig. 5.3 refers to two moles of gas with equations of state which can be written in the alternative forms

$$u_1 = f(\theta, v_1) \qquad u_2 = F(\theta, v_2)$$
$$p_1 = g(\theta, v_1) \qquad p_2 = G(\theta, v_2)$$

so that

$$dQ_1 = du_1 + p_1 dv_1 = \left(\left(\frac{\partial f}{\partial v_1}\right)_\theta + g\right)dv_1 + \left(\frac{\partial f}{\partial \theta}\right)_{v_1} d\theta$$

with a similar equation in F and G for dQ_2.
For an adiabatic change of the combined system

$$dQ = dQ_1 + dQ_2 = 0 = \left(\left(\frac{\partial f}{\partial \theta}\right)_{v_1} + \left(\frac{\partial F}{\partial \theta}\right)_{v_2}\right)d\theta + \left(\left(\frac{\partial f}{\partial v_1}\right)_\theta + g\right)dv_1 + \left(\left(\frac{\partial F}{\partial v_2}\right)_\theta + G\right)dv_2$$

which is in the required form since f, g, F, G are functions of the three independent variables θ, v_1, v_2 only.

In the case of two variables there is one adiabatic path through each point in the x, y plane. It is easily found since its slope $dy/dx = -X/Y$ is a definite function of the position x, y. Even if a simple algebraic solution is not possible, the adiabatic line can be constructed by graphical methods. For three (or more) variables the picture is much more complicated. There are now many adiabatic paths through each point in x, y, z space. Simple analogy leads one to suspect that these adiabatic paths will form an adiabatic surface so that any line lying in this surface is an adiabatic path. If this is so there is no difficulty in finding the integrating factor by a routine extension of the method which enabled one to find it for two variables and the whole problem is solved. Since the equation $dQ = 0$ defines a small plane surface in which all the adiabatic vectors must lie at each point in space it is natural to assume that all these locally defined plane elements must always join up to form the required adiabatic surface so that there is always an integrating factor. This is not so. A more complicated arrangement of the local planes is possible. They might not join up to form a surface, in which case there is no integrating factor. It is obvious that if all adiabatic paths lie in non-intersecting surfaces, paths normal to these surfaces cannot be adiabatics and there are always neighbouring points which cannot be joined by an adiabatic path.

Systems without an integrating factor thus have the remarkable property (from a molecular physicist's point of view) that all neighbouring points can be joined by an adiabatic path. Carathéodory proved the converse of this: If there are neighbouring points which cannot be joined by an adiabatic path then an integrating factor exists. Carathéodory's theorem is difficult to prove and the interested reader is referred to Born (1948) or Margenau and Murphy (1943) or Wilson (1966) for an outline of the proof.

The general reader need not however burden himself with the details of the proof: substances which behaved in this way could not be the subject of conventional thermodynamic arguments since they would not be represented by continuous thermodynamic surfaces. They are of mathematical interest only. Carathéodory's conclusions do however provide a physically satisfying explanation of the existence of entropy. For a perfect gas there is no problem. The existence of entropy and its properties have been demonstrated by relying on the simple result $pv = rT = \frac{2}{3}u$. When the gas is divided into two mechanically independent parts by a perfectly conducting partition, the composite system now has three independent degrees of freedom and its adiabatics lie in a set of non-intersecting surfaces. What happens if we gradually introduce arbitrary forces $(\mu f(r))$ between the molecules? As far as the undivided gas goes we know that there will be an integrating factor (not necessarily now $1/T$) and the adiabatics will be single curves which as the forces get smaller and smaller $(\mu \to 0)$ will get nearer and nearer to the adiabatic curve $(pv^{\gamma} = k)$ of a perfect gas. This is an example of the well-established principle of continuity; small changes produce small effects. Correspondingly in the composite system we expect to find small departures from the system of adiabatics which obtained when the gas was perfect. Now this can only happen by changes in the surfaces in which the adiabatics lie because the alternative is a sudden change in which the adiabatics wander through the whole of the space available to them, there being according to Carathéodory's theorem no half way house in which the adiabatics are confined to a small region; they are either all wholly within a set of non-intersecting surfaces or else they fill the whole space available to them. Thus, the expected continuous change in the equations of state produced by a continuous introduction of forces between the molecules includes the continued existence of an integrating factor for the composite state and this ensures that it still obeys the relationship $dQ = TdS$. While not a proof this argument strongly suggests that only in limited and very exceptional circumstances will the entropy theorem not be true. One can draw an analogy with a man walking across a field along a footpath on a windy day. If there is no wind he will follow the footpath closely. As the wind rises the random buffects will slightly deflect him from his path but he will never be found very far from the footpath. This will not however happen in the exceptional case that the footpath coincides with the edge of a cliff; then a small deflexion caused by the wind may sweep him over the edge and a small change in the forces will have precipitated a large change in the path. From the point of view of the student who must learn about entropy and complete differentials if he is going to make use of his formal thermodynamics the sooner he is introduced to these the better. For the student with more general interests the sooner he understands the broad molecular explanation of entropy the better.

The conclusions of this discussion may be summed up in the following way:

The answer to many important physical questions can be given by the molecular theory for the particular case of a perfect monatomic gas. Two examples are:

(a) equal temperature means equal average kinetic energy of the molecules

(b) the incomplete differential dQ is converted into the total differential dS by the integrating factor $1/T$ in which T is the perfect gas temperature

With the aid of these results many difficult problems can be solved for a perfect gas without further appeal to the molecular theory as shown in the following example.

Worked example

Helium at 40 atm and 15 K in a thin walled well-insulated vessel of negligible heat capacity is expanded adiabatically to atmospheric pressure. On the assumption that the entropy of helium gas is that of a perfect monatomic gas down to the boiling point T_B calculate the fraction that is liquified in terms of the latent heat of evaporation λ per mole.

Work in terms of one mole of helium. From section 5.1 the entropy of one mole of a perfect monatomic gas is given by $S = 3R/2 \log T + R \log V +$ a constant, which, since $V = RT/P$ can be expressed in terms of the pressure as $S = R(\frac{5}{2} \log T - \log P) + K$. During the expansion T and P will change so as to keep S constant until the liquid phase is formed; below this temperature P and T must follow the saturated vapour pressure curve. Since the expansion is adiabatic S the total entropy of the two phases remains constant. In the final state for which $P = 1$ atm $= P_2$ Pa, and $T = T_B$, let a fraction α have been liquified. Evaporate this liquid reversibly thereby adding $\alpha\lambda/T_B$ to the entropy which becomes $S + \alpha\lambda/T_B$. The entropy of the one mole of perfect gas so obtained is

$$R\left(\frac{5}{2} \log T_B - \log P_2\right) + K$$

$$= S + \frac{\alpha\lambda}{T_B}$$

$$= R\left(\frac{5}{2} \log T_1 - \log P_1\right) + K + \frac{\alpha\lambda}{T_B},$$

so that

$$\alpha = \frac{RT_B}{\lambda}\left(\frac{5}{2} \log \frac{T_B}{T_1} - \log \frac{P_2}{P_1}\right)$$

$$= \frac{RT_B}{\lambda}\left(\frac{5}{2} \log_e \frac{T_B}{15} - \log_e \frac{1}{40}\right)$$

From the tables $T_B = 4.22$ K and $\lambda = 80$ J/mole so that $\alpha = 0.23$. This is an example of a mixed calculation in which λ and T_B are obtained experimentally since we cannot as yet calculate them accurately from the known atomic structure of helium. The deviations from Boyle's law near the boiling point which have been neglected can be measured experimentally to give an estimate of the error to be expected in α.

If these two relationships remain true for real gases the same otherwise insoluble problems can be solved in the same way. It turns out that the results of such calculations are in agreement with experiment.

Quite remarkably, on the theoretical side, this procedure can be justified by a single additional assumption which can be made in many forms:

1 Carnot's Theorem (Carnot)
2 The second law of thermodynamics (Kelvin)

3 That the imperfect differential dQ always has an integrating factor $1/T$. (Sommerfeld)
4 That there are always neighbouring states of a real system which cannot be connected by an adiabatic path. (Carathéodory)

The same conclusion (i.e. that a single extra assumption is needed) will be reached in the next chapter in which an explanation of entropy will be given in terms of the molecular theory.

In which form the assumption should be made is not now of great importance. It is a matter of taste combined with Occam's razor. Left over are the more subtle questions:

Are the different assumptions precisely equivalent? Indeed, are they assumptions at all or can they be deduced from already agreed principles? These are indeed difficult questions which have teased the best mathematical physicists over the last 100 years.

5.5 Heat pumps and fuel conservation

The low efficiency of heat engines from which most of the world's power is derived poses serious problems to a world economy demanding more and more mechanical power from diminishing sources of fuel. There is little immediate prospect of increasing the efficiency of power stations beyond that of the best (~40 per cent) and it is galling to a fuel conscious community that more than half of the fuel consumed in a power station is used to heat rivers which would be better left unheated.

Worked example

What is greatest possible thermal efficiency of a power station prime mover if it takes in steam above the critical temperature at 400 °C and delivers it to a condenser cooled by a river at 25 °C?

The efficiency cannot exceed that of an engine working in a Carnot cycle between the maximum temperature T_1 and the minimum temperature T_2. The efficiency of a Carnot engine is

$$\frac{T_1 - T_2}{T_1} = \frac{375}{673.15} = 55.7 \text{ per cent.}$$

An elegant theoretical solution of the considerable part of the problem connected with keeping warm in winter and the general low temperature (process steam) activities of industry has been known for many years. It is usually referred to as the heat pump. A heat pump (so called because of a false analogy to the device for raising water from a well) is a heat engine following a Carnot cycle backwards. Such an engine will take in mechanical work W and deliver heat Q_1 at a temperature T_1: it is characterised by an efficiency $= 1 - T_2/T_1$ ($T_1 > T_2$) determined by the two temperatures T_1 and T_2 between which it is working. The efficiency is determined by these two temperatures only and does not depend upon the direction in which the cycle is traversed. A theoretical domestic heating scheme using a heat pump is illustrated in Fig. 5.6. If η_s is the efficiency of the Carnot engine at the power station $\eta_s = 1 - T_2/T_1$ and $W = \eta_s Q_1$. If $\eta_p = 1 - t_2/t_1$ is the efficiency of the heat pump, $W = \eta_p q_1$. Eliminating W the heat q_1 transferred to the building at temperature t_1 is given by

$$q_1 = \frac{Q_1 \eta_s}{\eta_p}$$

The attractions of this scheme are:

1 the convenience and flexibility of electrical heating has been retained

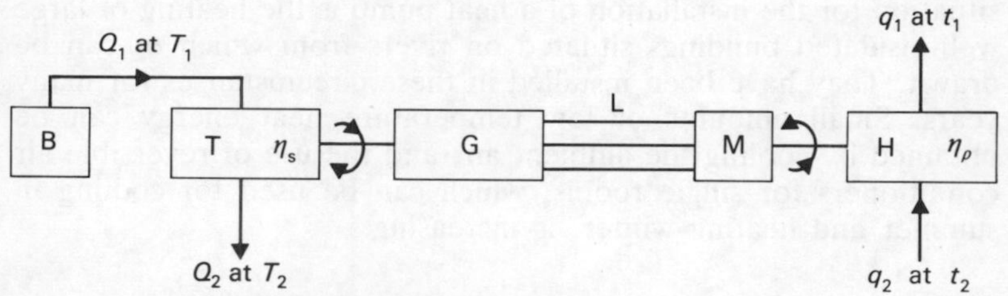

Fig. 5.6 Domestic heating with a heat pump.
B is the power station boiler. T is the turbine taking in high pressure steam at T_1 and delivering it as low pressure steam to a condensor whose temperature is T_2. The reason the combination of boiler and turbine has such a low efficiency η_s is that T_1 is determined by the thermodynamically irrelevant properties of steel. If T_1 were determined as it could be in theory by the maximum temperature which could be reached in the combustion of the fuel the efficiency η_s would be much higher. The generator G, transmission line L and motor M transfer the mechanical power from the power station to the domestic site. The losses in the system G, L, M, are remarkably low and could in theory be zero, i.e. superconducting leads etc. H is the heat pump taking in heat q_2 from a convenient source at temperature t_2 and delivering heat q_1 at temperature t_1 to the domestic heating system. The whole scheme depends for its success upon the heat pump behaving like a Carnot cycle and exchanging heat energy with its surroundings at two temperatures (t_1 and t_2) only.

2 because η_p can be chosen to be less than η_s more heat is transferred to the building than is released by the combustion of the fuel at the power station and $q_1 > Q_1$.

The theoretical gain can be quite large. For example if $T_1 = 350\,°C$ and $T_2 = 50\,°C$ at the power station

$$\eta_s = 48.1 \text{ per cent}$$

and for $t_1 = 70\,°C$, $t_2 = -10\,°C$ at the heat pump

$$\eta_p = 80/343 = 23.3 \text{ per cent}$$

The heat delivered to the building is therefore $48.1/23.3 = 2.06$ times the heat taken in from the boiler at the power station. The gain over straightforward electrical heating is $2.06/0.481 = 4.28$. Clearly gains of this order are worth striving for; to obtain them compressors must be developed whose performance approximates reasonably to that of an ideal Carnot engine. A source of heat must also be provided from which the heat pump can draw the heat q_2 which must flow into the Carnot engine at the low temperature t_2. These technical problems are more or less solved and the use of heat pumps is increasing. The most favourable situation for the installation of a heat pump is the heating of large well-insulated buildings situated on rivers from which q_2 can be drawn. They have been installed in these circumstances for many years. Small amounts of low temperature heat energy can be obtained by cooling the ambient air, and the use of reversible air conditioners for single rooms, which can be used for cooling in summer and heating winter, is increasing.

5.6 Irreversible changes

Processes such as State A → State B which actually occur in the laboratory do not proceed through a series of equilibrium states; the external conditions being biased in favour of the forward change. As has already been stated these changes are called 'irreversible' and strictly are outside the theory of states in equilibrium which has been discussed.

The term 'irreversible' has a very specialised meaning and does not mean that the reverse change State B → State A cannot take place. All it means is that the conditions which obtain when the change is going forward are different from the conditions when it is going backwards. In elementary physics one does not contemplate changes which cannot be reversed. In the classical mechanics of particle all changes will be strictly reversed if the velocities of all the particles are reversed.

In quantum mechanics the situation is not so simple because of particle diffraction, but these complications do not enter into the simple arguments being used here owing to the very large numbers of molecules involved. One of the reasons the study of reversible changes is important is that in an isolated system in which all the changes are reversible the entropy remains constant, the decrease in entropy Q/T_1 suffered by one part of the system being exactly compensated by the gain of entropy Q/T_2 in those parts of the system in which the heat Q is absorbed. Since the changes are reversible they proceed through systems in equilibrium so that $T_1 = T_2$ and the total change of entropy is zero. If the change is a real change going on at a finite rate the heat flow Q requires a temperature gradient to maintain it so that the loss of entropy Q/T is only partly compensated for by a gain of entropy $Q/(T - \Delta T)$ and there is always a net gain of entropy. There is also no change of entropy on the transfer of heat energy between bodies at quite different temperatures provided the change is carried out reversibly as is shown in the following example.

Worked example

Two bodies at temperatures T_1 and T_2 have the same heat capacity C. If heat is transferred between them reversibly show that they reach a common temperature $T = \sqrt{T_1 T_2}$ and that there is no change of entropy.

The heat must be transferred so that there is no finite temperature difference at any stage of the transfer. This can be done through the intermediary of a gas whose temperature can be changed by adiabatic expansion without changing its entropy. The gas at temperature T_1 and entropy S_1 is placed in contact with the hot body, and heat ΔQ_1 is absorbed by expanding the gas isothermally, thereby increasing its entropy by $\Delta Q_1/T_1$. The gas is then isolated and expanded reversibly without changing its entropy, until its temperature is T_2. It is now placed in thermal contact with the cold body and compressed absorbing heat ΔQ_2 until it has returned to its original entropy. Thermal contact with the cold body is now broken and the gas is compressed reversibly and adiabatically to its initial state with temperature T_1 and entropy S_1. In this imaginary process the temperature of the hot body has changed by $\Delta T_1 = -\Delta Q_1/C$ and that of the cold body by $\Delta T_2 = -\Delta Q_2/\mathrm{C}$ so that the difference of temperature has been reduced by the reversible transfer of the heat energy without change of entropy because $\Delta Q_1/T_1 + \Delta Q_2/T_2 = 0$. By repeating the process indefinitely the two bodies will be brought to a common temperature. The usual arguments of the calculus are used to show that if the steps are made small enough the temperature changes ΔT_1 and ΔT_2 caused by the heat transfers at each step can legitimately be neglected and ΔT_1, ΔT_2 replaced by dT_1, dT_2. Since the gas returned to its original entropy after each transfer

$$\frac{\Delta Q_1}{T_1} + \frac{\Delta Q_2}{T_2} = 0 = -C\left(\frac{dT_1}{T_1} + \frac{dT_2}{T_2}\right)$$

Integration shows that throughout the change $\log T_1 + \log T_2 =$ constant as $T_1 T_2 = T^2$ as stated.

A good example of the increase in entropy accompanying irreversible changes is afforded by the experiment devised by Joule with the object of measuring the internal potential energy of compressed gases. A cylinder containing $\tilde{n}$ moles of gas at a high pressure p is connected by means of a stopcock to a similar cylinder containing the gas at a very low pressure p'. The two cylinders are in an enclosure which completely insulates them from the surrounding matter. The heat capacity of the cylinders has a part to play in the theoretical discussion and is conveniently represented by a block of copper suspended in the enclosure. The cylinders can then be regarded as inert containers. The irreversible change is initiated by opening the stopcock and allowing the high pressure gas to expand into the empty cylinder. No external work is done and no heat enters or leaves the system, and the total internal energy of the gas is unchanged by the expansion. For a perfect gas there is no change of temperature on expansion. The internal energy of a perfect gas is wholly kinetic and the kinetic energy and so the temperature is unchanged. For a real gas a decrease in temperature is to be expected. After the expansion the molecules are on the average farther apart and so their potential energy has increased. This increase can only have come from the kinetic energy which has accordingly decreased and so there is a drop in temperature. The expansion can also be carried out reversibly (at least in imagination) by expanding the gas in the high pressure cylinder isothermally to the low pressure p'; the stopcock can now be opened and the two gases connected together under equilibrium conditions: the same pressure p' and the same temperature T. During the expansion work W_1 will have been done and heat Q_1 absorbed from the copper block which acts as a heat source. The gas at p' is now compressed into the combined cylinders isothermally and reversibly, work W_2 being absorbed and Q_2 given out and returned to the copper block which now acts as a heat sink. Finally, the gas is allowed to expand adiabatically doing work W_3 to bring its temperature to the final temperature which was reached in the free expansion. As a result the gas has gained entropy $(Q_1 - Q_2)/T$, the copper block has lost entropy $(Q_1 - Q_2)/T$, and work $W_1 + W_3 - W_2$ has been done.

Thus for the enclosure as a whole there has been no change of entropy, the gain of entropy by the gas being compensated for by the loss in the entropy of the copper. The work can be converted into potential energy within the system by raising the copper block against gravity. The increase in entropy which takes place when the gas expands irreversibly can be evaluated by a simple formal argument. The increase in entropy $(Q_1 - Q_2)/T$ of the gas is the same as before because entropy is a function of state and the final state of the gas is the same by whatever path, reversible or irreversible, it has been reached. For the irreversible expansion however there is no compensating decrease in the entropy of the copper block which remains unaffected. The free expansion has thus been accompanied by a net gain $(Q_1 - Q_2)/T$ in the entropy of the isolated system. That the terms $Q_1 - Q_2$ is always positive can be seen by considering a pv diagram of the expansion (e.g. example 5.10). The way in which this increase in entropy occurs can be explained by considering the irreversible expansion in more detail. When the stopcock is opened that part of the gas which will remain in its own cylinder will expand rapidly and approximately adiabatically and so drops considerably in temperature according to the relationship of section 2.4. The gas which rushes into the empty cylinder is initially at a low pressure and is compressed nearly adiabatically and so rises in temperature. To reach the final state in which all the gas is at the same temperature there has to be a large flow of heat by conduction through the gas. It is during this process that most of the increase in entropy occurs, since heat leaves a region of high temperature and is received in a region of low temperature. Incidentally, it was the time required for this heat flow which prevented Joule's experiment from achieving its objective. While he had massive thermometers sufficiently sensitive to record the small temperature change expected ($\sim 10^{-3}$ K) the temperature drift (due to the insulation of the calorimeter not being perfect) during the time required to reach equilibrium was large compared with the effect being looked for.

For a real gas it is quite difficult to calculate the change of entropy and the final temperature and pressure of the gas after the expansion from the equation of state, unless it is expressed in terms of the volume and internal energy $F(u, v, T) = 0$. If the gas is perfect gas all the quantities concerned can be easily calculated. The final state is now easily foreseen. No external work has been done by the gas and no heat has flowed into it from the enclosure. The internal energy and therefore the temperature of the gas will not have been changed by the expansion. Since its volume has been doubled and it obeys Boyle's law, the final pressure will be $p/2$. The trace of residual gas in the empty cylinder can be neglected; its pressure p' is a dummy introduced to avoid the absurdity of having to expand to zero pressure and infinite volume when the change is carried out reversibly.

Worked example

Find the change in temperature, $T_1 - T_2$, and the increase in entropy, $S_2 - S_1$, when a perfect monatomic gas changes its volume in a Joule expansion from v to $2v$.

In a Joule expansion the internal energy u is unchanged since the gas does no external work and no energy flows into it. If the equation of state is written in the form $T = f(u, v)$ $T_1 - T_2 = f(u, v) - f(u, 2v)$.

For a perfect monatomic gas $pv = rT = \frac{1}{3}nm\overline{c^2} = 2u/3$ and the internal energy u does not depend upon the volume v. Thus $T_1 - T_2 = 2u/3r - 2u/3r = 0$ and a Joule expansion produces no change in temperature. The entropy of 1 mole of a perfect monatomic gas is given by the expression in section 5.1 as

$$S = \frac{3}{2}R \log T + R \log V + \mathrm{K}$$

and since there is no change of temperature in the Joule expansion

$$S_2 - S_1 = R \log 2V - R \log V = R \log_e 2 \text{ per mole.}$$

Worked example

Make a similar approximate calculation for a monatomic gas which obeys van der Waals' equation.

The internal energy of the gas = kinetic energy + potential energy. The kinetic energy is $3RT/2$ per mole from the equi-partition theorem, and the potential energy is $-a/v$. The expression for the potential energy can be derived by the following approximate calculation. When a gas expands doing work $\int p\mathrm{d}v$ the part of the pressure due to the internal forces is, from the way in which van der Waals' equation has been derived, a/v^2. Thus the work done by the

$$\text{internal forces} = \int_{v_1}^{v_2} p\mathrm{d}v = \int_{v_1}^{v_2} -\frac{a}{v^2}\,\mathrm{d}v = \frac{a}{v_2} - \frac{a}{v_1}$$

$$= \text{decrease in the potential energy of the gas.}$$

This is equivalent to putting the potential energy $= \mathrm{K} - a/v$ with K chosen to be zero so that the p.e. $= 0$ when v is very large and the forces between the molecules can be neglected.

Thus for the Joule expansion of one mole of gas

$$U = \frac{3RT_1}{2} - \frac{a}{V_1} = \frac{3RT_2}{2} - \frac{a}{2V_1}$$

and

$$T_1 - T_2 = \frac{2}{3R}\left[\frac{a}{V_1} - \frac{a}{2V_1}\right] = \frac{a}{3RV_1}$$

The corresponding change in entropy can be calculated by carrying out the change reversibly so that

$$\mathrm{d}Q = T\mathrm{d}S = \mathrm{d}U + P\mathrm{d}V$$

$$\mathrm{d}U = \frac{3R}{2}\,\mathrm{d}T + \frac{a}{V^2}\,\mathrm{d}V$$

and

$$\mathrm{d}S = \frac{3R}{2}\frac{\mathrm{d}T}{T} + \frac{1}{T}\left(P + \frac{a}{V^2}\right)\mathrm{d}V$$

but $(P + a/V^2) = RT/(V - b)$ and $\mathrm{d}S$ can be integrated directly to give

$$S_2 - S_1 = \frac{3R}{2}\log\frac{T_2}{T_1} + R\log\frac{V_2 - b}{V_1 - b}$$

but $T_1 - T_2 = a/3RV_1$ so that the increase in entropy $S_2 - S_1$ per mole is given by

$$S_2 - S_1 = \frac{3R}{2}\log\left(1 - \frac{a}{3RT_1V_1}\right) + R\log\left(\frac{2V_\mathrm{I} - b}{V_\mathrm{I} - b}\right)$$

in which T_I and V_I are the initial temperature and volume of the gas.

One can generalise Joule's experiment to include all simple mechanical changes and say a non-reversible change always causes an increase in the entropy of an isolated system, that is a system contained by a rigid non-conducting wall.

A corollary of this is the important statement: an isolated system left to itself will reach a state of equilibrium in which the entropy is the maximum compatible with the constraints of the system. This is a useful statement since it can be used to formulate a principle analogous to that of virtual work in mechanics. Namely, that small changes in an isolated system in equilibrium will produce no change in its entropy.

The application of the principle turns out to be a very fruitful way of studying equilibria and one would like to be able to generalise it so that it was true for all systems of molecules, not only the rather simple ones for which it has just been demonstrated. If one attempts this one runs into serious difficulties of which it is worthwhile to give a very brief account.

Consider some electrolytic gas ($2H_2 + O_2$) in a strong cylinder fitted with a sparking plug and suspended in an enclosure with a copper block as described for Joule's experiment. This mixture, as is well known, is in a state of false equilibrium and if triggered off by a spark proceeds violently to a new state which consists of steam at a high temperature and pressure. When these violent changes have subsided, the final state will consist of a few drops of water in the cylinder at room temperature T and a copper block whose entropy has been raised by a considerable quantity Q/T in which Q is the heat energy released by the combination of the hydrogen and oxygen.

The change can also be carried out reversibly by means of a hydrogen-oxygen gas battery. In its operation the hydrogen and oxygen combine giving out very little heat q, which can be used to raise the entropy of the copper block by the small amount q/T. An electric motor driven by the cell will do a large amount of work Q the heat of combustion. Thus, it will be true to say that there is a much bigger increase ($\sim Q/T$) of entropy of the isolated system when the change is allowed to take place irreversibly by explosion, than when it takes place reversibly through a gas cell. But one cannot from this example justify the statement that an irreversible change always proceeds with an increase of entropy. On formulating the entropy balance one will have: initial entropy = entropy of 2 moles of H_2 + entropy of 1 mole of O_2 + initial entropy of one copper block and final entropy = entropy of 1 mole of water + initial entropy of 1 copper block + Q/T. The gain of entropy of the system is final entropy − initial entropy, which gives gain of entropy = (entropy of 1 mole H_2O − entropy of 2 moles of H_2 − entropy of 1 mole of O_2) + Q/T. Further progress is not possible without knowing the value of the bracketed term. This is not immediately available even in principle because as will be remembered from section 5.1 entropy as so far defined has an arbitrary constant in it. In Joule's experiment this difficulty did not arise because only differences of entropy of a single substance were involved and the integration constant did not appear in the calculations. This difficulty in generalising the statement 'irreversible changes always cause an increase in entropy' was first understood by Nernst; it is said in a flash of insight during a lecture (1905) to undergraduates: at any rate it so stated on a plaque in the lecture room which he used in the Berlin Physicochemical Institute. In his hands it led to the formulation of a new principle; Nernst's theorem or the Third Law of Thermodynamics. This although true, and a very useful 'aide memoire' turned out to be indeed a 'theorem' which could be deduced from acknowledged principles. Its proof experimental (see section 8.5) and theoretical is one of the most beautiful applications of quantum theory to the general description of matter in terms of the molecular theory.

5.7 Thermodynamic potential

The power of the principle of maximum entropy for the equilibrium state is illustrated by applying it to a simple example in which camphor is placed in an isolated evacuated enclosure. The camphor will then assume the maximum entropy allowed by the conditions.

With a little camphor it might be all vapour; with more camphor it might well be that camphor vapour, liquid camphor and solid camphor were all present. The mechanical theory of heat cannot of itself solve the problem of how the camphor will be divided between the three phrases. This is because the given energy of the closed system can be achieved in two independent ways. Raising the temperature increases the energy of all the phases since they all have positive heat capacities; increasing the proportion of the condensed phases decreases the distance between the molecules and so reduces the internal potential energy. As a result the given energy of the system can be achieved by a whole series of states and the actual state is not determined. Willard Gibbs in his general study of the equilibrium between phases (1875) showed that the principle, that from the series of states of an isolated system allowed by the conservation of energy one must select the one for which the entropy is a maximum, is sufficient to resolve the indeterminacy. For a pure substance a simpler demonstration (Birtwhistle 1925) is possible. It is of course Gibbs' general demonstration with n the number of components equal to one. Suppose a mass m of material with internal energy u is sealed in a rigid vessel of volume v which isolates it from its surroundings. For simplicity consider a state in which the matter divides itself between two phases. Let m_1, m_2 be the mass of the phases, u_1, u_2 their energy per unit mass, v_1, v_2 their volume per unit mass, and s_1, s_2 their entropy per unit mass. Then the following relationships must hold:

$m_1 + m_2 = m$ Conservation of mass

$m_1 u_1 + m_2 u_2 = u$ Conservation of energy

$m_1 v_1 + m_2 v_2 = v$ Rigid container of fixed volume

In addition the entropy is a maximum so that $s = m_1 s_1 + m_2 s_2$ is to have the greatest possible value consistent with the other conditions. For a maximum a small change in the proportion of the two phases will produce no change in the entropy so that $ds = m_1 ds_1 + s_1 dm_1 + m_2 ds_2 + s_2 dm_2 = 0$. This expression takes into account that although s_1 and s_2 are entropies per unit mass they are not constants but will depend upon the unknown temperature and pressure of the system which will change as the ratio m_1/m_2 of the two phases changes. The same consideration applies equally to u_1, u_2, v_1 and v_2 so that

$$du = m_1 du_1 + u_1 dm_1 + m_2 du_2 + u_2 dm_2 = 0$$

From the series of equations obtained in this way the small changes du_1 and du_2 can be eliminated by expressing the conservation of energy for each phase independently through the relationships

$$dQ_1 = T ds_1 = du_1 + p dv_1$$
$$dQ_2 = T ds_2 = du_2 + p dv_2$$

in which p and T are the unknown common pressure and temperature of both phases. Substituting for du_1 and du_2 in this way gives at once that

$$\begin{aligned} dm_1 + dm_2 &= 0 \\ m_1(T ds_1 - p dv_1) + m_2(T ds_2 - p dv_2) + u_1 dm_1 + u_2 dm_2 &= 0 \\ m_1 ds_1 + m_2 ds_2 + s_1 dm_1 + s_2 dm_2 &= 0 \\ m_1 dv_1 + m_2 dv_2 + v_1 dm_1 + v_2 dm_2 &= 0 \end{aligned}$$

These four equations limit the way in which the specific properties u_1, u_2 and v_1, v_2 of the material can vary. They express the fact, inherent in the assumption that entropy is a parameter of state, that the properties of matter cannot be chosen arbitrarily but are related to each other. The relationship between them can be obtained by combining the four restrictive equations into a single equation containing all the variables by Lagrange's method of undetermined multipliers. Multiply each equation by the arbitrary constants 1, λ, μ, ν, respectively and add the results to give the combined equation

$$\begin{aligned} & dm_1(1 + \lambda u_1 + \mu s_1 + \nu v_1) + dm_2(1 + \lambda u_2 + \mu s_2 + \nu v_2) \\ & \quad + m_1 ds_1(\lambda T + \mu) + m_2 ds_2(\lambda T + \mu) + m_1 dv_1(\nu - \lambda p) \\ & \quad + m_2 dv_2(\nu - \lambda p) = 0 \end{aligned}$$

The coefficients of this equation depend upon the physical properties (u, v, etc.) of the substance being considered which have so far been treated as if they could be chosen in an arbitrary way. The whole basis of the argument is that they are not arbitrary and there must be relationships between them which makes the solution of the composite equation not only possible but fully determined; otherwise there would not be an equilibrium between the two phases as assumed at the start of the investigation. The equation is obviously satisfied if all the coefficients of the six differentials are zero and this is the only way it can be satisfied. A formal proof of this can be given (Rushbrooke 1949) but it can also be inferred from general physical considerations. If the coefficients are not zero choose a value for one of the variables say dv_1; this can always be done since there are more differential variables than relations between them. This fixes the values of the other variables and the equation becomes a statement which cannot in general be true since the coefficients contain λ, μ, ν three arbitrary numbers. Therefore, the physical quantities $u_1, s_1, v_1, u_2, s_2, v_2$ must be related in such a way that the combinations which occur in the coefficients are zero whatever the values of λ, μ, ν. As will now be shown this on the face of it rather unlikely statement determines the relationship between them.

Accordingly equating the coefficients to zero we have

$$\lambda T + \mu = 0 \quad \text{therefore} \quad \mu = -\lambda T$$
$$\nu - \lambda p = 0 \quad \text{therefore} \quad \nu = \lambda p$$

and using these values

$$1 + \mu s_1 + \nu v_1 + \lambda \mathrm{u}_1 = 0 = 1 + \mu s_2 + \nu v_2 + \lambda u_2$$

becomes

$$u_1 - Ts_1 + pv_1 = u_2 - Ts_2 + pv_2$$

and these are the quantities which must be the same for the two phases if they are to be in equilibrium. The quantity

$$U - TS + PV = G$$

is known as the Gibbs free energy and is the same per unit mass for any two phases of a pure substance which are in equilibrium. This result can be obtained more simply without showing that the entropy is a maximum. If λ is the latent heat of the change per mole λ/T will be the change in entropy if (and only if) T is the temperature at which the two phases are in equilibrium, and

$$S_2 - S_1 = \frac{\lambda}{T} \quad \text{(definition of entropy)}$$

also

$$U_2 - U_1 = \lambda - P(V_2 - V_1) \quad \text{(conservation of energy)}$$

Eliminating λ gives at once that

$$U_1 - TS_1 + PV_1 = U_2 - TS_2 + PV_2$$

and $G_1/\text{mole} = G_2/\text{mole}$ at equilibrium.

It will be observed that the number of coexisting phases cannot exceed three. Suppose the value of G for the first phase is $G_{1/\text{mole}}$; then $G_{2/\text{mole}}$ for the second phase can be made equal to $G_{1/\text{mole}}$ by adjusting either the temperature T or the pressure P and for every P there will be a corresponding T. If there is a third phase the condition $G_{1/\text{mole}} = G_{2/\text{mole}} = G_{3/\text{mole}}$ can only be met at one value of T and one value of P. It is not possible to accommodate a fourth phase and if one is introduced (i.e. a second crystalline form) one of the four phases will be permanently unstable and disappear from the system. This is a rudimentary form of the Gibbs' phase rule. It is remarkable that so definite and important a prediction can be made from the molecular theory of heat and the assumption that there are states of the system which cannot be connected by adiabatic paths.

Problems

1 A heat engine is designed to work between an upper temperature $T_1 = 1\,°\text{C}$ and a lower temperature $T_2 = -1\,°\text{C}$. It uses water as the working substance and consists of a strong cylinder and a piston connected to a heavy load through a ratchet arrangement. The engine follows the following cycle. The cylinder is connected to the low temperature source; the water freezes and expansion causes the piston to move forward doing considerable work because although the motion is small the load is large. The cylinder is then connected to the high temperature source at 1 °C. The ice melts and the piston falls back under a negligible load because of the ratchet. Show that such an engine can have a thermal efficiency much greater than the 2/274.15 predicted by Carnot's theorem if water always freezes at 0 °C however great the pressure exerted on it by the piston. (It was the agreement between the predicted and measured value of the change in melting point with pressure which convinced the Thomson brothers that Carnot's theorem was true.)

2 Show that if entropy is a function of state that $\oint dQ/T$ round a closed cycle is zero. Use this to prove Carnot's theorem that the efficiency of an ideal heat engine taking in heat at T_1 and rejecting it at T_2 is $(T_1 - T_2)/T_1$.

3 A well-insulated house can be looked upon as a cube of side 10 m with walls of mineral wool ($k = 0.042$ W/mK) 20 mm thick. If the house is to be maintained at 22 °C when the outside temperature is −30 °C, calculate the minimum power to drive a heat pump drawing its heat from an underground river at 2 °C.

4 The Pfaffian equation $-y\,dx + x dy + k dz = dQ$ has no integrating factor. By rewriting it in cylindrical coordinates show that all radial paths are adiabatic and hence that all points can be connected by an adiabatic path.

5 Show that the Pfaffian $X dx + Y dy + Z dz = 0$ has a solution if $\mathbf{R}$. Curl $\mathbf{R} = 0$, in which X, Y, Z are the components of a vector $\mathbf{R}$. (Margenau & Murphy, p. 86).

6 It is easy enough in a school laboratory to show that $(1/p_0)(dp/dT)_v$ for air is about 1/273. Give a brief account of the difficulties which have to be overcome in converting this into a precise measurement for N_2 or He over the temperature range −50 °C–1000 °C.

7 Two bodies at temperature T_1 and T_2 and heat capacities C_1 and C_2 are brought to a common temperature T by a reversible process. Find the common temperature and show that there is no change of entropy.

8 Show that in the derivation of Gibbs' function in 5.7 it is not necessary to assume that the phases have a common temperature and pressure.

9 Verify that the condition

$$\left(\frac{\partial h}{\partial s_1}\right)_{s_2}\left(\frac{\partial s}{\partial s_2}\right)_{s_1} = \left(\frac{\partial h}{\partial s_2}\right)_{s_1}\left(\frac{\partial s}{\partial s_1}\right)_{s_2}$$

is fulfilled for the system of Fig. 5.3.

10 By drawing the pv diagram of the gas in Joule's experiment, justify the statement that $Q_1 - Q_2$ is positive made in section 5.5.

11 Calculate the change in temperature and entropy when a mole of Argon at 50 atm at room temperature doubles its volume in a Joule expansion. (Calculate a and b in van der Waals' equation from the critical data which are $p_c = 4.86$ MPa, $v_c = 75.2\ \text{cm}^3/\text{mole}$, $T_c = 150.7$ K).

Further reading

Born, M. (1949) *Natural Philosophy of Cause and Chance*, Clarendon Press, Oxford.

Gibbs, J.W. (1925) *The collected works of J.W. Gibbs*, Longman, Green & Co., London.

Margenau, H. and **Murphy, G.M.** (1946) *The Mathematics of Physics and Chemistry*, D. Van Nostrand and Co. New York.

Mendelssohn, K. (1973) *The World of Walther Nernst*, Macmillan Press Ltd., London.

Wilson A.H. (1966) *Thermodynamics and Statistical Mechanics*, C.U.P. London.

References

Birtwhistle, G. (1925) *The Principles of Thermodynamics*, University Press, Cambridge.

Born M. (1921) *Phys. Zeitschr.* **22,** 218, 249, 282.

Kipling R. (1910) *Action and Reactions*, p. 109, Macmillan and Co. London.

Metrologia (1969), **43,** 5.

Piaggio, H.T.H. (1931), *Differential Equations*, G. Bell and Sons, London.

Rushbrooke, G.S. (1949), *Introduction to Statistical Mechanics*, Clarendon Press, Oxford.

Weinhold, F. (1976) *Physics Today* **29**(3), 23.

Wilson, S.S. (1981) *Scientific American*, **245**(2), 102.

CHAPTER 6

Entropy

6.1 Entropy and the energy distribution function

Entropy is clearly an important variable of state. In the last chapter it was discussed in terms of variables like temperature, pressure and reversible heat flow which can be observed by instruments at the boundary of the gas. Very little use was made of arguments based upon the molecular structure. In the present chapter entropy will be discussed from the point of view that the gas is a collection of very weakly interacting molecules. The interaction is supposed to be so small that the gas can be treated as a perfect gas. If one knew the mass, position and velocity of every molecule in the gas as accurately as is allowed by the uncertainty principle one would know all that could be known by experiment about the system. Provided with this information about the state of a perfect gas at some definite time it should be possible to calculate the five variables of state in terms of which the properties of the gas can be described without having to refer to its past history. The first four are easily calculated. The volume can be written down at once. It is clearly the volume of the vessel in which the gas is confined. The pressure is $\frac{1}{3}nm\overline{c^2}$, the temperature $m\overline{c^2}/3k$ and the internal energy $\frac{1}{2}Nm\overline{c^2}$. The last three quantities depend only on the mean square velocity and all four are independent of the way in which the energy is divided among the molecules. A gas in which all the molecules had the same scalar velocity c could not be distinguished from a gas in which the energy was distributed according to Maxwell's law by measuring its equation of state $pv = rT = \frac{1}{3}Nm\overline{c^2}$.

This exhausts the possibility of expressing state variables in terms of mean quantities averaged over all the molecules. The remaining variable the entropy must depend upon the way in which the energy is distributed amongst the molecules. Boltzmann (1866) was the first physicist who not only realised this but put forward a way in which the entropy could be calculated from the primitive description of the molecular state in terms of the positions and velocities of the molecules.

It is intuitively clear that if one had a nearly perfect gas (say helium) in which all the molecules had the same scalar velocity c, this simple velocity distribution would soon be destroyed by molecular collisions, and replaced by successively more complicated distributions. However, this increase in the complexity of the distribution function cannot go on indefinitely as the gas would then always be changing and never reach a steady state. Clearly the steady state distribution may be recognised by the property that it is self-reproducing. In an element of time as many molecules leave the category in which their velocity lies between c and $c + \mathrm{d}c$ as enter this category by collisions between molecules which have other velocities. One notes in passing that one has abandoned the strictly ideal gas in which molecular collisions are ignored. It is molecular collisions which enable the molecular distribution function to be maintained. Boltzmann was able to show by considering a hard sphere model that Maxwell's distribution function and no other has just this self-reproducing property. One could describe the process of reaching the steady state in the following form of words. The original state in which all the helium molecules had the same velocity was a very artificial or improbable one and the effect of each collision was to produce a more probable state. The increasingly chaotic or disordered states produced by successive collisions were states of increasing probability and the final self-producing state was one with the greatest possible probability compatible with the constraints of the system. This state was self-reproducing, for otherwise the effect of molecular collisions would be to produce less probable states which is absurd.

6.2 Probability

In a sense this very plausible description is a play on words. It depends on the meaning of the word probable. The motion of the molecules after a collision is determined by the laws of mechanics which say nothing about collisions giving rise to more probable states. This is also true in quantum mechanics and the motion of the particles after a collision is determined by Schrödinger's equation as definitely as the uncertainty principle allows. A mathematical study of the way in which a gas composed a perfectly elastic spheres reached and maintained a Maxwell distribution enabled Boltzmann to put forward a practicable scheme for calculating the entropy of a gas whatever the law of force between the molecules. He did this by adopting ideas from the laws or probability developed by Pascal (1654).

Pascal, an austere theologian and no mean mathematician, was consulted by a gamester, the Chevalier de Méré, as to the correct odds in games of chance played with dice and cards. Pascal in a correspondence with Fermat was the first mathematician to give a precise, if limited, meaning to the term probability. The probability of an event in a game of chance is defined to be the number of ways in which an event can happen divided by the total number of ways in which the rules of the game can be fulfilled. It is thus a number between 0 and 1. This meaning is illustrated by the following very simple examples. The probability of throwing a six with a die is 1/6 because the die can fall in six different ways and for only one of these is the six uppermost. A card can be drawn from a pack by selecting any one of 52 cards. Of these 52 ways of drawing a card only four will correspond to drawing an ace. Accordingly the probability of drawing an ace is $4/52 = 1/13$. In general, this is a difficult branch of mathematics. Great care has to be taken neither to miss, nor count more than once, the ways in which an event can occur. When a correct value for the probability has been reached it is often in a very untidy form which it requires great technical ability to reduce.

Worked example

A has a coins, B has b coins. They play by spinning a true coin. If it falls heads A gives B one of his a coins; if it falls tails B gives A one of his b coins. The game is won by the first player to beggar his opponent and hold all the $a + b$ coins. What is A's chance of winning?

This is in essence the original problem posed to Pascal by de Mere, who wished to know how a stake was to be divided between equally skilled players who were forced to abandon a game before it was completed. The solution of the more difficult problem in which the coin is biassed was given sixty years later by Montfort and elaborated by Laplace.

Let $f(a)$ be A's chance of winning. Then, A has obtained a coins in one of two ways; either he had $(a + 1)$ coins and lost one, or he had $(a - 1)$ coins and gained one.

Thus, $f(a) = \frac{1}{2}f(a + 1) + \frac{1}{2}f(a - 1)$

From this it follows that $f(a)$ is a linear function of a and $f(a) = ka + c$, say.
By a similar argument $f(b) = kb + c$.
But either A or B must win so that

$$f(a) + f(b) = 1,\ k = \frac{1 - 2c}{a + b} \quad \text{and} \quad f(a) = \frac{(1 - 2c)}{a + b}\, a + c$$

But $f(1) = \frac{1}{2}f(2)$ so that $c = 0$ and $f(a) = a/a + b$. For a more detailed study of this type of problem the reader is referred to 'Theory of Stochastic Processes', Cox & Miller (1965).

Worked example

In how many different sequences can n_1 identical red cards and n_2 identical green cards be arranged?

Make the cards distinguishable by printing capital letters on the backs of the red cards and small letters on the backs of the green cards. The n_1 plus $n_2 = n$ distinguishable cards can be arranged in $n!$ different orders because the first position in the order can be occupied by any one of the n different cards, the second position by any one of $(n - 1)$ remaining cards etc. to give $n(n - 1)(n - 2) \dots 1 = n!$ different sequences. Imagine the $n!$ possible sequences set out in a matrix of n columns and $n!$ rows. This includes all the ways of ordering the $n_1 + n_2 = n$ cards. Now turn the cards face upwards to give a matrix which must include all the ways of ordering n_1 red and n_2 green cards. Many of the rows will be duplicates and the answer to the problem is the number of rows left when all the duplicates have been removed. For example the six different arrangements AbFdeC, FbAdeC, FbCdeA, CbAdeF, CbFdeA, will give the same arrangement red green red green green red when turned face upwards. All sequences of cards which face downwards differ only in the order of the small and capital letters will give the same sequence of red and green cards when turned face upwards. Since as has just been shown n different objects can be arranged in $n!$ different orders there will be $n_1!\ n_2!$ duplicate rows in the matrix and the number of different red and green sequences is $n!/n_1!n_2!$.

Worked example

A long lived radioactive element (e.g. ^{238}U) emits α particles at random so that the probability that a particle is registered by a particular detector in a short time Δt is $\lambda\Delta t$.

Find the probability W_t^n that n particles will be counted in t seconds. Hence find the mean counting rate and the mean square deviation from the average rate.

One expects W_t^n to be a continuous function of t and a discontinuous function of n. The probability that n particles are detected in a time $t + \Delta t$ is $W_{t+\Delta t}^n$. This event can happen in two ways: *either* $(n - 1)$ particles are counted in t seconds with a probability W_t^{n-1} followed by Δt seconds in which one particle is counted (probability $W_{\Delta t}^1$), *or* all the n particles are counted in the first t seconds (probability W_t^n) followed by an interval Δt in which no particle is counted (probability $W_{\Delta t}^0$). These two ways in which n particles are counted in $t + \Delta t$ seconds are independent, so that one can add their probabilites and write, $W_{t+\Delta t}^n = W_t^{n-1} \times W_{\Delta t}^1 + W_t^n \times W_{\Delta t}^0$. The approximate values of $W_{\Delta t}^1$ and $W_{\Delta t}^0$ can be found directly on the assumption that the probability of detecting two particles in this small interval is negligible. In other words it is certain (probability 1) that either no particle or one particle will be registered in Δt and $W_{\Delta t}^1 \times W_{\Delta t}^0 = 1$.

Since by definition $W_{\Delta t}^1 = \lambda\Delta t$, $W_{\Delta t}^0 = 1 - \lambda\Delta t$.

Accordingly

$$W^n_{t+\Delta t} = W^{n-1}_t \lambda\Delta t + W^n_t (1 - \lambda\Delta t)$$

$(n - 1)$ in t followed by 1 in Δt — n in t followed by 0 in Δt.

which can be rewritten as

$$\frac{W^n_{t+\Delta t} - W^n_t}{\Delta t} = \lambda(W^{n-1}_t - W^n_t) = \frac{dW^n_t}{dt} \quad \text{as} \quad \Delta t \to 0$$

From this difference equation the value of W^n_t can be built up step by step. W^0_t can be evaluated directly. Split the time t into N equal intervals t/N in which the probability of observing no particles is $(1 - \lambda t/N)$. The probability of observing no particles in N successive intervals is therefore

$$\left(1 - \frac{\lambda t}{N}\right)^N \text{ and } W^0_t = \underset{N\to\infty}{\text{Lt}} \left(1 - \frac{\lambda t}{N}\right)^N = e^{-\lambda t}$$

by definition of the exponential function. W^1_t is evaluated by putting $n = 1$ in the difference equation to give

$$\frac{dW^1_t}{dt} = \lambda(e^{-\lambda t} - W^1_t)$$

This is a straightforward first order differential equation for W^1_t whose solution is $W^1_t = \lambda t\ e^{-\lambda t}$. By successively putting $n = 2$, $n = 3$ etc. the general result $W^n_t = ((\lambda t)^n/n!)\ e^{-\lambda t}$ can be built up. The mean rate $\bar{n}$ can be obtained directly. The probability of observing n particles in a time t is W^n_t, so that the total number of particles observed in a large number N of successive observations of duration t is $N\Sigma n W^n_t$ and the average rate $\bar{n}$ is

$$\Sigma n W^n_t = e^{-\lambda t}\Sigma n \frac{(\lambda t)^n}{n!} = e^{-\lambda t} \cdot \lambda t \Sigma \frac{(\lambda t)^{n-1}}{(n-1)!}$$

$$= \lambda t e^{-\lambda t} e^{\lambda t} = \lambda t = \bar{n}$$

The average number of particles observed in a time t is therefore proportional to t as it should be. The mean square deviation $\overline{(n - \bar{n})^2} = \overline{(\Delta n)^2}$ can be obtained indirectly as follows:- By a similar argument to that used for finding $\bar{n}$, the mean value of $n(n - 1) = \overline{n^2} - \bar{n}$ is given by

$$\Sigma n(n-1) W^n_t = e^{-\lambda t}\Sigma \frac{n(n-1)}{n!} (\lambda t)^n$$

$$= (\lambda t)^2 e^{-\lambda t} \frac{(\lambda t)^{n-2}}{(n-2)!}$$

$$= (\lambda t)^2 = (\bar{n})^2$$

so that $\overline{n^2} - (\bar{n})^2 = \bar{n}$
But $\overline{(\Delta n)^2} = \overline{n^2} - 2\bar{n}\bar{n} - (\bar{n})^2 = (\overline{n^2}) - (\bar{n})^2$
so that $(\overline{\Delta n^2}) = \bar{n}$.

Worked example

In how many different orders can 600 passengers disembarking from a steamer go down a single gangway?

The number of different orders is clearly 600! This is not a useful answer until evaluated. The factorials of large numbers can be approximated to (very accurately for $n > 100$) by making use of Stirling's theorem. A rough approximation to $n!$ can be obtained by giving a geometrical interpretation to the identity $\int \log x \, dx = x \log x - x$. If a graph of $\log x$ against x as shown in Fig. 6.1 is divided up into unit steps it is clear that the area under the logarithmic curve is approximately equal to the sum of the strips of area $\log 2 + \log 3 \ldots \log n = \log n!$ out of which it can approximately be made up.

Thus,

$$\log n! \approx \int_{3/2}^{n+1/2} \log x \, dx = (n + \tfrac{1}{2}) \log (n + \tfrac{1}{2}) - n + 0.3918$$

$$\text{and} \quad n! \approx \sqrt{5{\cdot}94n} \; e^{-n} \, n^n \left(1 + \frac{1}{8n} \ldots\right)$$

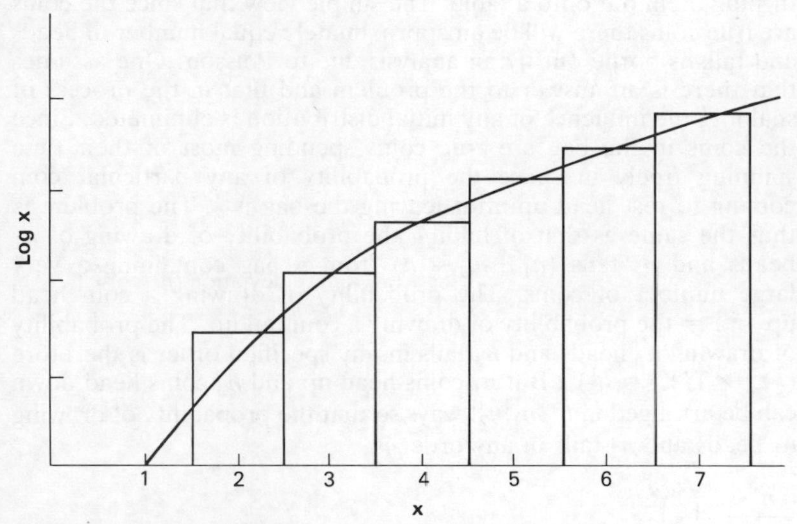

Fig. 6.1 Stirling's theorem.

This may be compared with Stirling's correct approximation

$$n! = \sqrt{2\pi n}\ e^{-n}\ n^n\left(1 + \frac{1}{12n}\ldots\right)$$

For the $n \approx 600$ the difference between the two expressions can usually be neglected and it is sufficient to write $n! = n^n\ e^{-n}$. It is left to the reader to show that $600! = 1.27 \cdot 10^{1408}$.

These large numbers have no physical meaning as they wildly exceed any estimates of the total number of protons in the universe. Stirling's theorem provides a way of treating an expression like $n!/n_1!n_2!$ when all the n's are large numbers as if it were a function of continuous variable amenable to the methods of the calculus. This is an immense mathematical simplification. For molecular systems for which $n \approx \mathrm{L} \approx 10^{23}$ the difference between the approximate and the true answer is quite beyond the limits of direct measurement.

6.3 Molecular distribution functions

Boltzmann thought that the problem of finding a distribution function could be looked upon as a complicated example of the well-known problem of the distribution of heads and tails in a pile of coins obtained by violently shaking a bag of n true coins and then turning them out onto a table. The simple view that since the coins are true coins there will be an approximately equal number of heads and tails is borne out by an analysis due to Poisson. One assumes that there is an answer to the problem and that in the process of shaking, the influence of any initial distribution is eliminated. Since the coins in the bag are true coins spending most of their time spinning freely in space the probability of any particular coin coming to rest head up after leaving the bag is $\frac{1}{2}$. The problem is thus the same as that of finding the probability of drawing of n_1 heads and n_2 tails ($n_1 + n_2 = n$) from a bag containing a very large number of coins. The probability of drawing a coin head up $= \frac{1}{2} =$ the probability of drawing a coin tail up. The probability of drawing n_1 heads and n_2 tails in any specified order is therefore $(\frac{1}{2})^{n_1} \times (\frac{1}{2})^{n_2} = (\frac{1}{2})^n$. But n_1 coins head up and n_2 coins head down can be arranged in $n!/n_1!n_2!$ ways so that the propability of drawing n_1 heads and n_2 tails in any order is

$$\frac{n!}{n_1!n_2!}\cdot\left(\frac{1}{2}\right)^n$$

This expression has the property that it only has an appreciable value for large n if $n_1 \approx n_2$ and can be represented by a Gaussian distribution (cf. problem 6.7) centred about $n_1 = n_2$. The features of this calculation which are of interest in the present discussion are

1 the conversion of an apparently insoluble dynamical problem (the shaking of coins in a bag) into a comparatively simple problem in mathematical probability.

2 the result that $n_1 \approx n_2$ is due to the overwhelmingly large number of ways in which (for large n) the distribution for which $n_1 \approx n_2$ can be reached compared with the number of ways of reaching any other distribution.

In order to bring the laws of probability to bear on molecular problems one must invent a procedure in which they can be described in terms analogous to the drawing of coins out of a bag. A suitably simple procedure is to represent the energy levels of the elements of the system $E_1, E_2, E_3 \ldots$ by a set of parallel grooves cut in a board. These can be equally spaced, the scale to which they are drawn not entering into the representation. The material system being considered is thought of as a collection of elementary parts (the elements of the system) which differ only in their energy.

Beads representing the elements of the system are then drawn from a bag and placed in the grooves. If the system contains n distinct elements n beads will have to be distributed. There is only one rule to be followed, namely, the conservation of energy. The total energy represented by the beads on the board must be the constant energy of the system. A given state of the gas is then represented by a pattern of beads on the board. If the elements are molecules a pattern in which all the beads are in one groove E_n will represent a perfect gas in which all the molecules have the same velocity c given by $E_\mathrm{n} = \frac{1}{2}mc^2$. As the molecules in the gas collide other energy levels will be occupied and the beads representing them will spread out until, when a steady state is reached, there is no further change in the representative pattern. So far this is no more than a pictorial way of conveying the information which is conveyed more succinctly and accurately by the relationship.

$$\mathrm{d}N = \frac{2N}{\sqrt{\pi k^3 T^3}} \mathrm{e}^{-E/kT} E^{1/2}\, \mathrm{d}E$$

which represents a Maxwell distribution in terms of the energy.

Boltzmann's representation however draws attention to the fact that in the steady state the distribution function remains constant, although the velocities of the individual molecules are continually changing and individual beads move from one groove to another. Boltzmann showed that the pattern corresponding to Maxwell's distribution function could be obtained by applying the same probability arguments to molecular motion as Poisson had applied to the distribution of heads and tails in a pile of coins. At this point the investigation into molecular distribution functions divides into two parts:-

1 the detailed justification of Boltzmann's methods from the laws of mechanics. The arguments are necessarily abstract and depend upon general theorems which are difficult to prove. This branch of the subject is usually called Statistical Mechanics and it took two generations of physicists and the introduction of quantum mechanics to complete it. Its study will not be attempted here.

2 The consequences of Boltzmann's method which allow a relatively simple calculation of the entropy of molecular systems to be made. These were important, far reaching and to a great extent unexpected and are the subject of this chapter.

6.4 Entropy and probability

Boltzmann introduced an enormous simplification by transferring the calculations from the actual molecular collisions which set up the pattern to the pattern itself. To do this he introduced a method of representing patterns by a single variable Ω. It represents the total number of distinct ways in which the pattern representing the system can be built up by placing beads representing the elements of the system into grooves representing the energy levels of an element. The evaluation of Ω will be considered later but it clearly exists and for small number of beads can be evaluated by simple enumeration. For example, one 'way' of building up a particular pattern would be to complete the bottom (low energy) groove first and proceed systematically placing the correct number of beads in the grooves in increasing order of energy. The whole process would count as one 'way'.

Another way would be to complete the top groove first and proceed systematically downwards to the bottom groove. If Ω_i is the total number of ways of building up the ith pattern then $\Omega = \Sigma\Omega_i$ the summation being taken over all the possible patterns of beads. The probability of finding the gas in a state represented by the ith pattern is Ω_i/Ω. This may be interpreted as meaning that in the course of the changes produced by molecular collisions the gas will spend a fraction of its time Ω_i/Ω in a state having a distribution function represented by the ith pattern.

Worked example

Find Ω_i and Ω by enumeration in the very simple case for which there are 3 beads, the energy levels are $0, \varepsilon, 2\varepsilon, 3\varepsilon, 4\varepsilon$, and the total energy is (a) 12ε, (b) 0, and (c) 3ε.

(a) All the beads must be in level 4. This state can only be reached by putting each bead as it is withdrawn from the bag into this level; Ω_i is therefore 1 and $\Omega = \Omega_i = 1$ since there are no other possible patterns.

(b) All the beads must be in level 0 and as before $\Omega_i = \Omega = 1$.

(c) Three possible patterns are possible:
(i) one bead in level 3ε and 2 beads in level 0. This can be achieved by putting either the first, second or third bead drawn from the bag into level 3ε and $\Omega_i = 3$.
(ii) all three beads in level ε. Each bead must be put into this level after drawing from the bag and $\Omega_i = 1$.
(iii) one bead in each of the three levels 0, ε, and 2ε. The first bead can go into any one of the three levels; the second bead into either of the remaining levels and for the third bead there is then no choice $\therefore \Omega_i = 3 \times 2 \times 1 = 6$.
The value of $\Omega = \Sigma\Omega_i = 3 + 1 + 6 = 10$.

It is clear from this example that enumeration is not a practical method of calculating Ω when the numbers are large.

Boltzmann's next suggestion was that S the entropy of the state was simply related to Ω so that

$$S = F(\Omega)$$

This puts into a manageable mathematical form the suggestion that the entropy depends directly upon the distribution function. The pure number Ω is often represented in the literature and many textbooks by the letter W. Boltzmann's work was outstandingly original and he modified his notation as the subject developed. The first use of Boltzmann's expression for the entropy of a system outside the field of molecular physics was made by Planck (1905) in his classical researches into black body radiation. He introduced a different notation replacing Ω by W derived from the German word GesamtWahrscheinlichkeit meaning total probability. This notation was generally adopted and the triumphant expansion of the subject (1905–1923) was made using this notation. This expansion was primarily concerned with the experimental verification of the theoretical expressions for the entropy of gases. The return to the modified form of Boltzmann's notation (he wrote Ω when one now writes $\log \Omega$) corresponds to a revived interest in the mathematical foundations of the subject.

Curiously enough if Boltzmann's suggestion is well founded the unknown function $F(\Omega) = F(W)$ is easily found.

Suppose one has a gas in a vessel divided into two portions by an isolating partition, and that the entropies of the gases in the two sections are S_1 and S_2. Let Ω_1 and Ω_2 be the corresponding values of Ω. Then if the system is regarded as one system with entropy S and "GesamtWahrscheinlichkeit" Ω we know from our previous results (section 5.3) that

$$S_1 + S_2 = S.$$

Also, it is clear that any one of the Ω_1 ways of representing state 1 can be combined with any one of the Ω_2 ways of representing state 2 so that Ω the number of ways of representing the combined state is $\Omega_1 \cdot \Omega_2$.

If as Boltzmann supposed

$$S_1 = F(\Omega_1)$$
$$S_2 = F(\Omega_2)$$
$$S = F(\Omega) = F(\Omega_1 \Omega_2)$$

so one has

$$F(\Omega_1) + F(\Omega_2) = F(\Omega_1 \Omega_2).$$

This functional equation is clearly satisfied by $F(\Omega) = k \log \Omega$ and it can be shown (Hardy, 1955) that this is the only solution.

The arbitrary constant k is universally known as Boltzmann's constant and is related to the gas constant R by the relationship $R = N_A k = Lk$ ($N_A = L$ = number of molecules in a mole).

The adoption of Boltzmann's suggestions thus effectively solves the problem of calculating the entropy of a gas from first principles if the energy levels are known. Boltzmann's ideas were not generally accepted when he put them forward in 1867. They gave rise to a great deal of controversy. A brief account of the matters in dispute, which played a considerable part in the development of theoretical physics, is given in section 6.14 at the end of the chapter. It is a classic example of how easy science is when one has the key and knows the answer and how difficult when one does not. The method of calculating Ω is described in section 6.6 after the bearing of Willard Gibbs contribution to the subject has been described.

6.5 Gibbs' paradox

It was more than fifty years before Boltzmann's ideas were generally accepted. Broadly speaking his ideas on the calculation of entropy have turned out to be correct and can be applied in a straightforward way to those simple systems for which the discrete energy levels of the elements of the system are known.

These systems were not known about in Boltzmann's time and everybody thought (quite erroneously) of a perfect gas as the simplest system on which to try out a new method of calculating entropy. Any calculation of the entropy of a collection of molecules must be divided into two parts: a purely quantum mechanical calculation of the energy levels of the elements of the system and a subsequent calculation of Ω and so of the entropy through the relationship $S = k \log \Omega$. Now it so happens that although a perfect gas is easy to describe in molecular terms, it is not a system whose quantum mechanics is correspondingly simple. Its energy levels are so closely spaced that for purposes of calculation they can be treated as forming a continuous spectrum which is described in terms of a density of levels $\mathrm{d}n/\mathrm{d}E$, the number of discrete levels whose energy lies between E and $E + \mathrm{d}E$. The density of levels of a perfect gas was not correctly known until quite late in the development of quantum mechanics. Boltzmann who, of course, knew nothing about quantum mechanics was able to calculate the entropy of a perfect gas by starting from a continuous system or spectrum of energy levels based upon Maxwell's distribution function and the continuously variable energy $\frac{1}{2}mc^2$ of a single molecule of the gas. Boltzmann's spectrum although wrong in principle was not very different in form from the correct one; his expression for the entropy of a perfect gas was nearly right and at the time was not seriously challenged. There was then no way of checking it experimentally, (vide section 8.6) as the erroneous part was in the constant term which was in any case thought to be arbitrary in agreement with the arbitrary integration constant which appears in the thermodynamic treatment of entropy. That something was wrong could have been deduced from a *reductio ad absurdum* argument based on an extension of the paradox noticed by Gibbs and Rayleigh in their work on the diffusion into each other of gases like CO and N_2 which have the same molecular weight. Two equal vessels connected by a tap each contain N/2 molecules at the same temperature of two different perfect gases, say hydrogen and helium respectively. The tap is then turned so that the vessels communicate and the two gases diffuse into each other. A straightforward classical calculation using semi-permeable membranes shows that the increase in entropy is $Nk \log_e 2$.

Boltzmann's method leads to the same result, since in calculating Ω for the mixed gases each molecule may be assigned to either vessel, each represented by its own set of grooves on its own board, while for the separated gas no such choice exists, each molecule having to be put into its own vessel. So (Ω after diffusion) = $2^N \times$ (Ω before diffusion) and the entropy gained by diffusion in $Nk \log_e 2$, as before. This agreement is very satisfactory.

Worked example

Find the change in entropy when a mixture of one mole of a perfect gas A and one mole of a perfect gas B at pressure P and temperature T is separated into its component parts.

To calculate the change in entropy the separation must be carried out through a series of equilibrium states. This can be done (in imagination) by putting the mixed gases in a cylinder of volume $2V$ fitted with two pistons of negligible thickness moving in opposite directions as shown in Fig. 6.2. The left-hand piston contains a membrane permeable to A and impermeable to B; the right-hand piston contains a membrane permeable to B and impermeable to A. Initially the pistons are withdrawn to opposite ends of the cylinder and the space between them is filled with a volume $2V$ of the mixed gases. Displace the left hand piston isothermally until it reaches the centre of the cylinder. The gas A is unaffected by the motion but gas B is compressed isothermally and reversibly into the right hand section of volume V which now contains one mole of gas B. The work done on the

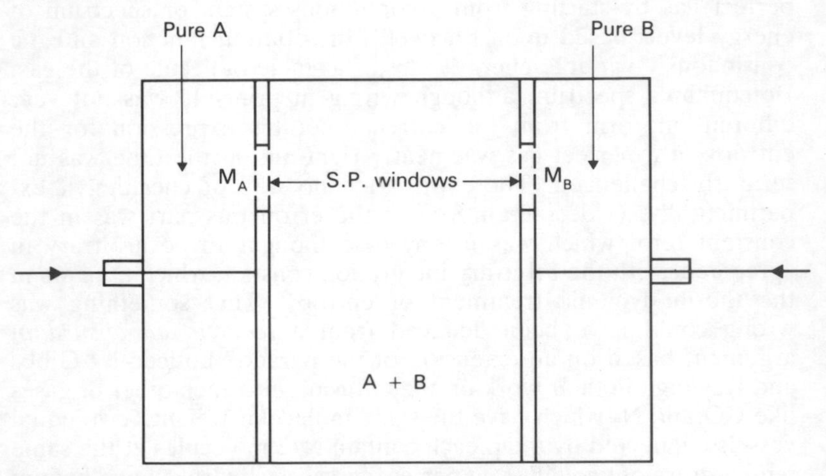

Fig. 6.2 Separation of gases by semi-permeable membranes.
M_A is a membrane permeable to A only
M_B is a membrane permeable to B only
Practical examples of these membranes are warm palladium which is very permeable to hydrogen and Pyrex glass which is permeable to helium.

system $= \int_{2v}^{V} P\mathrm{d}V = RT \log_e 2$. For a perfect gas an equal amount of heat energy Q flows out of the system to keep the temperature constant and the entropy has decreased by $Q/T = R \log_e 2$. Now move the right-hand piston to the centre of the cylinder. The gas B is unaffected by the motion and gas A is compressed isothermally and reversibly into the left-hand section and as before entropy $R \log_e 2$ leaves the system, which now consists of one mole of A and one mole of B at pressure P and temperature T. The entropy of the separated gases is thus $2R \log_e 2$ less than the entropy of the mixed gases.

If however the two vessels are filled with the same gas (say He) nothing happens when the tap is turned and there is no change in entropy. However, if the entropy is calculated by evaluating Ω, the factor 2^N still remains and the difference of entropy is still $Nk \log_e 2$, which is absurd. This paradox requires a sophisticated application of quantum theory to resolve it so that it is no wonder that the entropy of a perfect gas caused trouble to the pioneers in this difficult branch of theoretical physics. The physical basis of this paradox is that for the purpose of calculating its entropy, a gas must be treated as a whole and not as a collection of separate non-interacting molecules. This as has already been mentioned in section 6.1 is because the entropy depends upon the distribution function and for a gas to have any entropy there must be some interaction between the molecules by which the distribution function can be set up and maintained. This is the reason the rather pedantic phrase 'elements of the system' has been used instead of the term molecule which Boltzmann supposed meant the same thing. There is a sort of 'Cheshire cat effect' (Carroll 1866) in which the effect of the intermolecular forces upon the calculated entropy of a gas remains long after the terms depending directly upon their magnitude (ε and σ of a Lennard–Jones potential) have become quite negligible. This effect must somehow be taken into account. This can be done in a simple way by adopting a method of describing a collection of molecules introduced by Gibbs. He regarded a gas not as a collection of independent molecules but as an 'ensemble' of systems each containing many molecules.

6.6 The ensemble

Willard Gibbs, a man of excessive modesty, developed in his lectures at Yale the statistical mechanics underlying Boltzmann's method of calculating entropy. Gibbs' characteristically austere work (1902), firmly based on Newtonian mechanics, showed that

Boltzmann's work if not right was at least not wrong as his critics maintained. This in his modest way he thought was as far as a mathematical physicist could hope to take the subject. Gibbs advanced two methods of calculating the entropy of a perfect gas. In the first method, which was physically simple, he followed Boltzmann in considering a gas as a collection of a large number of interacting molecules which he referred to as a system. In his second method he imagined a collection of such systems (in an isolating enclosure) which could exchange energy but not molecules and calculated the average energy of each system. He referred to the collection of systems as an ensemble. To Gibbs the two methods differed only in mathematical convenience, but it has turned out that his second method, based on the ensemble, is a very suitable vehicle for transferring the basis of the calculation from Newtonian to quantum mechanics. Although the merits of Gibbs' work was generally admitted it was usually looked upon as unsuitable for an introductory course; making science easy for the young men being no part of Gibbs' philosophy. This objection is no longer true. The simplification of Gibbs' method to make it suitable for a first reading is due to Schrödinger (1944, 1952).

The molecular explanation of entropy is closely linked with the replacement of Newton's laws of motion, by Schrödinger's equation, as the fountain head from which flows the understanding of all physical phenomena. One of the merits of Schrödinger's equation is that it leads naturally to a difference between the behaviour of systems of distinguishable particles (e.g. $He - H_2$) and systems of indistinguishable particles (e.g. $He - He$) which has no counterpart in classical mechanics. Now Gibb's paradox has already suggested that distinguishable and indistinguishable molecules must be treated differently if absurd results are to be avoided in calculating the change of entropy which occurs when two gases mix by diffusion. Quantum mechanics provides a rational way of doing this. A combination of Boltzmann's ideas and Gibbs' treatment of a gas as an ensemble leads to an unambiguous way of calculating Ω which will now be described. Boltzmann's central idea of representing a distribution function by a single number Ω is retained. All that is changed is the meaning of the phrase 'elements of the system' which is no longer synonymous with the molecules of which the system is composed. A large number of examples of the system under consideration are imagined to be available. These are weakly coupled together into a composite system which is referred to as the ensemble. This ensemble is supposed to be isolated from its surroundings in an isolating adiabatic enclosure. In other words it can be described as uncoupled from its surroundings. The term weakly coupled is supposed to have a very definite meaning. The individual members of the ensemble will slowly exchange energy by means of the weak coupling so that in the long run they will reach a state of equilibrium in which they have as nearly as possible a common temperature. The mechanism by which the interchange of energy is accomplished is so 'weak' that it can be neglected in calculating the energy levels of the individual systems. These remain the E_1, E_2, $E_3 \ldots E_n$ of the individual systems whether they are considered in isolation or joined together to form the ensemble. An example of what is meant is afforded by a series of identical independently supported heavy pendulums joined together by a fine torsion thread running along their common axis of oscillation. The fine torsion thread ensures (by resonance) that all the pendulums have on the average the same energy. However, if it is fine enough its effect on the restoring force ($mg \sin \theta$) can be neglected in calculating the frequency of oscillation of each individual pendulum in the chain. The frequency ν of the coupled pendulum, and therefore its energy levels $(n + \frac{1}{2})h\nu$ is the same as that of a single pendulum (Table 1.1) considered in isolation. While each member of the ensemble will have the same energy levels E_1, E_2, E_3, . . . E_n they will not all have the same energy since they are continually exchanging quanta of energy. A given state of the ensemble can be described by saying that n_1 members are in the energy level E_1, n_2 members in an energy level E_2, etc. We now set up the description of the ensemble on our grooved board. The grooves represent as before E_1, E_2, $E_3, \ldots E_n \ldots$ the energy levels of each individual member of the ensemble. The beads in a groove however no longer represent molecules but of the members of the ensemble which happen to be in that energy level. With this understanding Ω for the ensemble is calculated as follows: to build up a pattern, the ith pattern say, one has to be provided with a box containing as many beads as the number of systems in the ensemble and a set of Ω_i programmes telling one what to do with the n beads as they are taken one by one from the box. For example, if the state of the ensemble happened to be one in which all the individual systems were in the same energy level E_m, Ω_i would be one, and the instruction tape would monotonously repeat, n times 'Put the bead in the mth groove corresponding to the energy level E_m'. If the pattern consisted of all but one of the members in the state E_m and one member in the state E_q, Ω_i for that arrangement would be equal to n, the number of systems in the ensemble; the n instruction tapes would each carry n − 1 instructions to put the bead in the mth groove and one instruction to put the bead in the qth groove. This instruction could be in any one of the n possible positions in the chain giving n separate ways, and no more, of building up the pattern. The order in which the instruction tapes are used does not affect Ω_i. This way of describing the ensemble and so calculating Ω can reasonably be objected to by the student who points out that quantum mechanics does not require a system to be in a pure state represented by a simple energy level. This is so, but it can and has been shown by the pundits that a more correct representation makes a very small difference to the answer.

The essential qualifications for membership of the ensemble are manifestly to have energy levels and to be recognisable. If the energy levels are very numerous or overlap and form a continuum this may be mathematically very tiresome but such a system can still, so to speak, join the entropy club. If, however, you cannot be identified you cannot join the club. In these circumstances it would not be possible to calculate Ω. An imaginary secretary going round the club to compile a list of those in energy level E_m would have no way of knowing whom he was addressing and so ticking him off on his list as having declared his energy. In the case of an idealised crystalline solid a single atom can represent a unit in an ensemble. It has a distinct set of energy levels, $(n + \frac{3}{2})h\nu$ in which ν is the frequency with which it vibrates under the influence of the interatomic forces. Although all atoms are alike, so that it cannot be identified by inspection, it can be identified by its position in the lattice which it cannot leave.

It is thus in principle possible to count up the number of atoms with a given energy in the enormous assembly of atoms which constitutes the lattice. These are the simple systems to which Boltzmann's ideas are most easily applied. At the other extreme is a gas in which single molecules cannot be identified because the molecules are all identical and have no fixed position because they are wandering about in the gas. The more complicated calculation of the entropy of a gas is described in section 6.10.

6.7 Calculation of Ω and S for systems with discrete energy levels

The simplest system for which Ω can be calculated is one in which each element of the ensemble has only one energy level. It is represented by a board with a single groove on which all the beads representing the elements of the assembly must be placed. As has already been noted $\Omega = 1$ and $S = 0$; it is a dead system, incapable of change.

The simplest active system is one in which each member of the ensemble has two energy levels E_1 and E_2, with $E_2 - E_1 = \varepsilon$. It is represented by a board with two grooves. It is an interesting system which exhibits most of the features of more complicated systems. It is not as artificial as would at first appear and there are plenty of real systems in which changes between only two levels are effective over a limited temperature range.

If there are n members of the ensemble n_1 will be in the lower energy level E_1 and n_2 in the upper level E_2. The system will therefore be represented by placing n_1 beads in the lower groove and n_2 beads in the upper groove. For a fixed energy n_1 and n_2 are determined and only one pattern is possible.

A programme for setting up this state of affairs is easily made by taking n_1 white cards and n_2 black cards and combining them into a well-shuffled pack of $n_1 + n_2 = n$ cards. Each time a bead is drawn from the box a card is drawn from the pack. If it is a white card the bead is placed in the lower groove. This must produce an arrangement with n_1 beads in the lower groove and n_2 beads in the upper groove because the number of white cards was n_1 and the number of black cards was n_2. The numbers of possible programmes is the number of distinct ways of arranging the cards. This problem is the same as the one solved in example 2 of section 6.2. There are $n!/n_1!n_2!$ different programmes and $S = k \log \Omega$ is given by

$$S = k \log \frac{n!}{n_1!n_2!}$$

This is a definite quantity since n_1 and n_2 are definite numbers determined by the relationships

$$n_1 + n_2 = n$$
$$n_1E_1 + n_2E_2 = E = \text{total energy of the system.}$$

The problem is thus completely solved but not in a form easily compared with experiment. A more suitable form is obtained by expressing the factorials as continuous functions by Stirling's theorem which can be used in its simplest form, $\log n! = n \log n - n$, the term, $\log \sqrt{2\pi n}$ being negligible if n is large.

Thus, $S = k \log \Omega = k(n \log n - n_1 \log n_1 - n_2 \log n_2)$.

In this expression n_1 and n_2 are not independent variables and can be expressed in terms of the single variable α by means of the relationships

$$n_1 = \frac{n}{2}(1 + e^{\alpha})$$

$$n_2 = \frac{n}{2}(1 - e^{\alpha})$$

One readily finds by substitution that

$$S = k \log \Omega = nk\left[\log n - \frac{1}{2}\left\{(1 + e^{\alpha}) \log (1 + e^{\alpha}) + (1 - e^{\alpha}) \log(1 - e^{\alpha})\right\}\right]$$

so that

$$\frac{dS}{d\alpha} = -\frac{nk}{2} e^{\alpha} \log \frac{1 + e^{\alpha}}{1 - e^{\alpha}} = -\frac{nk}{2} e^{\alpha} \log \frac{n_1}{n_2}$$

The energy of the system E is given by

$$E = n_1E_1 + n_2E_2 = nE_1 + n_2(E_2 - E_1) = nE_1 + n_2\varepsilon$$

and

$$\frac{dE}{d\alpha} = \varepsilon \frac{dn_2}{d\alpha} = -\frac{n\varepsilon}{2} e^{\alpha}.$$

The system is not capable of doing external work so that if heat dQ is added to the system it all goes to raising the energy from E to $E + dE$ and

$$dQ = dE = TdS$$

in which T is by definition the absolute temperature. Eliminating α one has

$$kT \log \frac{n_1}{n_2} = \varepsilon \quad \text{and} \quad \frac{n_2}{n_1} = e^{-\varepsilon/kT}$$

The expression $n_2/n_1 = e^{-\varepsilon/kT}$ is a very important one known as a Boltzmann distribution and lies at the heart of the subject. It is essentially correct and turns up in the consideration of more complicated systems however the subject is approached.

The entropy S is given by

$$S = nk\left[\log(1 + e^{-\varepsilon/kT}) + \frac{\varepsilon}{kT} \frac{e^{-\varepsilon/kT}}{1 + e^{-\varepsilon/kT}}\right]$$

a quantity which varies from 0 when $T = 0$ to $nk \log_e 2$ when, T is very large.

The specific heat

$$C_v = \frac{dE}{dT} = \frac{nk\left(\frac{\varepsilon}{kT}\right)^2 e^{-\varepsilon/kT}}{(1 + e^{-\varepsilon/kT})^2}$$

a function which peaks sharply when $\varepsilon/kT \approx 2.4$. The 'total entropy under the peak', i.e.

$$\int \frac{dQ}{T} = \int \frac{C_p dt}{T}$$

is $nk \log_e 2$ as is to be expected as the two levels are then equally populated.

The next case to consider is clearly the one in which each member of the ensemble has three energy levels E_1, E_2, E_3. As before a particular state of the assembly is described by three numbers n_1, n_2, n_3 which are the numbers of members in energy levels E_1, E_2, E_3 respectively. The ensemble can be represented on a board with three grooves with n_1, n_2 and n_3 beads in them. By an argument exactly analogous to that used previously the number of ways Ω_i in which this particular arrangement can be constructed is $n!/n_1!n_2!n_3!$ and Ω is

$$\Sigma\Omega_i = \Sigma \frac{n!}{n_1!n_2!n_3!}$$

the summation being carried out over all triplets of numbers n_1, n_2, n_3 compatible with the conservation of energy, and number of particles, so that $n_1 + n_2 + n_3 = n$. The presentation of Ω in a useful form is a much more difficult mathematical problem than in the example with two energy levels. This is because n_1 and n_2 are independent variables and the numbers n_1, n_2, n_3 can no longer be expressed in terms of a single variable α. It is important to realise that Ω is a quite definite quantity which can be evaluated by enumeration for small numbers. If $n = 4$ it is quite easy to see that $\Omega = 12$ if $E = E_1 + 2E_2 + E_3$. Such a calculation will also convince that direct enumeration is not a practicable method of finding Ω for even moderate values of n and is out of the question for $n \sim$ Avogadro's number $= 6.02 \times 10^{23}$.

To evaluate Ω two courses can be followed. The first, a counsel of perfection is to evaluate it as exactly as required by a highly ingenious method invented for the purpose by Darwin and Fowler and to be found in the latter's work on statistical mechanics (1929). The only objection to this course is that it requires a thorough understanding of contour integration. The results when obtained are in fact simple and memorable and it is quite reasonable to make use of them without worrying about the details of the mathematics. For those physicists who like to attempt their own mathematics one resuscitates the method originally used by Boltzmann. Effectively, the method comes to saying that the final summation $\Omega = \Sigma\Omega_i$ over the Ω_i's of all possible states of the ensemble with n_1 elements with energy E_1, n_2 with energy E_2, etc., is unnecessary, because the expression for Ω_i has such a sharp maximum for a particular distribution that all other states can be neglected. Substantially the same values for the entropy will be obtained whether we calculate it from $S = k \log \Omega$ or from $S = k \log \Omega_{i_{max}}$ which is much easier. An idea of the way this comes about can be seen by considering an analagous calculation for a Poisson distribution in the following example.

Worked example

Calculate the number of distinct distributions of heads and tails when n (assumed even) coins are tipped from a bag. How great must n be before one can neglect the distributions in which the numbers of heads and tails are not the same?

The distribution of heads and tails will follow a Poisson distribution (section 6.3) and the probability of a distribution with n_1 heads and n_2 tails is $(\frac{1}{2})^n \; n!/n_1!n_2!$

Since the probability of all possible distributions is unity, one has that

$$\left(\frac{1}{2}\right)^n \Sigma \frac{n!}{n_1!n_2!} = 1$$

the summation being taken over all values of n_1 from $0 - n$. Accordingly the total number of distributions is $\Sigma(n!/n_1!n_2!) = 2^n$. This answers the first part of the question.

To answer the second part one wants to know what difference it makes to the sum if all the terms are omitted except the central term for which $n_1 = n_2 = n/2$. Evaluating the factorials by Stirling's theorem one has that the approximate sum, that is the value of the central term, is given by (approximate sum)

$$= \frac{n!}{\frac{n}{2}!\,\frac{n}{2}!} = \frac{\sqrt{2\pi n}\,e^{-n}n^n}{\pi n e^{-n}\left(\frac{n}{2}\right)^n} = \sqrt{\frac{2}{\pi n}}\,2^n = \sqrt{\frac{2}{\pi n}} \quad \text{(true sum)}$$

and the difference between the two sums increases steadily with n.

If as in an entropy calculation one is only interested in the logarithm of the number of distributions one has log (approximate sum) = log (true sum) + log $\sqrt{2/\pi n} = n \log 2 - \frac{1}{2}\log n - \frac{1}{2}\log \pi + \frac{1}{2}\log 2$. For $n = 10^6$ say, the difference between the two logarithms is 7.13, quite small compared with the true value 693 146 and would only have to be taken into account if one were working to 0.01 per cent. For $n = 10^{23}$, the value to be expected in molecular calculations with molar quantities of material, the difference between 6.93×10^{22} and $6.93 \times 10^{22} - 26.7$ is beyond the possibility of direct measurement.

If Ω_i is a maximum log Ω_i is a maximum and d log $\Omega_1 = 0$. Now Ω_i by a simple extension of the argument used for two levels, is clearly given by

$$\Omega_i = \frac{n!}{n_1!n_2!n_3!}$$

in which n_1, n_2, n_3 are the number of beads in the first second or third levels respectively in the ith pattern and log $\Omega_i = n \log n - n - n_1 \log n_1 + n_1 - n_2 \log n_2 + n_2 - n_3 \log n_3 + n_3$ and $\mathrm{d}\log \Omega_i = -\mathrm{d}n_1 \log n_1 - \mathrm{d}n_1 - \mathrm{d}n_2 \log n_2 - \mathrm{d}n_2 - dn_3 \log n_3 - \mathrm{d}n_3$, setting $d\Omega = 0$ for a maximum gives $\mathrm{d}n_1(\log n_1 + 1) + \mathrm{d}n_2(\log n_2 + 1) + \mathrm{d}n_3(\log n_3 + 1) = 0$. Thus entropy is to be calculated from that value of

$$\Omega_i = \frac{n!}{n_1!n_2!n_3!}$$

for which

$$1 \sum_1^3 \mathrm{d}n_i(1 + \log n_i) = 0 \qquad \Omega \approx \Omega_i \text{ a maximum}$$

$$\lambda \sum_1^3 E_i \mathrm{d}n_i = 0 \qquad \text{Total energy } E \text{ constant}$$

$$\mu \sum_1^3 \mathrm{d}n_i = 0 \qquad \text{Total number } n \text{ constant}$$

Multiplying by the arbitrary multipliers 1, λ, μ shown in the margin, and adding, one gets

$$\sum_1^3 \mathrm{d}n_i(1 + \log n_i + \lambda E_i + \mu) = 0$$

Following Lagrange, since dn_1 and dn_2 are arbitrary and can have any small unrelated values it follows that their coefficients must be zero since otherwise any relationship true for one pair of values of dn_1 and dn_2 would not be true for a different pair.
So

$$1 + \log n_1 + \lambda E_1 + \mu = 0$$

and

$$n_1 = \mathrm{e}^{-(1+\mu)}\,\mathrm{e}^{-\lambda E_1} = \mathrm{A}\mathrm{e}^{-\lambda E_1}$$

similarly

$$n_2 = \mathrm{A}\mathrm{e}^{-\lambda E_2}$$

$$n_3 = \mathrm{A}\mathrm{e}^{-\lambda E_3}$$

in which A and λ are arbitrary constants available to meet the particular circumstances of the ensemble being considered.

6.8 Partition functions

The constants A and λ may be identified as follows:
Introduce a quantity

$$Z = \mathrm{e}^{-\lambda E_1} + \mathrm{e}^{-\lambda E_2} + \mathrm{e}^{-\lambda E_3} \quad \text{so that}$$

$$n = n_1 + n_2 + n_3 = \mathrm{A}\sum_{1}^{3} \mathrm{e}^{-\lambda E_i} = \mathrm{A}Z \text{ and } \mathrm{A} = \frac{n}{Z}$$

Then using Stirling's theorem to evaluate the factorials

$$\begin{aligned} S = k \log \Omega &= k[n \log n - n - n_1 \log n_1 + n_1 \\ &\quad - n_2 \log n_2 + n_2 - n_3 \log n_3 + n_3] \\ &= nk\left[\log n - \frac{1}{n}\sum_{1}^{3} n_i \log n_i\right] \end{aligned}$$

but

$$\sum_{1}^{3} n_i \log n_i = \sum_{1}^{3} n_i(\log \mathrm{A} - \lambda E_i) = n \log \mathrm{A} - \lambda E$$

and

$$S = nk\left[\log \mathrm{A} + \log Z - \log \mathrm{A} + \frac{\lambda E}{n}\right] = nk \log Z + k\lambda E.$$

As λ varies the particular state of the ensemble being studied will vary. The relationship between λ and the temperature of the ensemble can be found by a method due to Sommerfeld. Suppose there is a small general change in the system in which λ changes to $\lambda + \mathrm{d}\lambda$ and all the energy levels change so that $E_{\mathrm{i}} \rightarrow E_{\mathrm{i}} + \mathrm{d}E_{\mathrm{i}}$ etc. Then

$$\begin{aligned}\mathrm{d}S &= nk\frac{\mathrm{d}Z}{Z} + k\lambda\mathrm{d}E + kE\mathrm{d}\lambda \\ &= -\frac{nk}{Z}\left[\sum_1^3 \mathrm{e}^{-\lambda E_{\mathrm{i}}}(\lambda\mathrm{d}E_{\mathrm{i}} + E_{\mathrm{i}}\mathrm{d}\lambda)\right] + k\lambda\mathrm{d}E + kE\mathrm{d}\lambda\end{aligned}$$

Now

$$\sum_1^3 E_{\mathrm{i}}\,\mathrm{e}^{-\lambda E_{\mathrm{i}}} = \frac{1}{\mathrm{A}}\sum_1^3 n_{\mathrm{i}}E_{\mathrm{i}} = \frac{E}{\mathrm{A}} = \frac{ZE}{n}$$

and

$$\sum_1^3 \mathrm{e}^{-\lambda E_{\mathrm{i}}}\,\mathrm{d}E_{\mathrm{i}} = \frac{1}{\mathrm{A}}\sum_1^3 n_{\mathrm{i}}\mathrm{d}E_{\mathrm{i}}$$

so that

$$\begin{aligned}\mathrm{d}S &= -\frac{nk}{Z}\left[\frac{ZE}{n}\mathrm{d}\lambda + \frac{\lambda}{\mathrm{A}}\sum_1^3 n_{\mathrm{i}}\mathrm{d}E_1\right] + k\lambda\mathrm{d}E + kE\mathrm{d}\lambda \\ &= k\lambda\left[\mathrm{d}E - \sum_1^3 n_{\mathrm{i}}\mathrm{d}E_{\mathrm{i}}\right]\end{aligned}$$

In this general change all the characteristics including the energy levels of the members of the ensemble have been allowed to change by a small amount. In the case of a gas the position of the walls determines the wave functions and so the energy levels directly. For single atoms in a solid the wave function is determined by the fact that it must meet certain boundary conditions at the neighbouring atoms and so, as in the case of a gas, an expansion of the solid is accompanied by a change in the energy levels.

The total energy E of the ensemble is $\Sigma_1^3 n_i E_i$ so that the change in $E = dE$ is given by

$$dE = \sum_1^3 n_i dE_i + \sum_1^3 E_i dn_i.$$

The first term on the right hand side represents the work done by the external forces in effecting the change in energy levels. This is not equal to the increase in energy dE by the term $\Sigma_1^3 E_i dn_i$ which by conservation of energy must represent the heat dQ which flows into the system from outside, and one can write

$$dQ = dE - \sum_1^3 n_i dE_i$$

Comparing this with the expression for the change in entropy

$$dS = k\lambda \left[dE - \sum_1^3 n_i dE_i \right]$$

one sees that $k\lambda$ is the integrating factor which converts the incomplete differential dQ into the exact differential dS. This is by definition $1/T$ so that $\lambda = 1/kT$ in which T is the absolute thermodynamic temperature of the system.

Thus,

$$n_1 = Ae^{-E_1/kT} = \frac{n}{Z} e^{-E_1/kT}$$

$$Z = \sum_1^3 e^{-E_i/kT}$$

Once again in this more general case one sees that it is the Boltsmann factor $e^{-E_i/kT}$ which dominates the situation.

The expressions for the energy E and the entropy S become

$$E = \sum_1^3 n_i E_i = nkT^2 \frac{\partial \log Z}{\partial T}$$

and

$$S = nk \log Z + \frac{E}{T}$$

From these expressions the whole thermodynamic behaviour of the system can be deduced.

6.9 Degenerate levels and weight functions

The general case in which there are an indefinite number of energy levels $E_1, E_2, E_3 \ldots E_n$ can be dealt with in exactly the same way. No new matters of principle are involved. The only difference is that the auxiliary function Z now becomes

$$Z = \Sigma g_i \, e^{-E_i/kT}$$

The factors g_i, the weight factors, take into account that some of the levels may be 'degenerate' so that quite different wave functions have the same energy level. Thus if three different wave functions had the same energy E_n (like the three p states of a hydrogen atom) one would have to include three terms $e^{-E_n/kT}$, one for each wave function in Z, and g_n would be three. The function Z was introduced by Planck and descriptively called 'Zustandssumme' or sum over states; it is usually anglicised as 'partition function'; it forms a very convenient link between the properties of a system in terms of the atoms and molecules from which it is made up and the properties of the substance in bulk measured in the laboratory. The theoretical physicist calculates the energy levels of the system and this information is so to speak stored in the partition function. The large scale thermodynamic properties are then calculated in terms of the partition function and compared with experiment by means of the theoretical formulae which are given below for reference:

$$Z = \sum_{1}^{\infty} g_i \, e^{-E_i/kT}$$

$$S = nk \log Z + \frac{E}{T}$$

$$E = nkT^2 \frac{\partial \log Z}{\partial T}$$

It should be noted that n is the number of systems in the ensemble and E the total internal energy of the ensemble. All the thermodynamic expressions are ultimately proportional to the total number of atoms in the ensemble and can be so written if necessary.

A few examples illustrate the use of these formulae.

(a) The two-level system of section 6.6

$$Z = e^{-E_1/kT} + e^{-E_2/kT} = e^{-E_1/kT}(1 + e^{-\varepsilon/kT})$$

$$S = nk\left[\log Z + T\frac{\partial \log Z}{\partial T}\right]$$

$$= nk\left[-\frac{E_1}{kT} + \log(1 + e^{-\varepsilon/kT}) + \frac{1}{kT}\frac{(E_1 e^{-E_1/kT} + E_2 e^{-E_2/kT})}{e^{-E_1/kT} + e^{-E_2/kT}}\right]$$

which after some reduction becomes

$$nk\left[\log(1 + e^{-\varepsilon/kT}) + \frac{\varepsilon}{kT}\frac{e^{-\varepsilon/kT}}{1 + e^{-\varepsilon/kT}}\right]$$

the result previously obtained.

(b) The energy levels are those of a linear oscillator and each member of the ensemble contains only one atom

$$E_n = (n + \tfrac{1}{2})h\nu = (n + \tfrac{1}{2})\hbar\omega$$

so

$$Z = \sum_0^\infty e^{-(n+1/2)h\nu} = e^{-h\nu}(1 + e^{-h\nu/kT} + e^{-2h\nu/kT} + \cdots)$$

$$= e^{-h\nu/2kT}\frac{1}{1 - e^{-h\nu/2kT}}$$

and E the total internal energy is given by

$$nkT^2\frac{\partial \log Z}{\partial T} = \frac{nh\nu}{2} + \frac{nh\nu e^{-h\nu/kT}}{1 - e^{-h\nu/kT}}$$

and

$$E = n\frac{h\nu}{2} + \frac{nh\nu}{e^{h\nu/kT} - 1}$$

The result for a solid which has three degrees of freedom can be obtained by multiplying by three. It is more instructive to calculate it from first principles. The fundamental oscillator is a simple harmonic oscillator in three dimensions. The energy levels are well known to be $(3h\nu/2) + nh\nu$ in which the first term represents the zero point energy above which the energy levels are spaced at equal intervals of $h\nu$. These energy levels are degenerate, several quite different modes of oscillation having the same energy, and one must take account of this in the relationship $Z = \Sigma g_i e^{-E_i/kT}$. The weights can be calculated by working in a cartesian system in which the three dimensional motion of the atomic oscillator is split up into that of three quantised oscillators vibrating along the three axes. An oscillator with n quanta can split these up between the three axes in $\frac{1}{2}(n + 1)(n + 2)$ ways, the first few of which are illustrated in Fig. 6.3. So Z is given by

$$Z = \Sigma \frac{1}{2} (n + 1)(n + 2) \, e^{-(n+(3/2))h\nu/kT}$$

Writing $e^{-h\nu/kT} = x$ the series can be summed as follows:

$$Z = \frac{1}{2} x^{3/2} \Sigma (n + 1)(n + 2) x^n$$

$$= \frac{1}{2} x^{3/2} \Sigma \frac{d^2}{dx^2} x^n = \frac{1}{2} x^{3/2} \frac{d^2}{dx^2} \Sigma x^n$$

since $x < 1$ and the series is absolutely convergent. After some reduction this gives without difficulty

$$Z = x^{3/2} \frac{1}{(1 - x)^3} = e^{-3h\nu/2kT} \frac{1}{(1 - e^{-h\nu/kT})^3}$$

and

$$E = nkT^2 \frac{\partial \log Z}{\partial T} = \frac{3nh\nu}{2} + \frac{3h\nu}{e^{h\nu/kT} - 1}$$

in which the first term is the zero point energy and the second term is the celebrated expression first found by Einstein (1906) for the internal energy of a monatomic solid.

(c) The rigid rotator

The energy levels of a rigid diatomic molecule of moment of inertia I, due to its rotation about the centre of gravity, are well known to

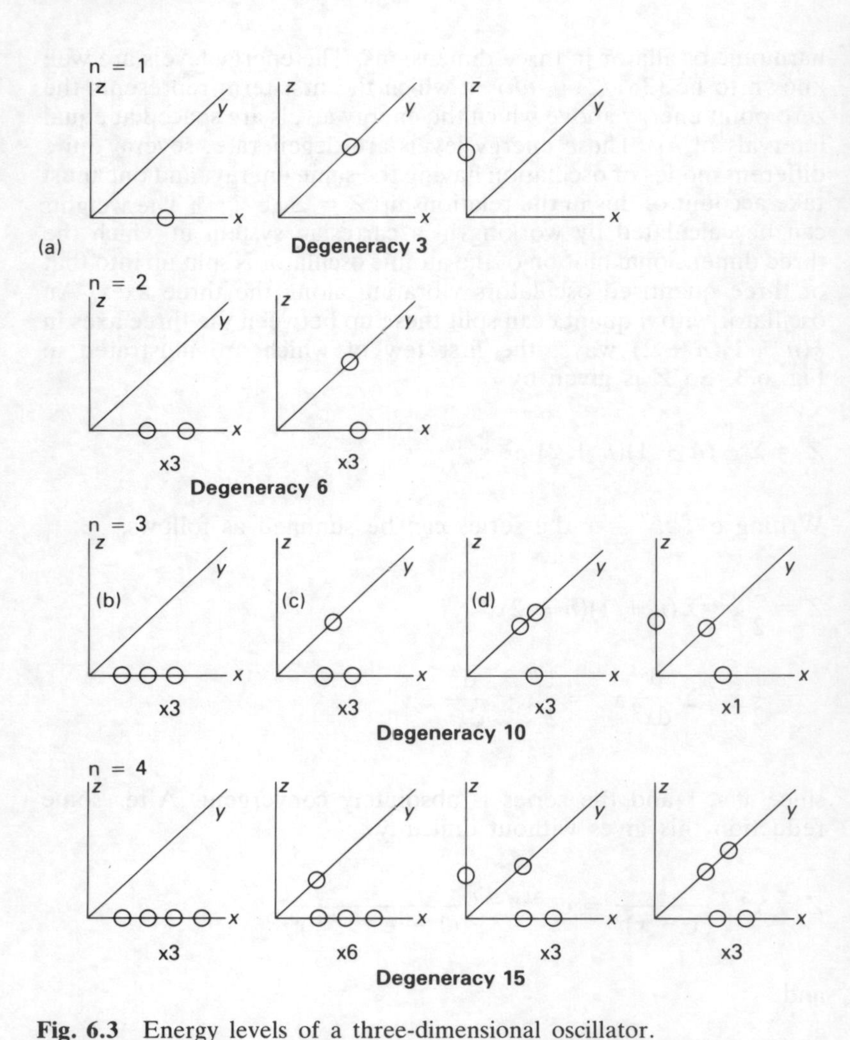

Fig. 6.3 Energy levels of a three-dimensional oscillator.
Each ○ represents a quantum of energy for the axis on which it is placed

(a) corresponds to an oscillator with one quantum of energy. For oscillations along the x axis the wave function is $Axe^{-(x^2+y^2+z^2)/2}$ (cf Pauling & Wilson (1935) p. 100).

(b) the state in which all three quanta are in oscillations along the x axis is three-fold degenerate since there are clearly similar states for the y and z axes.

(c) this state is six-fold degenerate. The unexcited axis can be either the x, the y or the z axis; for each choice there are then two possibilities as shown in (d).

The general formula for the degeneracy is $\dfrac{(n+1)(n+2)}{2}$.

be

$$\frac{K(K+1)\hbar^2}{2I} \quad \text{(Schiff Chapter XI, Pauling \& Wilson Chapter X).}$$

Since the angular momentum $\sqrt{K(K+1)}\hbar$ and the component of angular momentum $K\hbar$(K = 0, 1, 2, . . .) about a fixed axis are both quantised there are $2K + 1$ separate wave functions for each value of K and the energy levels are $(2K + 1)$ fold degenerate. The partition function Z_{rot} is accordingly

$$Z_{\text{rot}} = \Sigma g_i \, e^{-E_i/kT} = \Sigma(2K+1)\, e^{\frac{-K(K+1)\hbar^2}{2IkT}}$$

If the temperature is high enough (as it is for most substances above the boiling point), the steps are so small that the summation can be replaced by an integration and

$$Z_{\text{rot}} = \int (2K+1)\, e^{\frac{-K(K+1)\hbar^2}{2IkT}}\, dK = \frac{2IkT}{\hbar^2\sigma}$$

The factor σ is 2 or 1 according to whether the molecule is a symmetric one like $^{16}O.\,^{16}O$ with identical atoms or a molecule like $O^{16}.\,O^{18}$ in which the two atoms in the molecule are different. The factor σ takes into account the fact that in the former case only even values of K are allowed.

6.10 Entropy of a perfect gas

The individual molecules of a perfect gas cannot be accepted as members of the ensemble because they cannot be identified and because their energy cannot be assigned to them alone but is part of the collective energy of the gas. For a gas the smallest unit which can be considered as an element of the system is a specimen large enough to be recognised by the vessel in which it is contained. The ensemble is a collection of weakly interacting specimens of the gas each containing many molecules. This representation can also be looked upon as turning the gas into an artificial crystal by dividing it up into a large number of equal cells by means of partitions. Each cell contains a very large number of molecules. The walls of the partitions are imagined impermeable to molecules but must allow a slow exchange of energy between the cells. It is to be understood that this process is carried out in imagination only so as to make it possible to calculate Ω. It will be found that the number of cells, provided it is large enough, does not appear in the final result for Ω. To calculate Ω for a perfect gas the energy levels $E_1, E_2, E_3 \ldots E_n$ of each member of the ensemble must be calculated on the basis that it is a collection of identical particles. Like the three-dimensional oscillator (section 6.8) the energy levels of a single particle in a box of volume v are highly degenerate, but it is not difficult to find the weight factors. The difficulty lies in finding a mathematical method of allowing for the fact that the molecules are indistinguishable from each other. In the previous examples this difficulty was avoided by making each member of the ensemble a single atom which, for the purposes of the calculation, was taken to be a recognisable particle; recognisable by its position and in the case of degenerate levels by their orientation in space. Now that each member of the ensemble is a volume of gas containing very many molecules the problem of the correct calculation of the energy levels in terms of the known energy levels of a single molecule cannot be avoided. For perfect gases at low pressure the calculations can be greatly simplified by making use of the fact that the number of energy levels available to the molecules of the gas far exceed the number of molecules available to occupy them. Most of the energy levels of the gas are thus unoccupied and the probability of a level being occupied by two molecules is so small that its consequence can be neglected in the calculations. Systems of this type are said to be highly undegenerate: the term degenerate being used to describe systems in which the number of molecules is comparable with the number of levels (always limited by the conservation of energy) available to receive them. The reader should note that the word degenerate in the phrase 'degenerate level' has a different meaning from that in the phrase 'degenerate system'.

Bearing this meaning of the term degenerate system in mind one can calculate the partition function of a perfect undegenerate gas in the following way.

Let each of the n members of the ensemble consist of a large number q of weakly interacting point molecules. The energy levels available to a single molecule are determined by the box in which it is confined and are $\varepsilon_1, \varepsilon_2, \varepsilon_3 \ldots \varepsilon_n$. For simplicity assume that each level is distinct so that the weight factors are unity. This assumption does not affect the argument but makes the meaning of the expressions easier to grasp. The weight factors can easily be put in at the end of the calculation. The partition function for a single molecule in the box is

$$z = \sum_{\text{all } i} e^{-\varepsilon_i/kT}$$

The system of energy levels, ε_1, ε_2, ε_3 . . . of a single molecule is not changed when other molecules are put in the box because the mutual potential energy of the molecules is negligible in a perfect gas. If, when the box contains q molecules, l_1 have energy ε_1, l_2 energy ε_2 etc., the total energy of the gas E will be given by $E = l_1\,\varepsilon_1 + l_2\,\varepsilon_2 + l_3\,\varepsilon_3 \ldots$ with $\Sigma_i\, l_i = q$. While this statement is true not every combination of the l_i's which satisfies the conservation condition $\Sigma_i\, l_i = q$ will in fact occur, if E is one of the energy levels E_i of the elements of the ensemble. This is because the wave function associated with the energy E_i must solve Schrödinger's equation for a system of q identical particles.

Thus Z the partition function of the ensemble is given by

$$Z = \Sigma e^{-E_i/kT} = \Sigma e^{-(l_1\varepsilon_1 + l_2\varepsilon_2 \ldots)/kT}$$

the summation being taken over all the values of l permitted by the laws of quantum mechanics. This summation is in general difficult but can be carried out quite easily if the gas is undegenerate so that the l_i's are either 0 or 1. The summation is then over the number of ways in which the very large number of levels can be occupied by the much smaller (but still large) number q of molecules.

A comparision of Z with the expansion of $(z)^q$ by the multinominal theorem shows that

$$Z = \frac{(z)^q}{q!} - \frac{1}{q!}\Sigma_i \quad (\text{terms in which some } l_i > 1)$$

The second sum can be neglected for the conditions under consideration because there are relatively so few terms in it. For ordinary gases at n.t.p. only about one energy level in a million is occupied and terms corresponding to multiple occupation can legitimately be ignored in calculation Z. Thus for a completely undegenerate perfect gas $Z = (z)^q/q!$. Exactly the same result would have been obtained had the levels ε_1, ε_2 . . . been degenerate and the partition function been $z = \Sigma g_i\, e^{-\varepsilon_i/kT}$ and the expression $Z = z^q/q!$ can always be used to calculate the partition function of an undegenerate perfect gas. That ordinary gases are fully undegenerate is easily shown by calculating the number of levels available to each molecule. The number of energy levels available for occupation is very large. For vessels of laboratory size most of the occupied levels have very large quantum numbers and the difference in energy between adjacent levels is so small that the energy spectrum can be described in terms of a 'density of states' $\mathrm{d}n/\mathrm{d}E$ so that the number of levels Δn whose energy lies between E and $E + \Delta E$ is given by $\Delta n = (\mathrm{d}n/\mathrm{d}E)\Delta E$. When the levels are very closely spaced $\mathrm{d}n/\mathrm{d}E$ can be treated as though it were a continuous function. This procedure is analogous to the description of a molecular distribution function $\mathrm{d}n = f(c)\mathrm{d}c$ for a very large number of molecules in terms of a continuous function $f(c)$ in spite of the fact that $\mathrm{d}n$ must always be a whole number. In dealing with systems of free particles (gases) it is often convenient to express the density of states $\mathrm{d}n/\mathrm{d}E$ in terms of the momentum $p = mv$ instead of the energy $E = p^2/2m$ so that $\Delta n = (\mathrm{d}n/\mathrm{d}p)\,\Delta p$ is the number of levels which lie between the momentum limits p and $p + \Delta p$.

Worked example

The energy levels of the bound electron in a hydrogen atom are $-R\hbar/n^2$ (R = Rydberg's constant, h = Planck's constant, n = integral quantum number, and the sign is negative because the electron is bound).

Find the number of levels between $E = -R\hbar$ and $E = -R\hbar/20$, between $E = -Rh/10^4$ and $E = -Rh/1.1 \cdot 10^4$ and between $E = -Rh/10^8$ and $-Rh/1.1{\cdot}10^8$
Compare the result with the density of states approximation

$$\Delta n = \frac{\mathrm{d}n}{\mathrm{d}E}\,\Delta E \text{ with } \frac{\mathrm{d}n}{\mathrm{d}E} = \frac{n^3}{2Rh}.$$

Since $E = -Rh$ for $n = 1$ and $-Rh/20 = -Rh/(4.47)^2$ the levels $n = 1, 2, 3, 4,$ are within the specified energy range and $\Delta n = 4$. Similarly since $100^2 = 10^4$ and $(104.88)^2 = 1.1 \times 10^4 n = 100, 101, 102, 103, 104$ are within the range and $\Delta n = 5$. For the last energy range Δn is $10\,488 - 10\,000 = 488$. In the density of states approximation $\Delta n = n^3/2Rh\ \Delta E$, n^3 must be taken as the mean value over the energy range $\Delta\varepsilon$ and is $(n_2{}^4 - n_1{}^4/4(n_2 - n_1))$ in which n_1 and n_2 are effective values determined from E and $E + \Delta E$.

$$\text{Thus}\quad \Delta n = \frac{n_2{}^4 - n_1{}^4}{8(n_2 - n_1)}\left(\frac{1}{n_1{}^2} - \frac{1}{n_2{}^2}\right)$$

The values of n_1 and n_2 corresponding to the three energy ranges are (1, 4.47), (100, 104.88) and (10 000, 10 488) respectively. The corresponding values of Δn are easily calculated to be 13(.6), 4(.89) and 488(.95). Not until quantum numbers greater than 10^4 is the density of states formula correct to $\frac{1}{4}$ per cent.

The number of energy levels available for occupation by a single particle of mass m in a box of volume v is derived in the appendix where it is shown that the number of states in the momentum range between p and $p + \mathrm{d}p$ is $4\pi v/h^3\, p^2\, \mathrm{d}p$. The density of states is a very large number because of the small value of Planck's constant h.

The number of molecules in this momentum range available to occupy these levels is given by Maxwell's distribution function which expressed in terms of the molecular momentum p is

$$\mathrm{d}N = 4\pi N \frac{1}{(2\pi mkT)^{3/2}} \mathrm{e}^{-p^2/2mkT}\, p^2\, \mathrm{d}p$$

so that

$$\frac{\text{number of energy levels between } p \text{ and } p + \mathrm{d}p}{\text{number of molecules between } p \text{ and } p + \mathrm{d}p}$$

$$= \frac{v}{Nh^3}(2\pi mkT)^{3/2}\, \mathrm{e}^{p^2/2mkT} \approx 10^6$$

for oxygen at s.t.p. since $\mathrm{e}^{p^2/2mkT}$ is approximately unity and cannot be less than one.

Using the expression for the density of energy levels the value of z can be found by integration;

$$z = \Sigma g_i\, \mathrm{e}^{-\varepsilon_i/kT} = \frac{4\pi v}{h^3}\int_0^\infty p^2\, \mathrm{e}^{-p^2/2mkT}\, \mathrm{d}p$$

$$= \frac{v}{h^3}(2\pi mkT)^{3/2}$$

and

$$Z = \frac{1}{q!}\left[\frac{v}{h^3}(2\pi mkT)^{3/2}\right]^q$$

S can be calculated from $S = nk \log Z + E/T$

In the second term E is the total energy of all the nq molecules of the ensemble: since the mean energy of the molecules of a perfect gas is $3kT/2$, the value of the second term is $3nqk/2$.

Thus,

$$S = nk\left[q\log\left(\frac{2\pi mkT}{h^2}\right)^{3/2} - q\log q + q + q\log v\right] + \frac{3nqk}{2}$$

$$= nqk\left[\log\frac{v}{q} + \frac{3}{2}\log T + \log\left(\frac{2\pi mk}{h^2}\right)^{3/2} + \frac{5}{2}\right].$$

The entropy can be expressed in terms of the pressure by using the relationship $pv = qkT$ to give

$$S = Nk\left[\frac{5}{2}\log T - \log p + \frac{3}{2}\log\frac{2\pi mk^{5/3}}{h^2} + \frac{5}{2}\right].$$

On comparing this with the classical expression $S = Nk(\frac{5}{2}\log T - \log p) + \mathrm{K}$ one sees that they are the same except that the constant of integration K is no longer arbitrary but has the definite value

$$Nk\left[\frac{3}{2}\log\frac{2\pi mk^{5/3}}{h^2} + \frac{5}{2}\right].$$

The correctness of this expression depends upon having used the correct value for the density of states. As will appear at the end of section 6.11 the constant term may have to be modified for some monatomic gases like sodium and potassium vapours but the dependence of S upon $\log v/q$ is unaffected. Any doubts about the correctness of a theoretical formula for S which contains a term in $\log p$ (which is clearly impossible since the argument of a logarithm must be a number) can be set at rest by noting that the term $m^{3/2}k^{5/2}/h^3$ has the dimensions of a pressure p_0, so that the logarithmic term is $\log p_0/p$ and has a dimensionless argument.

6.11 Quantum mechanical resolution of Gibbs' paradox

The expression for the entropy of a perfect gas although obtained by a direct calculation of Ω is free from the objections raised in the discussion of Gibb's paradox. The entropy depends upon the total number of molecules nq in the system and upon the density of the particles q/v. It does not change if each member of the ensemble is divided into two systems of equal volume. This satisfactory resolution of Gibb's paradox depends upon a fundamental philosophical difference between classical and quantum mechanics. When molecules collide they behave differently from the way in which a study of the collisions between Newtonian point masses would lead one to expect. If the molecules differ in mass only (i.e. isotopes) the differences in behaviour although important in a detailed study are not large enough to affect the behaviour of a perfect gas in which the interaction between the molecules is first of all averaged out into a mean free path and then eliminated by working at a sufficiently low pressure. If the molecules are identical (e.g. collision between two ${}^{4}_{2}He$ molecules) fundamental differences in behaviour occur which cannot be neglected even in the limiting case of a perfect gas. While no one would deny that all the molecules of a pure gas are identical and indistinguishable, it is nevertheless difficult to discuss molecular collisions mathematically without labelling the molecules 1 and 2. Having done this, in classical mechanics it is difficult to unlabel them again, because each point particle although featureless and indistinguishable can still be identified by the definite trajectory which it follows, always remaining a finite distance from the particle with which it is colliding. In classical mechanics the molecules are identical in the same sense that well graded steel ball bearings are identical. They have the same mass and have been put through a series of selective tests which ensure that they all behave in the same way in an elastic collision. Nevertheless, carefully examined under a microscope they will all show minute surface scratches, too small to influence their behaviour in a collision, but which will enable the ball bearing initially labelled No. 1 to be identified and distinguished from No. 2.

Molecules are identical in a more strict sense of the word and it appears that many calculations about identical molecules in classical mechanics incorporate a concealed logical fallacy. It is a salutary commentary on the fallibility of human judgements that in spite of the determined attacks on Boltzmann's work the one point on which he was philosophically vulnerable seems never to have been raised until Bose and Einstein drew attention to it.

In a quantum mechanical calculation the situation is more satisfactory owing to the uncertainty principle, conveniently expressed in the wavelike representation of matter. The simple trajectory of the classical particle is now replaced by a finite band or tube within which there is a high probability of finding the particle. In a close collision these bands overlap and the particles are inextricably mixed so that it is not possible to say after such a collision which particle is to be found in which of the broad tubelike trajectories by which they leave the region in which the collision took place. This result can be expressed mathematically by arranging that every expression (such as the square of the wave amplitude Ψ^2) which can be directly compared with experiment, is symmetrical in the positions r_1, r_2 of the identical molecules and so independent of the way in which the particles are labelled. The squared amplitude is then interpreted as the probability of finding one particle at r_1 and another particle at r_2. In such a scheme the identity of the particles is lost and they are truly indistinguishable. Questions about which particle is in a particular place have no physical meaning and cannot be answered either theoretically or experimentally. The result of this mathematical procedure is to eliminate from consideration many wave functions which formally solve Schrödinger's equation. The rejection of some solutions of an equation is not a new idea in mathematical physics. No one has any difficulty in deciding, when faced with the answer that the number of sheep in a field is $+5$ or -3, that there are five sheep in the field and that -3 although a solution to the quadratic equation, is not a solution to the problem about the number of sheep in the field.

Direct and decisive experimental evidence for the correctness of the quantum mechanical point of view is found by measuring the scattering of slow α particles in helium gas. Rutherford's scattering law based upon the electrostatic interaction between the incident α particle and the scattering nucleus can be deduced from either system of mechanics and has been well verified experimentally for many different scattering nuclei. However, if the α particles are scattered in ${}^4_2$He whose nucleus is identical with the α particle, classical and quantum mechanics predict different scattering laws. The classical expressions for the number of particles arriving at the detector has two terms. The first represents the scattered α particles. The second the ${}^4_2$He nuclei of the target atoms projected in the direction of the detector by particles scattered in a direction at right angles. Quantum mechanics yields a symmetrical expression which gives the rate at which α particles arrive at the detector. It is impossible in this expression to say whether a given α particle is scattered from the original beam or is an originally stationary ${}^4_2$He nucleus projected towards the detector after collision with the original α particle beam. This quantum mechanical scattering is usually known as Mott scattering after its discoverer (Mott and Sneddon 1948). The difference between the two scattering laws is quite large and experiment is decisively in favour of the expression derived from quantum mechanics as shown in Fig. 6.4. The kinetic energy of the molecules of a gas is not large enough to enable them to be detected molecule by molecule by the experimental methods of nuclear physics. It is therefore not possible to obtain direct experimental evidence, as one could for α particle scattering, that the molecular motion is not the same as that predicted by classical

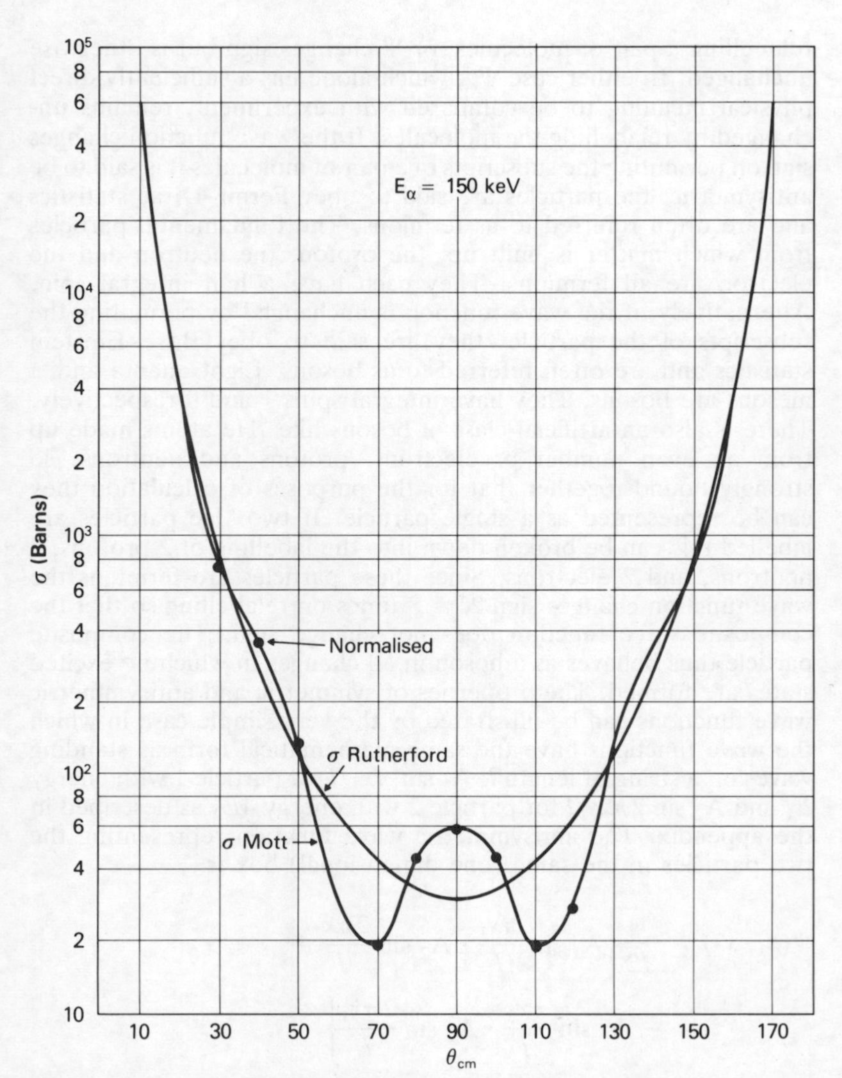

Fig. 6.4 Comparison of Rutherford and Mott scattering of α particles in helium (Heydenberg and Temmer 1956).

mechanics. Nevertheless, the same principles are expected to apply and should be taken into account in calculating Z the partition function of the ensemble.

The condition that Ψ^2 the squared amplitude of the wave function must be independent of the way in which the molecules are labelled can be fulfilled in two ways. Either Ψ itself is unchanged by relabelling a pair of molecules, or Ψ changes sign but is otherwise unchanged. In either case Ψ^2, which alone has a sufficiently direct physical meaning to be compared with experiment, remains unchanged by relabelling the molecules. If the wave function changes sign on permuting the subscripts of a pair of molecules it is said to be antisymetric, the particles are said to obey Fermi–Dirac statistics and are often referred to as fermions. The fundamental particles from which matter is built up, the proton, the neutron and the electron are all fermions. They each have a half integral spin. Alternatively, if the wave function is unchanged by permuting the subscripts of the particles they are said to obey Bose–Einstein statistics and are often referred to as bosons. Light quanta and π mesons are bosons. They have integral spins 1 and 0 respectively. There is also an artificial class of bosons like $^4_2$He atoms made up from an even number of electrons, protons and neutrons, so strongly bound together that for the purposes of calculation they can be represented as a single particle. If two $^4_2$He particles are labelled this can be broken down into the labelling of 2 protons, 2 neutrons, and 2 electrons. Since these particles are fermions the wave function changes sign $2^3 = 8$ times on relabelling so that the composite wave function does not change sign. The composite particle thus behaves as a boson in all changes in which no excited states are formed. The properties of symmetric and antisymmetric wave functions can be illustrated by the very simple case in which the wave functions have the same mathematical form as standing waves on a string of length l; $A_1 \sin \pi x_1/l$ for particle 1 with energy E_1 and $A_2 \sin 2\pi x_2/l$ for particle 2 with energy $4E_1$ as described in the appendix. The antisymmetric wave function representing the two particles in the same (one dimensional) box is

$$\Psi(x_1, x_2) = \frac{1}{\sqrt{2}}\left\{A_1 \sin\frac{\pi x_1}{l} \cdot A_2 \sin\frac{2\pi x_2}{l} - A_1 \sin\frac{\pi x_2}{l} \cdot A_2 \sin\frac{2\pi x_1}{l}\right\}$$

Simple trial shows that Ψ changes sign if the subscripts 1 and 2 are interchanged. The wave function Ψ can be used to calculate the probability of finding one particle on the l.h.s. of the box and the other in the r.h.s. of the box but can give no information as to which particle it will be. If the particles have the same energy E_1 and so the same wave functions $A_1 \sin \pi x_1/l$ and $A_1 \sin \pi x_2/l$ it will be observed that $\Psi = 0$ and that no such state exists. The corresponding symmetric function does not, of course, vanish. The difference between the two systems is illustrated in Fig. 6.5. This difference in

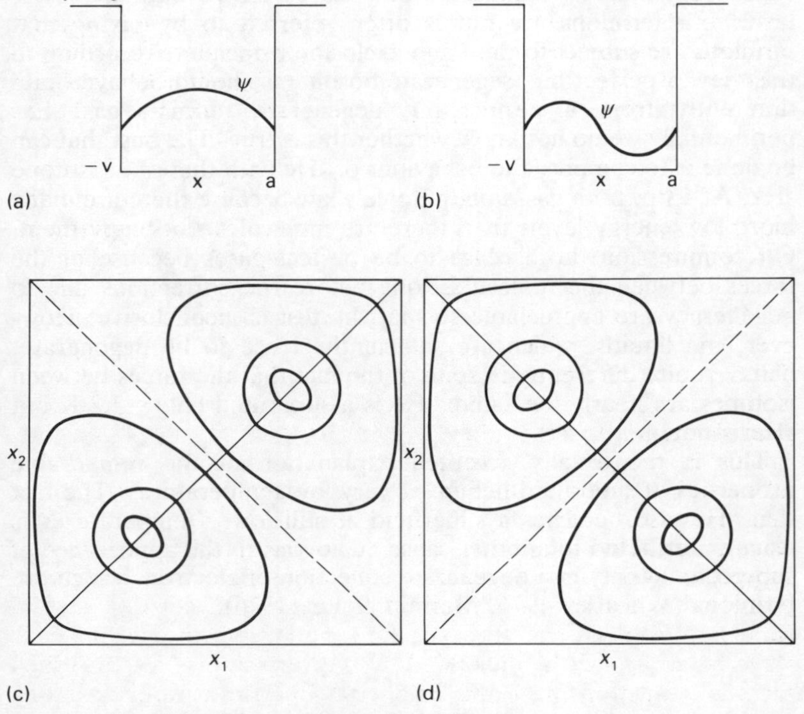

Fig. 6.5 Particles in a one-dimensional box

(a) shows the wave function of particle 1 with energy ε in a one-dimensional well.

(b) shows the wave function of particle 2 with energy 4ε in the same well. If both particles are in the well the probability of finding the particles at positions x_1 and x_2 are shown by the contours;

(c) if the particles are bosons

(d) if they are fermions.

Bosons are most likely to be found near to each other and $x_1 \sim x_2$. Fermions are never found together ($x_1 = x_2$) and are most likely to be found about 0.4 of the width of the well apart. An uninformed observer would deduce that the walls of the well strongly repelled the particles. This correct deduction is represented in the diagram by the large depth $-V$ of the potential well. The corresponding deduction that bosons attract each other and fermions repel each other is contrary to the initial conditions that the particles do not interact with each other and there is no potential term $V(x_1 - x_2)$.

It is this difference between the motions of fermions and Newtonian particles which gives rise to the valency force of chemistry.

behaviour whereby bosons are allowed to crowd into the lower levels but fermions are not is often referred to by saying that fermions are subject to the Pauli exclusion principle. According to this view a perfect but degenerate boson gas should behave quite differently from a perfect but degenerate fermion gas. Experimentally we do not know whether this is true. The best that can be done is to compare the behaviour of $^4_2$He with that of its isotope $^3_2$He. At s.t.p. both gases obey Boyle's law because there are many more low energy levels than there are molecules to occupy them. On compression, both cease to be perfect gases because of the forces between the molecules long before the corrections due to degeneracy are appreciable so that the test is inconclusive. However, the liquids, which are sufficiently dense to be degenerate, behave quite differently in spite of the fact that the forces between isotopes are nearly the same: $^4_2$He is a superfluid below 2.2 K but $^3_2$He is not.

This is the usually accepted explanation of the remarkable properties of superfluid helium at very low temperatures. The fact that $^3_2$He also becomes a superfluid at still lower temperatures is usually attributed to another cause analogous to the appearance of superconductivity in a degenerate collection of electrons which are fermions (Wheatley 1976, Mermin & Lee 1976).

6.12 Entropy and spin

Strictly speaking the expression for the entropy of a perfect gas arrived at in section 6.10 is that for a gas whose molecules are bosons without spin e.g. $^4_2$He. If the particles of the gas are fermions each level E_i is doubled, there are 2^N more possible representations so that $\Omega_{\text{spin }1/2} = 2^N\Omega_{\text{no spin}}$ and a term $k \log 2^N = Nk \log_e 2$ must be added to the entropy constant. The doubling of the levels can be understood by considering the first two (short) periods of the periodic table of the elements. The atomic structure of the light elements (up to Mg, $Z = 12$ say) is based upon the levels of a single electron (charge $-e$) bound in the field of a point charge $+Ze$. These are the levels

$$E_n = -\frac{Z'^2(Ie)}{n^2}$$

discovered by Bohr (1911) for a hydrogen-like atom. The constant Ie is the energy required to ionise a hydrogen atom = 13.598 eV (1 eV = $1.602 \cdot 10^{-19}$ J) and Z' is the effective nuclear charge.

Owing to the dependence of E_n the energy of the nth level upon $1/n^2$ the lower levels are well separated into the familiar K, L and M shells for which $n = 1$, $n = 2$, $n = 3$ respectively. An atom with atomic number Z is represented by Z electrons placed in the Z lowest energy levels. The mutual repulsion of the electrons is taken into account by assuming that since the shells are well separated in space the outer shells have a Z appropriate to the total charge inside them.

To make this scheme work there must be some law of nature which prevents all the Z electrons crowding into the lowest level E_1. In the early history of the subject (1911–1926) the rule was that each level could only take one electron. This scheme although successful in predicting a periodic table failed in that it predicted just half the elements in a period of the table than were found in practice. The rule that two electrons could be placed in each hydrogen-like level gives excellent agreement between the predicted and observed periodic table of the elements.

The explanation of the new rule 'the Pauli exclusion principle' starts with a revival of a long discarded suggestion (Parsons, 1910) that the electron is a spinning particle with an associated magnetic field. Each energy level can take two electrons with spins and associated magnetic fields pointing in opposite directions, usually called up and down. In this descriptive picture elements like Mg ($Z = 12$), which have two valency electrons, one up and one down in the same level, have no resultant spin and can be treated as bosons. Atoms like Na ($Z = 11$) with one valency electron have one unbalanced spin and have a magnetic moment $\mu_B = eh/4\pi m = 9.27 \cdot 10^{-24}$ JT^{-1} (subscript B for Bohr). If a sodium atom is placed in a weak magnetic field B its energy level ε_i due to its motion as a point particle in a box is changed to $\varepsilon_i \pm B\mu_B$ according to whether the spin is up or down with respect to the magnetic field. This statement can be verified by direct experiment. A parallel beam of neutral sodium atoms is split into two beams when it is passed through an inhomogeneous magnetic field. Experiments of this type (Stern–Gehrlach experiments) are direct evidence for electron spin. Measurements on beams of free electrons are frustrated by electron diffraction and the much larger electromagnetic deflection. Thus in working out Ω for sodium vapour in a magnetic field each level ε_i is replaced by two energy levels $\varepsilon_i \pm B\mu_B$, since the splitting is very small (about 10^{-7} kT in the earth's field) $z' = 2z$ in which z' is the partition function as calculated in section 6.9 for a point atom. Proceeding as before one has

$$Z_{\text{magnetic field}} = \frac{(z')^q}{q!} = \frac{2^q z^q}{q!}$$

and

$$S_{\text{spin } 1/2} = nk \log Z_{\text{spin } 1/2} = nk \log Z + nkq \log_e 2$$

and

$$S_{\text{spin } 1/2} = S_{\text{no spin}} + Nk \log_e 2$$

as previously stated. The external weak magnetic field is now seen to have been a dummy since it does not appear in the result and can be allowed to tend to zero without changing Ω.

If one does not like this argument one can rely on the pragmatic argument that all measurements on sodium vapour will have been made in a magnetic field (the earth's) so that the spin term must be included in a comparison between the calculated and the measured values of the entropy of sodium vapour such as are discussed in section 8.6. There is a full quantum theoretical theory of electron spin upon which the $NK \log_e 2$ terms can be based but it is difficult and is not usually studied at the level of this text. The subject is difficult because 'electron spin' is not to be explained in terms of a small but finite sized electron actually spinning round on an axis. As Dirac discovered, the 'spin' appears as a property of a structureless point electron if one insists that the wave equation it obeys must be consistent with Einstein's relativistic mechanics. The change from Schrödinger's to Dirac's wave equation is often regarded as the dividing line between advanced and elementary theoretical physics.

6.13 Degenerate gases

If one tries to evaluate the partition function in the general case when the gas is degenerate one finds that the behaviour of a boson gas is quite different from the behaviour of a gas composed of fermions. Of all the large number of wave functions corresponding to the state with energy $E_i = l_1\varepsilon_1 + l_2\varepsilon_2 + l_3\varepsilon_3 \ldots$ there is only one which is completely symmetric in the coordinates of all the q particles, (i.e. unchanged on relabelling any pair of particles) and only one which is antisymmetric in all the particles, i.e. changes sign when any two particles are relabelled. All the other states with energy E_i have a mixed symmetry and behave differently according to which pair of particles are relabelled. As only completely symmetric or completely antisymmetric functions can represent identical particles, the degenerate boson gas behaves differently from the degenerate gas composed of fermions because the restrictions placed upon the l_i's are different for the two gases. For bosons all values on l_i consistent with the condition $\Sigma l_i = q$ are allowed. For fermions the condition that the wave functions are antisymmetric imposes the extra condition that all the l_i's are either 0 or 1.

The evaluation of the partition function of a degenerate gas would be relatively easy if the l_i's were not restricted by the condition $\Sigma l_i = q$. This restriction does not apply to the gas whose particles are photons as these are created and destroyed at the walls and the number of photons in any volume of black body radiation is an indefinite quantity. There is no a priori reason to try and define the system by prescribing the number of photons at the outset. Their number can be allowed to adjust itself to the temperature of the system which is an initially undetermined number in the partition function. For such a photon gas the partition function can be rewritten as

$$Z = \Sigma \mathrm{e}^{-(l_1\varepsilon_1 + l_2\varepsilon_2 \ldots)/kT}$$

where the summation is over all values of l_1, l_2 etc. without the restriction that they add up to q and

$$Z = (1 + \mathrm{e}^{-\varepsilon_1/kT} + \mathrm{e}^{-2\varepsilon_1/kT} + \mathrm{e}^{-3\varepsilon_1/kT} \ldots)(1 + \mathrm{e}^{-\varepsilon_2/kT} + \mathrm{e}^{-2\varepsilon_2/kT \ldots}) \times (1 + \mathrm{e}^{-\varepsilon_3/kT} + \mathrm{e}^{-2\varepsilon_3/kT \ldots}) \ldots$$

as is easily verified by multiplying out the first few terms. Each bracketed term is an infinite geometrical progression (g.p.) with a common ratio $e^{-\varepsilon_i/kT}$.

Summing each g.p. before multiplication gives at once that the partition function is given by the continued product

$$Z = \frac{1}{1 - \mathrm{e}^{-\varepsilon_1/kT}} \times \frac{1}{1 - \mathrm{e}^{-\varepsilon_2/kT}} \times \frac{1}{1 - \mathrm{e}^{-\varepsilon_3/kT}} \ldots$$

The total energy E of the ensemble is given by

$$E = nkT^2 \frac{\partial \log Z}{\partial T} = n\Sigma \frac{\varepsilon_i}{\mathrm{e}^{\varepsilon_i/kT} - 1}$$

in which the summation is taken over all the energy levels $\varepsilon_1, \varepsilon_2, \varepsilon_3 \ldots$ of a single particle in a box of volume V. The energy of the photon is always $h\nu$ and the number of possible standing waves of frequency ν to $\nu + d\nu$ is $2V\, 4\pi^2/c^3\, d\nu$ as shown in the appendix.

The summation can now be replaced by an integration and

$$E = \frac{8\pi nVh}{c^3} \int_0^\infty \frac{\nu^3\, d\nu}{e^{h\nu/kT} - 1} = \frac{8\pi\nu Vk^4}{c^3h^3} \cdot T^4 \int_0^\infty \frac{x^3\, dx}{e^x - 1}$$

The definite integral is of course a number and is well known to be $\pi^4/15$, giving the density of black body radiation to be

$$\frac{8\pi^5k^4}{15h^3c^3} \cdot T^4$$

and the corresponding value of Stefan's constant

$$\sigma = \frac{2\pi^5k^4}{15h^3c^2}.$$

The form of the integral shows that the density radiation of frequency ν is given by

$$\rho\nu = \frac{8\pi\nu^2}{c^3} \cdot \frac{h\nu}{e^{h\nu/kT} - 1}$$

which is the famous law first found empirically by Planck (1900) to account for the experimental results obtained by Rubens in the near infrared. It is the foundation on which present theoretical physics has been built.

For gases composed of a finite number (q) of particles the calculation of Z is technically more difficult.

If one tries to evaluate the partition function of a degenerate gas whose molecules are bosons of mass m ($\varepsilon = p^2/2m$) in the same way that was used for a photon gas ($\varepsilon = cp$), the resulting value of Z will clearly be too large because the integrand includes the extra terms which make $\Sigma l_i > q$. What is needed is a procedure which excludes the unwanted terms from the final integration. This in fact is just the situation which the method of Darwin and Fowler was invented to deal with. The detailed calculation of the partition function is only difficult in the sense that it depends upon the theory of contour integration and can be accepted without shame by those who have a limited amount of time to devote to pure mathematics.

The details are given by Schrödinger (1944) and Born (1949) but they are only intelligible to those who have a good understanding of the integration of complex variables. The result of applying these principles to the calculation of the partition function leads to a somewhat complicated expression for Z which in the limit of non-degenerate gases agrees with the value obtained by the elementary method of section 6.10.

The effect of increasing gas degeneration is much more simply expressed in terms of $\bar{l}_i$ the average value of l_i taken over all the available energy levels ε_i. In the undegenerate gas as explained in section 6.5 the usual value of l_i is zero; it is occasionally 1 and the incidence of other values can be neglected. The mean value of $l_i = \bar{l}_i$ is therefore very small for all undegenerate gases. As the number of molecules per unit volume increases and the gas becomes degenerate the number of empty levels ($l_1 = 0$) falls, the number singly occupied levels ($l_1 = 1$) rises and one may even have to consider the contribution of multiply occupied levels ($l_i = 2$, $l_i = 3$, etc.) to the average. Thus, as the gas becomes degenerate the value of l_i rises.

The exact expressions (given without proof) are for fermions

$$\bar{l}_i = \frac{1}{e^{\alpha} e^{\varepsilon_i/kT} + 1} \text{ (F.D. statistics)}$$

and for bosons

$$\bar{l}_i = \frac{1}{e^{\alpha} e^{\varepsilon_i/kT} - 1} \text{ (B.E. statistics)}$$

The parameter α expresses the degree of degeneracy of the gas. Large α corresponds to an undegenerate gas and lies between 0 and ∞ for bosons and between $-\infty$ and $+\infty$ for fermions. Maxwell's distribution function corresponds to $\bar{l}_i \propto e^{-\varepsilon_i/kT}$ and both expressions tend to this limit when α is large. The cause of the deviations from Maxwell's distribution which occur in a degenerate gas can be understood in terms of the difference between the treatment of molecular collisions in classical and quantum mechanics. In classical mechanics the changes in molecular motions which are caused by a molecular collision are determined entirely by the motions of the molecules before the collision. In quantum mechanical collisions both the initial and the final states have to be considered. Collisions in which there are very few final states compatible with the conservation of energy and momentum are very unlikely to occur and the distribution function in distorted from the Maxwell distribution which is set up if all collisions are allowed. Colloquially one can say that classical particles leap before they look; quantum particles look very carefully before they leap and never leap until they know there is a place for them to leap to.

The theory of the perfect degenerate gas is not of much practical value in discussing the behaviour of real gases at high pressures. Perfect degenerate gases do not of course obey Boyle's law but the deviations are much smaller than those due to molecular forces between the molecules. Below the critical temperature real gases liquify due to attraction between the molecules long before deviations due to degeneracy are noticeable. However, the concept of degeneracy is of great importance in the general explanation of the properties of matter. For example, it had long been suspected (Drude 1913) that the electrical properties of metals were to be accounted for in terms of free electrons ejected from the metallic atoms and wandering freely about the metal. Owing to the large electrical forces between the electrons it seemed improbable that these could be treated as forming a perfect gas. Indeed there was poor agreement between experiment and conclusions based on this assumption. However, it was pointed out by Sommerfeld (1927) that these electrons would form a highly degenerate gas; the mean energy of the electrons would be several electron volts ($\sim$1.6 aJ) instead of the 1/40th eV (4×10^{-21} J) corresponding to a Maxwell distribution. The high kinetic energy of the electrons made the assumption that they could be treated as a perfect but highly degenerate gas much more plausible and on this assumption Sommerfeld was able to give a very satisfactory account of the electrical properties of simple metals.

6.14 Historical note

When Boltzmann first put forward his views Western Europe was in a turmoil, political, military and intellectual. A centre of power and culture was moving from ancient Vienna to upstart Berlin. Even in physics there was an uneasy feeling that all was not well with the Newtonian foundations. We of course know that relativity and quantum mechanics were on the way in, but the physicists of the day did not, and looked to a rather nebulous distinction between gravitational and inertial mass to calm their anxieties. Any suggestion that the behaviour of matter depended upon statistical averages or probability was very unpalatable to the majority of physical scientists who professed a strict determinism based upon Newtonian mechanics. It was thus an unfortunate time for a young Viennese of 22 to tell his distinguished elders that they should not pursue the Will o' the Wisp of the second law of thermodynamics because all that was necessary was to be found in his treatment of molecular motion in gases. Boltzmann's views were looked at very critically and accepted by few. The new views could not at that time be tested by experiment since absolute entropy free from an arbitrary constant could not be calculated until the value of the as yet undiscovered Planck's constant h was known. The subject was therefore wide open to uncontrolled speculation. The then (as now?) rather fashionable philosphers of science attacked him because his methods depended upon the frank admission that matter was composed of atoms and molecules. While nobody actually denied this the direct evidence for their existence was not very convincing and it was thought prudent to frame all physical theories in neutral language so that proofs and arguments would not collapse if it turned out that atoms and molecules did not really exist after all. For example, chemistry was by law presented not in terms of 'atoms' but of 'equivalents' in French schools until 1895. Even Ostwald's great text book of physical chemistry, the backbone of any continental course on elementary physical science, was written in this vein. Not until 1908, after Perrin's determination of Avogadro's number, did Ostwald in his preface admit that physical chemistry should without question be presented in terms of atoms. A similar attitude prevented the straightforward facts of electrolysis elucidated by Faraday being interpreted in terms of an atomic theory of electricity. It must be admitted that this rather cautious attitude has in a sense been justified. The atoms have not disappeared but they enter into the calculations less as particles than as systems of energy levels. As far as this branch of physics is concerned any theory of matter which gave the right energy levels would give agreement with experiment.

Boltzmann had the difficult task of expressing what was really discontinuous mathematics, dealing with whole numbers only, in the language of the calculus which operates on a continuum. The rough and ready intuitive methods of the physicist have never been very acceptable in this field and his methods were judged to be incapable of a rigorous presentation. Even the meticulous Gibbs was not spared. Zermelo, a mathematician expert in number theory, thought the whole development of the subject wrong and based upon a concealed logical fallacy. Zermelo's case was essentially that according to a strict interpretation of Newtonian mechanics a system of particles must in time return infinitely near to its initial state and therefore entropy could not be calculated from a knowledge of the velocities of all the molecules because entropy always approached a maximum and never decreased. Actually this particular attack was based upon a misunderstanding; after much controversy conducted at a rarefied mathematical level it was beaten off. However, it played a useful part in clarifying the not inconsiderable philosophical issues at stake. The interested reader may consult Born's account in his Waynflete lectures (Born 1949).

Apart from these almost metaphysical difficulties Boltzmann had to face the all too real one that there was no experimental way of testing his results. Without this possibility the whole enterprise was little more than empty verbiage. At first sight since Ω is a large but definite number $S = k \log \Omega$ should have a definite value. To illustrate his methods Boltzmann actually solved in detail the problem of seven molecules in a system with eight equally spaced energy levels $0, \varepsilon, 2\varepsilon \ldots 7\varepsilon$. His solution would have got him nearly full marks in a present day degree examination in elementary theoretical physics. His value of Ω was of course a number and so $S = k \log \Omega$ was a definite quantity. At that time it occurred to nobody that this was anything more than a ludicrously simple example. It was generally assumed (quite wrongly) that the simplest substance to be considered was a perfect gas. Thus to have any meaning at the time Boltzmann had to express his ideas in a form which could be applied to systems containing a very large number of particles and having a continuous energy spectrum. To do this he imagined the sharp levels to have a finite width so that they overlapped to form a continuous spectrum (an almost exact analogy is the way in which the line spectrum of a nitrogen molecule is converted into a band spectrum by a spectroscope of sufficient resolving power but used with too wide a slit).

The continuous energy spectrum was thus divided into finite sections of equal width with intensities corresponding to the weight functions introduced in the calculation of Z. The summation involved in the calculation of Z was then replaced by an integration. This is in effect the procedure of section 6.10 except that the width of the sections is not numerically determined by the factor $1/h^3$. The relative width of the sections at different parts of the energy spectrum (analogous to the dispersion term in an optical spectrum) is not arbitrary and is determined in the process of converting the summation into an integration. Equal sections are those which correspond to equal increments of the momentum of the particles. This procedure can be stated in a more sophisticated way by saying that molecular motion must be expressed in the canonical variables x, p_x of Hamiltonian mechanics. As this was the notation of choice of most workers in this field the physical importance of this necessity went on the whole unnoticed. The width of the sections (or bins) into which the energy spectrum was divided remained arbitrary, so long as they were equal in the sense which has just been explained. It was this arbitrary width which gave rise to the arbitrary constant in Boltzmann's expression for the entropy of a perfect gas: halving the width doubled the number of sections and thus increased Ω by the factor 2^n and added $nk \log_e 2$ to the entropy. The elimination of this unwanted arbitrary constant was achieved by combining three lines of thought being developed in Berlin by groups which, if not actually under the same roof, attended a common colloqium and were for the most part in walking distance of each other. The leaders were: Nernst whose chemical researches had convinced him that entropy was absolute and could have no floating constant because chemical equilibria could be predicted in terms of the entropies of the participating molecules. Planck who had discovered in his work on black body radiation a natural constant 'h' whose dimensions were action = momentum × distance and were therefore of the same dimensions as the width of the sections into which Boltzmann had divided his energy spectrum. Planck favoured a deterministic view of physics and was a little alarmed at the way his discovery was being used to support statistical explanations of phenomena in other branches of physics. Einstein who knew that physics was on the brink of great changes and favoured the statistical approach to physics made possible by Planck's discoveries.

In 1912 Sackur saw how to kill two if not three birds with one stone. He proposed that the arbitrary width of the sections into which Boltzmann had divided his energy spectrum should be replaced by h^3 (cubed because the integration was carried out over three dimensions) since this had a natural basis in Planck's work and had the right dimensions. This suggestion gave a definite value to the entropy of a perfect gas and satisfied Nernst's need for a well defined entropy; it had the further advantage that it was capable of being tested experimentally. When this was done by comparing the measured and predicted value of the saturated vapour pressure of mercury (section 8.6) it was at once obvious that Sackur's expression was correct except for a semi constant term $\log N! = N \log N - N$. This term was clearly wrong since it prevented entropy being proportional to the total number of molecules N being considered. Various ways of eliminating the unwanted term were considered but it was some time (and a devastating war) before the eminently satisfactory one of combining the notation of Gibbs' ensemble with the Pauli exclusion principle used in the text was gradually discovered. Boltzmann himself did not live to see the rather grandiloquent opening sentence of his first publication on the subject, 'The object of this paper is to furnish a purely analytical and perfectly general proof of the second law of thermodynamics as well as to investigate the corresponding principle in mechanics' triumphantly justified.

He died by his own hand in a fit of depression brought on it was said by the cold reception of his views. The pious conclusion of his obituary makes strange reading today: 'Mathematical research is a dangerous occupation if carried too far and the consequences that may have been the result of his intense concentration of thought should prove a warning to others not to allow themselves to be too deeply absorbed in any particular investigation'.

There is however no doubt that the single inscription on his memorial tablet

$$S = k \log W$$

has passed into the heritage of science.

Problems

1 In a raffle, n tickets are sold and $(n - 1)$ tickets and a winning ticket are placed in a hat from which each of the participants draws in turn. Is this a procedure which gives all the participants an equal chance of winning?

2 Write out the multinominal theorem and verify that it is true for $(a + b + c + d)^5$ by direct multiplication. Five pigs are placed at random into four sties. Show that the probability that l pigs are in the first sty, m pigs in the second sty, etc., is the coefficient of $(a^l b^m c^n d^o)$ in $(\frac{1}{4})^5(a + b + c + d)^5$.

3 Direct calculation shows that $\log_{10} 25! = 25.190\,65$, $\log_{10} 50! = 64.483\,07$, and $\log_{10} 100! = 157.970\,00$. Compare these results with the approximate values given by Stirling's theorem.

4 Verify that $A = A_o \sin 2\pi lx/\lambda \cdot \sin 2\pi my/\lambda \cdot \sin \omega t$ is a standing wave solution of the two dimensional wave equation

$$\frac{1}{c^2}\frac{\partial^2 A}{\partial t^2} = \frac{\partial^2 A}{\partial x^2} + \frac{\partial^2 A}{\partial y^2}$$

for waves of velocity c if $l^2 + m^2 = 1$. Sketch the profile of the waves if the amplitude is zero along the edges of a square frame of side a and calculate the number of standing waves which are possible between the lowest possible frequency and its second harmonic.

5 State the theorem of the equipartition of energy amongst the molecules of a system. Show that in the systems for which the

partition function has been found in section 6.7 that the mean energy of each particle tends to the equipartition value $kT/2$ per degree of freedom at high temperatures.

6 Show that Boltzmann's expression $n_i = Ae^{-\varepsilon_i/kT}$ for the number of molecules n_i with energy ε_i leads to equipartition of a continuous energy distribution if one interprets dn_i as the number of molecules whose momentum p lies between $p + dp$.

7 Show that for large n the Poisson distribution of section 6.3 approximates to a normal Gaussian distribution

$$\frac{1}{\sigma\sqrt{2\pi}}\,e^{-(x-\mu)^2/2\sigma^2}$$

about the mean value $\mu = n/2$. How does σ depend upon n? (Use Stirling's theorem including the $\sqrt{2\pi n}$ term).

8 The rotational energy of the hydrogen molecule is given by $J(J+1)h^2/8\pi^2I$ with $J = 0, 2, 4, 6 \ldots$ for para hydrogen and $J = 1, 3, 5 \ldots$ for orthohydrogen. Calculate the heat energy liberated in the ortho-para conversion at liquid hydrogen temperatures (~20 K) when only molecules with $J = 0$ and $J = 1$ occur to any appreciable extent and the initial mixture contains 25% para hydrogen.

9 Obtain the barometer formula $p = p_0e^{-mgh/kT}$ by a direct application of Boltzmann's theorem $n_i = Ae^{-\varepsilon_i/kT}$.

10 Calculate the entropy of one mole of helium (looked upon as a perfect gas) in J/K units.

11 Tin has an anomalous heat capacity with a maximum at about 47 K as shown in Fig. 8.6. On the assumption that this is due to two closely spaced levels (Schottky anomaly) calculate the separation ε of the levels and compare the theoretical shape of the anomaly with the experimental value given in Fig. 8.6.

12 Verify that the expression for the heat capacity of a two-level system obtained in section 6.7 from $C_v = dE/dT$ can also be obtained from the alternative relationship $C_v = T\,dS/dT$.

Further reading

Boltzmann, L. Boltzmann's celebrated 'Lectures' are of historical interest and are now available in English translation published in 1962 by the California University Press, Berkley.

Born, M. (1962) *Atomic Physics*, Blackie and Son Ltd., London.

Coulson, C.A. (1952) *Waves*, Oliver and Boyd, Edinburgh.

Cox, P.R. and **Miller H.P.** (1967) *Theory of Stochastic Processes*, Methuen and Co., London.

Schroedinger, E. (1944) *Statistical Thermodynamics*, Dublin Institute of Advanced Studies, Dublin and (1952) Cambridge University Press, London.

References

Born, M. (1949) *Natural Philosophy of Cause and Chance*, Clarendon Press, Oxford.

Carroll, L. (1866) *Alice's Adventures in Wonderland*, Macmillan, London.

Hardy, G.H. (1955) *Introduction to Pure Mathematics* p. 401 Clarendon Press, Oxford.

Heydenberg, N.P. and **Temmer G.M.** (1956) *Phys. Rev.* **104**, 123.

Mayer, J.E. and **Mayer, M.G.** (1940 & 1977) *Statistical Mechanics*, Wiley and Son, N.Y.

Mermin, M.D. and **Lee, D.M.** (1976) *Sci., Am.*, **235**(6), 56.

Mott, N.F. and **Sneddon, I.N.** (1948) *Wave Mechanics and its Applications*, Clarendon Press, Oxford.

Pauling L. and **Wilson, E.B. Jr.** (1935) *Introduction to Quantum Mechanics*, McGraw Hill Book Co., Inc., N.Y.

Schiff, L.I. (1946) *Quantum Mechanics*, McGraw Hill Book Co. Inc. N.Y.

Wheatley, J.C. (1976) *Physics Today*, **29** (2), 32.

CHAPTER 7

Formal thermodynamics

7.1 Entropy as a variable state

The object of formal thermodynamics is to obtain general relationships between the measurable properties of a substance by mathematical methods. The axioms from which the deductions are made are that all pure substances can be represented in terms of five variables P, V, T, U, S, only two of which are independent. There are thus ten related equations of state which a pure substance obeys. If the equations of state were all known explicitly any required relationship could very easily be found. Usually the opposite is the case and none of the equations of state are known, so that any relationships which are deduced are true for all substances. While it is true that the axioms on which the subject is founded are deducible from the general assumptions of a molecular explanation of the properties of matter, the converse is not necessarily true. The experimental verification of a predicted thermodynamical relationship does not imply that the system is composed of molecules. The continuous systems of electric and magnetic fields assumed by Faraday and Maxwell can be introduced without destroying the gratifying agreement between theory and experiment which is usually found in this branch of physics. In this divorce from any particular model lies the strength and weakness of formal thermodynamics as a research tool. The weakness is clearly that the experimental verification of one of the predicted relationships tells us nothing about what is going on; it merely confirms that a molecular explanation is possible. The experimental physicist may think this to be a meagre return for completing what may well have been a very difficult set of measurements. On the other hand a disagreement between theory and experiment is a serious matter which cannot be swept under the carpet, or explained away as being a 'special case'. There is a head-on collision between experiment and the whole body of established physical science. The experiment at once becomes a 'crucial experiment' so dear to the historian of science. Thermodynamics thus acts as a watchdog of experimental work ready to give the alarm if something is amiss. Usually this turns out to be a faulty experimental technique, which when rectified adds another confirmed relationship to the impressive number already in the scientific cupboard. This is not always so and this approach to experimental work has had its triumphs.

Early in the history of the subject W. Thomson (Lord Kelvin) noticed that the thermoelectric e.m.f. (Seebeck effect) was not proportional to the temperature difference in the circuit as thermodynamics seemed to predict. He correctly concluded that an unknown phenomenon (the Thomson effect) had been overlooked in the theoretical treatment, deduced its magnitude and demonstrated its existence experimentally. Curiously enough in the odd way in which theory and experiment are intertwined in physics this rather obscure phenomenon is important to telephone engineers as the main cause of 'pitting' in relays. More recently Meissner and Ochsenfeld (1933) noticed the converse situation in the field of superconductivity. The theoretical thermodynamics relationship between the specific heats in a normal and superconducting state of a metal in a magnetic field was well obeyed. However, thermodynamic arguments are only applicable to a system in thermodynamic equilibrium, that is an equilibrium which is independent of the way in which it is approached. Meissner noticed that the contemporary explanation of superconductivity led to the opposite conclusion. If a superconductor were inserted into a magnetic field the magnetic field was excluded from the superconductor by the electromagnetically induced superconducting currents. If however the same metal were inserted into the magnetic field above the temperature at which it became a superconductor, the induced currents soon died down and the field penetrated into the specimen. On now cooling below the superconducting threshold the field remained. The equilibrium between a magnetic field and a superconductor thus was supposed to depend upon the way in which the state was approached and so was not a state of thermodynamic equilibrium to which thermodynamic arguments could properly be applied. Meissner was thus led to discover a new fundamental property of superconductors, the 'Meissner effect' which had been unnoticed by experimenters for twenty years.

Meissner and Ochsenfeld observed that no magnetic field can exist in a superconductor. If a metal in which there is a static magnetic field is cooled, as soon as it becomes a superconductor appropriate superconducting currents spontaneously start up and neutralise the previously existing field. Thus contrary to previous belief the state of a superconductor in a magnetic field does not depend upon the way in which it is reached and the door is open to a thermodynamic treatment of the superconducting state.

The formal relationships of thermodynamics can also be used to find by indirect means the value of quantities which it is almost impossible to measure directly. The outstanding example of such a quantity is C_v the heat capacity at constant volume. Even for gases this is difficult to measure while for solids it is next to impossible. Nevertheless it is an important quantity since it (and not the easily measured C_p) is one of the few properties which can be accurately calculated for simple solids; it is thus a link between theory and experiment in this important field.

It is inevitable that many students find thermodynamics an uncongenial subject since many of the derivations seem to be 'trick proofs' without reference to the physical meaning of the symbols involved. This is because all the physics is included in and confined to the axioms, particularly the inclusion of the entropy S as a variable of state. Moreover many of the derivations go back to Maxwell, a notable 2nd Wrangler up to every trick of the mathematical trade. The cynic may express the view that the subject, although important, occupies too much time in an undergraduate course and keeps itself alive by the ease with which it lends itself to the setting of examination questions. There certainly seems no reason why students without mathematical interests should not use the principle results quoted in this chapter without bothering much about their mathematical derivation.

7.2 Mathematical background

Physical measurements are made under well defined conditions so that the most useful relationships are those which relate to changes under restrictive conditions imposed from outside. For example, for a perfect gas the general relationship $pv = rt$ is converted in the special relationship $pv^\gamma = k$ by the restrictive condition that the entropy S is to be constant during the expansion. If one tries to extend this relationship to real gases one at once finds that one needs to know the way in which the internal energy U and the entropy S depend upon the external conditions. Both these inaccessible parameters are usually measured from arbitrary zeros, so that the simplest initial theoretical statement of a problem is in the form of a relationship between partial differential coefficients such as $(\partial U/\partial V)_T$ or $(\partial S/\partial P)_V$ in which the arbitrary constant has been eliminated by differentiation and the restriction imposed by partial differentiation corresponds to the carefully controlled experimental conditions under which the agreement between theory and experiment is to be investigated. Since U and S are not variables which can be measured directly one needs a systematic mathematical way in which they can be eliminated from the initial expressions and replaced by the variables P, V, T which can be measured directly in the laboratory. This can be done in two ways:

1 The principal heat capacities C_p and C_v are admitted as explicit variables in their own right and not regarded as theoretical quantities which must be derived from the equation of state. In this way S can be eliminated through the relationship $\mathrm{d}S = \mathrm{d}Q/T = C\mathrm{d}T/T$ in which $\mathrm{d}Q$ is the heat energy absorbed during a particular reversible change and C is the corresponding heat capacity. The two changes usually considered are changes at constant pressure and changes at constant volume corresponding to the relationships

$$C_p = T\left(\frac{\partial S}{\partial T}\right)_p \quad \text{and} \quad C_v = T\left(\frac{\partial S}{\partial T}\right)_v$$

2 Great use is made of the fact that the surfaces defined by the equations of state and their derivatives are continuous and very well behaved in a mathematical sense. The worst that can happen, mathematically speaking, is that the derivatives in the neighbourhood of a phase change can be discontinuous.

This simple mathematical structure implies relationships between the derivatives which can be used to eliminate the concealed parameters U and S and replace them by the accessible parameters P, V and T. The relationships to be derived depend upon the continuity of the surfaces and not upon the physical meaning attached to the coordinates, so that it is convenient to use a neutral notation in which x, y are the two independent variables and z the dependent variable so that any of the ten possible equations of state are of the form $z = Z(x, y)$. Most thermodynamic arguments start by expressing the change in the dependent variable z caused by small changes in independent variables x and y. The change in z in going from the point x, y to the neighbouring point in the surface at $x + \mathrm{d}x$, $y + \mathrm{d}y$, is clearly a fixed quantity independent of the particular path by which they are joined since z is a function of x and y only; it can be calculated by any convenient method and the result will not be any the less true because the method was convenient. A very convenient method is to first of all make the required change in x keeping y constant, giving a change $\mathrm{d}z' = (\partial z/\partial x)_y\,\mathrm{d}x$ and then make the required change in y keeping x constant giving a change

$dz'' = (\partial z/\partial y)_x\, dy$. The total change in z, $= dz = dz' + dz''$ and

$$dz = \left(\frac{\partial z}{\partial x}\right)_y dx + \left(\frac{\partial z}{\partial y}\right)_x dy$$

a relationship which can be used to find the change in some other related variable w so that

$$\frac{dz}{dw} = \left(\frac{\partial z}{\partial x}\right)_y \frac{dx}{dw} + \left(\frac{\partial z}{\partial y}\right)_x \frac{dy}{dw}$$

It is important to note that dz/dw, dx/dw, dy/dw are full derivatives and the change dw is only restricted by the condition that the coordinates x and y remain in the surface $Z(x, y) = z$. Each element of the surface $z = Z(xy)$ defines a tangent plane $px + qy + rz =$ constant. This plane cuts the coordinate plane $z = 0$ in the line $px + qy =$ constant whose slope dy/dx is $-p/q$. This value of dy/dx is just the value of $(\partial y/\partial x)_z$ at the point where the tangent plane touches the surface $z = Z(x, y)$. Applying this argument to the other coordinate planes $x = 0$ and $y = 0$, one finds that

$$\left(\frac{\partial y}{\partial z}\right)_x = -\frac{r}{q} \text{ and } \left(\frac{\partial x}{\partial z}\right)_y = -\frac{r}{p}$$

so that

$$\left(\frac{\partial x}{\partial y}\right)_z \left(\frac{\partial y}{\partial z}\right)_x \left(\frac{\partial z}{\partial x}\right)_y = (-)^3 \frac{q}{p} \cdot \frac{r}{q} \cdot \frac{p}{r} = -1$$

An alternative proof is obtained by considering a path in the surface as generated by a point moving along some path $F(x, y)$ in the x, y plane, so that (dy/dx) is known at each point in the path. Then the general relationship

$$dz = \left(\frac{\partial z}{\partial x}\right)_y dx + \left(\frac{\partial z}{\partial y}\right)_x dy$$

becomes

$$\left(\frac{dz}{dx}\right)_F = \left(\frac{dz}{dx}\right)_y + \left(\frac{\partial z}{\partial y}\right)_x \left(\frac{\partial y}{\partial x}\right)_F$$

This result will remain true if $F(x, y)$ is chosen so that the curve in the surface is a contour with $z = K$, and $dz = 0$. In this case $(\partial y/\partial x)_F = (\partial y/\partial x)_z$ and

$$\left(\frac{\partial z}{\partial x}\right)_F = 0 = \left(\frac{\partial z}{\partial x}\right)_y + \left(\frac{\partial z}{\partial y}\right)_x\left(\frac{\partial y}{\partial x}\right)_z$$

which is the required result.
Another important relationship is that

$$\frac{\partial}{\partial y}\left(\frac{\partial Z}{\partial x}\right)_y = \frac{\partial^2 Z}{\partial y \partial x} = \frac{\partial^2 Z}{\partial x \partial y} = \frac{\partial}{\partial x}\left(\frac{\partial Z}{\partial y}\right)_x$$

This relationship is only true if the surface $z = Z(x, y)$ is continuous in a strict mathematical sense. It is however easily seen to be true if the surface can be expressed as a power series so that

$$Z = \Sigma a_{mn}\, x^m y^n$$

since it is clear that

$$\frac{\partial^2}{\partial x \partial y}\, x^m y^n = mnx^{m-1}y^{n-1} = \frac{\partial^2}{\partial y \partial x}\, x^m y^n.$$

It can safely be left to the professional mathematician to show that all experimentally realisable thermodynamic surfaces are included in the above derivation.

These fundamental results are true for surfaces which at all points have a unique tangent plane. These are just what the untutored physicist thinks of as a continuous surface and are the surfaces of thermodynamics. For a proper introductory discussion see *Differential and Integral Calculus*, by Courant (1936) or any other good text.

Finally, a word must be said about so called ‘cycle proofs’ in which the substance is taken through a ‘Carnot cycle’ with a small temperature difference ΔT. As the student of mathematical physics must by now have realised, all mathematical proofs are essentially the same and connect the result with the agreed axioms, which alone have any physical content. The merit of a cycle proof is that its mathematical steps dovetail smoothly with the physics of the problem. Its disadvantage is that the transition to the limit when $\Delta T \rightarrow$ zero cannot be avoided and the full arguments of the differential calculus have to be deployed in the open. These are difficult arguments, and are not made easier when the symbols are tied to physical quantities which distract the mind from the mathematical arguments. The result is only too often that the student with little interest in mathematics is left with an uneasy suspicion that the final answer is only an approximation, which is not so. In selected examples the cycle method is excellent, but it is difficult to apply to all cases. The interested student should master both methods. It is very easy to make a logical mistake in dealing with partial derivatives (i.e. treat them as full derivatives or to use relationships only true in a metrical space) and an alternative method of checking the result is highly desirable.

7.3 Maxwell's relations

By the conservation of energy for any small reversible change

$$\mathrm{d}Q = \mathrm{d}u + p\mathrm{d}v = T\mathrm{d}S$$

from the definition of entropy. This equation of course implies that the temperature T is measured on the absolute Kelvin thermodynamic scale. Rearranging this relationship to give

$$\mathrm{d}u = \mathrm{Td}S - p\mathrm{d}v$$

one has that

$$\left(\frac{\partial u}{\partial v}\right)_S = -p \text{ and } \left(\frac{\partial p}{\partial S}\right)_v = -\frac{\partial}{\partial S}\left[\left(\frac{\partial u}{\partial v}\right)_S\right]_v = -\frac{\partial^2 u}{\partial S \partial v}$$

Similarly

$$\left(\frac{\partial u}{\partial S}\right)_v = T \text{ and } \frac{\partial^2 u}{\partial v \partial S} = \left(\frac{\partial T}{\partial v}\right)_S$$

but since

$$\frac{\partial^2 u}{\partial S \partial v} = \frac{\partial^2 u}{\partial v \partial S}$$

equating the two partial differentials gives

$$\left(\frac{\partial p}{\partial S}\right)_v = -\left(\frac{\partial T}{\partial v}\right)_S$$

Other relationships of this type can be obtained by adding to u other quantities with the dimensions of energy such as pv and $-TS$ and so defining new composite variables of state. The most generally useful of these are the Gibbs free energy G already introduced (section 5.6), the Helmholz free energy $F = u - TS$ and the enthalpy $H = u + pv$. The symbol H comes from the old name, total heat, G is in honour of Gibbs and F is the original free energy introduced by Helmholz. The Helmholz free energy F is convenient for discussion of gaseous systems for which the volume is easily kept constant and G plays a similar role in condensed systems in which the (atmospheric) pressure is constant.

Starting with F one has

$$\mathrm{d}F = \mathrm{d}u - T\mathrm{d}S - S\mathrm{d}T = -p\mathrm{d}v - S\mathrm{d}T$$

since $\mathrm{d}Q = T\mathrm{d}S = \mathrm{d}u + p\mathrm{d}v$.

Since F is a function of state

$$\left(\frac{\partial F}{\partial v}\right)_T = -p \text{ and } \frac{\partial}{\partial T}\left[\left(\frac{\partial F}{\partial v}\right)_T\right]_v = \frac{\partial^2 F}{\partial T \partial v} = -\left(\frac{\partial p}{\partial T}\right)_v$$

$$\left(\frac{\partial F}{\partial T}\right)_v = -S \text{ and } \frac{\partial}{\partial v}\left[\left(\frac{\partial F}{\partial T}\right)_v\right]_T = \frac{\partial^2 F}{\partial v \partial T} = -\left(\frac{\partial S}{\partial v}\right)_T$$

so that

$$\left(\frac{\partial p}{\partial T}\right)_v = \left(\frac{\partial S}{\partial v}\right)_T$$

Similarly starting with G one finds that

$$\left(\frac{\partial S}{\partial p}\right)_T = -\left(\frac{\partial v}{\partial T}\right)_p$$

Starting with H one obtains

$$\mathrm{d}H = \mathrm{d}u + p\mathrm{d}v + v\mathrm{d}p = T\mathrm{d}S + v\mathrm{d}p$$

so that

$$\left(\frac{\partial T}{\partial p}\right)_S = \left(\frac{\partial v}{\partial S}\right)_p.$$

The four relationships obtained in this way,

$$\left(\frac{\partial p}{\partial S}\right)_v = -\left(\frac{\partial T}{\partial v}\right)_S$$

$$\left(\frac{\partial v}{\partial S}\right)_p = \left(\frac{\partial T}{\partial p}\right)_S$$

$$\left(\frac{\partial v}{\partial T}\right)_p = -\left(\frac{\partial S}{\partial p}\right)_T$$

$$\left(\frac{\partial p}{\partial T}\right)_v = \left(\frac{\partial S}{\partial v}\right)_T$$

are known as Maxwell's four thermodynamic relationships. They can be extended by adding further quantities of the dimensions of energy to U such as $-BM$ when the system has a magnetic moment M interacting with a magnetic field B or $-ED$ if the system is in an electric field. Maxwell's relations are used in thermodynamic arguments to remove differentials of S and replace them by differentials of directly measurable quantities. Examples of their use will be found in the sections which follow.

7.4 Heat capacity

In spite of the demise of caloric, calorimetry is still an active branch of experimental physics. The molar heat capacity C, namely the amount of energy (in Joules) required to raise the temperature of one mole by 1 K depends upon the conditions under which the change is carried out. The two principal specific heats are C_p when the pressure is kept constant and C_v when the volume is kept constant. Only $C_v = (\partial u/\partial T)_v$ is an intrinsic property which depends directly on the properties of the molecules. It is usually the quantity which can be most easily calculated from a molecular model and is thus the quantity through which the theoretical and experimental aspects of a molecular system are compared. It is an unsatisfactory quantity for this purpose because it is difficult to measure for gases and almost unmeasurable for solids.

Worked example

Show that C_v for a van der Waals gas is a function of temperature only. The equation of state is a relationship between any three of the variables of state. Choose these to be S, v, T so that $S = f(v, T)$ and $dS = (\partial S/\partial v)_T dv + (\partial S/\partial T)_v dT$. Using Maxwell's relationship $(\partial S/\partial v)_T = (\partial p/\partial T)_v$ this becomes

$$dS = \left(\frac{\partial p}{\partial T}\right)_v dv + \left(\frac{\partial S}{\partial T}\right)_v dT = \left(\frac{\partial p}{\partial T}\right)_v dv + \frac{1}{T} C_v dT$$

since $C_v = (\partial Q/\partial T)_v = T(\partial S/\partial T)_v$ by definition.
So

$$\left(\frac{\partial S}{\partial T}\right)_v = \frac{1}{T} C_v \quad \text{and} \quad \left(\frac{\partial^2 S}{\partial v dT}\right) = \frac{1}{T}\left(\frac{\partial C_v}{\partial v}\right)_T$$

$$\left(\frac{\partial S}{\partial v}\right)_T = \left(\frac{\partial p}{\partial T}\right)_v \quad \text{and} \quad \left(\frac{\partial^2 S}{\partial T \partial v}\right) = \left(\frac{\partial^2 p}{\partial T^2}\right)_v$$

Equating the two derivatives gives that

$$\left(\frac{\partial C_v}{\partial v}\right)_T = T\left(\frac{\partial^2 p}{\partial T^2}\right)$$

For a gas obeying van der Waals' equation direct differentiation shows that $(\partial^2 p/\partial_{T^2})_v = 0$ and so $(\partial C_v/\partial v)_T = 0$. Thus C_v is a function of T only since $(\partial F(T)/\partial v)_T = 0$.

Worked example

Show that $(\partial U/\partial V)_T = T(\partial P/\partial T)_V - P$ and that

$$dU = C_V dT + \left\{T\left(\frac{\partial P}{\partial T}\right)_V - P\right\} dV$$

Write U as a function $U(V, T)$ of the independent variables V and T so that

$$dU = \left(\frac{\partial U}{\partial V}\right)_T dV + \left(\frac{\partial U}{\partial T}\right)_V dT$$

Then from the conservation of energy

$$dQ = TdS = dU + PdV = \left(\frac{\partial U}{\partial V}\right)_T dV + \left(\frac{\partial U}{\partial T}\right)_V dT + PdV$$

and $T(\partial S/\partial V)_T = (\partial U/\partial V)_T + P$. Combining the relationship

$$\left(\frac{\partial S}{\partial V}\right)_T = -\left(\frac{\partial T}{\partial V}\right)_S \left(\frac{\partial S}{\partial T}\right)_V$$

with Maxwell's relationship

$$\left(\frac{\partial T}{\partial V}\right)_S = -\left(\frac{\partial P}{\partial S}\right)_V$$

gives

$$\left(\frac{\partial S}{\partial V}\right)_T = \left(\frac{\partial P}{\partial S}\right)_V\left(\frac{\partial S}{\partial T}\right)_V = \left(\frac{\partial P}{\partial T}\right)_V$$

and $(\partial U/\partial T)_V = T(\partial P/\partial T)_V - P$ as stated.
Substituting this value in the expression for $\mathrm{d}U$ gives at once that

$$\mathrm{d}U = \left(T\left(\frac{\partial P}{\partial T}\right)_V - P\right)\mathrm{d}V + \left(\frac{\partial U}{\partial T}\right)_V \mathrm{d}T$$

which is the required expression since $(\partial U/\partial T)_V = C_V$ by definition.

The measurement of C_p is however comparatively simple and for this reason the thermodynamic relationships between the principal molecular heats are important in experimental physics. The ratio $C_p/C_v = \gamma$ is an important quantity in the theory of sound and shock waves. Since acoustical measurements are both accurate and convenient the direct measurement of γ which they afford is often the only path by which a value of the theoretically important but experimentally inaccessible C_v can be reached. A third quantity C_s, the molecular heat of a vapour under conditions which ensure that it remains saturated, is important in the theory of steam and other vapour driven engines.

(a) $\dfrac{C_p}{C_v} = \gamma$

The molecular heat capacity $C = \mathrm{d}Q/\mathrm{d}T = T(\mathrm{d}S/\mathrm{d}T)$ by definition of S so that $C_p = T(\partial S/\partial T)_p$. This can be transformed to $-T(\partial p/\partial T)_S\ (\partial S/\partial p)_T$ by the product formula of section 7.2.

$$\text{Thus } \gamma = \frac{\left(\frac{\partial p}{\partial T}\right)_S\left(\frac{\partial S}{\partial p}\right)_T}{\left(\frac{\partial v}{\partial T}\right)_S\left(\frac{\partial S}{\partial v}\right)_T} = \left(\frac{\partial p}{\partial v}\right)_S\left(\frac{\partial v}{\partial p}\right)_T$$

since it is permissible to treat products of partial derivatives as if they were full derivatives provided that the variable which is being maintained constant is the same in both derivatives. The relationship $(\partial p/\partial v)_S = \gamma(\partial p/\partial v)_T$ is thus true for all substances however complicated their equations of state and is of considerable importance in experimental physics. Newton's formula for the velocity of sound V

$$V^2 = \frac{\text{elastic modulus}}{\text{density}} = \frac{E}{\rho}$$

is accurately obeyed for sound intensities well above that necessary for accurate measurement. A measurement of the velocity of sound thus gives a good value of the elastic modulus since the density can be measured, or obtained without difficulty from the equation of state. Since there is no time for much heat transfer in the rapid compressions and rarefactions in a train of sound waves the appropriate modulus is the adiabatic one

$$E_S = -v\left(\frac{\partial p}{\partial v}\right)_S = -\gamma v\left(\frac{\partial p}{\partial v}\right)_T = \gamma E_T.$$

Since the isothermal modulus E_T can be measured in relatively simple static measurements (or calculated from the equation of state $p = F(v, T)$) a measurement of the velocity of sound can be used to give an accurate value of γ and so by combining it with a measurement of C_p of the elusive C_v. It is worth remarking that γ is an important property of matter which can in principle be calculated for a gas whose molecular structure is known.

(b) $C_p - C_v$

There is usually a considerable difference between C_p and C_v. Roughly speaking it represents the work done by an expanding substance against external restraints. For gases which have a large expansion coefficient it is approximately R joules. For solids, although the expansion is small, the forces required to prevent expansion are large and $C_p - C_v$ has a value approximately $1/10 - 1/20$ of the value for gases. A direct calculation based on the physics of the problem is frustrated because there is no simple way of knowing how much of the work done in compressing a substance isothermally remains as an increase in internal energy and how much flows out again as heat. For any substance $S = f(p, T)$ so that in general

$$\mathrm{d}S = \left(\frac{\partial S}{\partial T}\right)_p \mathrm{d}T + \left(\frac{\partial S}{\partial p}\right)_T \mathrm{d}p.$$

By definition $C_p = T(\partial S/\partial T)_p$ so that using Maxwell's relation to eliminate $(\partial S/\partial p)_T$ one obtains

$$T\frac{dS}{dT} = C_p - T\left(\frac{\partial v}{\partial T}\right)_p \frac{dp}{dT}$$

a relationship which gives the heat capacity $C = T(dS/dT)$ for any conditions specified by dp/dT the way in which the pressure is to vary when the temperature is increased. If the volume is to be kept constant C becomes C_v and dp/dT becomes $(\partial p/\partial T)_v$ giving

$$C_p - C_v = T\left(\frac{\partial v}{\partial T}\right)_p \left(\frac{\partial p}{\partial T}\right)_v$$

In this expression the right hand side depends only on accessible parameters which can be measured or calculated without difficulty. For an ideal solid with temperature independent coefficients

$$v = v_0(1 + 3\alpha T) \quad \text{and} \quad dp = -K\frac{dv}{v_0}$$

$$\text{so that} \left(\frac{\partial v}{\partial T}\right)_p = 3\alpha v_0 \quad \text{and} \quad \left(\frac{\partial p}{\partial T}\right)_v = 3\alpha K;\ C_p - C_v = 9T\alpha^2 K v_0$$

in which K is the bulk modulus and α the coefficient of linear expansion of the isotropic solid. It is not practicable to test this relationship experimentally owing to the difficulty of measuring C_v. At low temperatures the expansion coefficient of solids is negligible so that $C_p - C_v$ is zero.

At normal temperatures the value of C_v obtained in this way is not more than 1 − 3 per cent from the theoretical value. This is as good agreement as can be expected when it is remembered that K is usually measured on a polycrystalline specimen and the theoretical calculation is made for a single crystal.

(c) Saturated vapours

The heat which must be added to a vapour to raise its temperature by 1 K and at the same time keep it saturated is the heat capacity of the saturated vapour C_s and may be related to the other properties of the system in the following way. The entropies per mole S^l of the liquid and S^g of the vapour in equilibrium with it are related by

$$S^g - S^l = \frac{\lambda}{T}$$

so that

$$\frac{dS^g}{dT} - \frac{dS^l}{dT} = \frac{1}{T}\frac{d\lambda}{dT} - \frac{\lambda}{T^2}$$

and

$$T\frac{dS^g}{dT} - T\frac{dS^l}{dT} = C_s^g - C_s^l = \frac{d\lambda}{dT} - \frac{\lambda}{T}$$

Since on general grounds C_s^l will not differ much from the ordinary heat capacity C_p one can write that

$$C_s^g = \frac{d\lambda}{dT} - \frac{\lambda}{T} + C_p^l.$$

If a more accurate expression is required the difference between C_p and C_s can easily be evaluated using the relationship

$$C = C_p - T\left(\frac{\partial v}{\partial T}\right)_p \frac{dp}{dT}$$

derived in section 7.4b. If the liquid and vapour are to remain in equilibrium while the temperature rises dp/dT must be the slope of the vapour pressure curve given by the Clausius–Clapeyron equation derived in the next section and

$$C_s^l = C_p^l - T\left(\frac{\partial v^l}{\partial T}\right)_p \frac{\lambda}{T(v^g - v^l)}$$

The last term is nearly always small because the coefficient of expansion of the liquid $(\partial v^l/\partial T)_p$ is small ($\sim 10^{-3}$) and v^g the volume of the vapour is large.

Calculated in this way the heat capacity per mole of a saturated vapour is often negative (-80/deg C for steam). Raising the temperature of a saturated vapour can be thought of as taking place in two stages. First the vapour is heated at constant pressure absorbing heat $C_p dT$; the vapour is now compressed isothermally to bring its pressure up to the new higher saturated vapour pressure. In this compression heat energy is released and it is not possible to tell without calculation whether the net result will be an absorption of heat (positive C_s) or an emission of heat (negative C_s). The occurrence of negative heat capacities need therefore cause no surprise. Not all saturated vapours have negative heat capacities. In ether, vapour C_s is positive and it has been reported (Hirn 1862) that it is a sudden compression of saturated ether vapour and not a sudden expansion which causes the formation of a cloud of liquid ether droplets. For an interesting review of this field see Ewing (1920).

7.5 The Clausius–Clapeyron equation

Assume as shown in section 5.6 that when two phases 1 and 2 of the same simple substance are in equilibrium not only must their temperatures and pressures be the same, but they must also have the same value for their Gibbs' functions G_1 and G_2 per mole. If at a slightly higher temperature they are still in equilibrium their Gibb's functions must still have the same value so that $\mathrm{d}G_1/\mathrm{d}T = \mathrm{d}G_2/\mathrm{d}T$, the differentiation being along the equilibrium curve.

Now

$$G_1 = U_1 - TS_1 + PV_1$$

and

$$\mathrm{d}G_1 = \mathrm{d}U_1 - T\mathrm{d}S_1 - S_1\mathrm{d}T + P\mathrm{d}V_1 + V_1\mathrm{d}P = V_1\mathrm{d}P - S_1\mathrm{d}T$$

since from the conservation of energy $\mathrm{d}U_1 + P\mathrm{d}V_1 - T\mathrm{d}S_1 = 0$
Thus

$$\left(\frac{\partial G_1}{\partial P}\right)_T = V_1 \quad \text{and} \quad \left(\frac{\partial G_1}{\partial T}\right)_P = -S_1$$

Similarly

$$\left(\frac{\partial G_2}{\partial P}\right)_T = V_2 \quad \text{and} \quad \left(\frac{\partial G_2}{\partial T}\right)_P = -S_2$$

and the equilibrium condition $\mathrm{d}G_1/\mathrm{d}T = \mathrm{d}G_2/\mathrm{d}T$ becomes

$$\frac{\mathrm{d}P}{\mathrm{d}T}(V_2 - V_1) = S_2 - S_1 = \frac{\lambda}{T}.$$

The relationship

$$\frac{dP}{dT} = \frac{\lambda}{T(V_2 - V_1)}$$

in which dP/dT represents the variation of the saturated vapour pressure of a liquid with temperature, first enunciated by Clapeyron, using the caloric theory of heat, (1834), is usually referred to as the Clausius–Clapeyron equation or more simply as the vapour pressure equation. It applies to any two phases in equilibrium and thus governs sublimation (vapour/solid equilibrium), variation of melting point with pressure (liquid/solid) and the change in transition temperature with pressure when the two phases are both solids as in the transition between rhombic and monoclinic sulphur, or diamond and graphite. In all these examples the factor $V_2 - V_1$ ensures that le Chatelier's principle is obeyed and that increasing pressure favours the phase with the greater density. The wide application of the equation warrants the inclusion of a second less abstract derivation based directly on Carnot's theorem. Consider a liquid of volume v_1 taken through a Carnot cycle as shown in the pv diagram of Fig. 7.1. From A to B the liquid is evaporated at constant temperature T and pressure p, absorbing latent heat λ and expanding to a volume v_2. From B to C it is expanded adiabatically until at C its temperature is $T - \Delta dT$ and the pressure has fallen to $p - \Delta p$. During this process an unknown fraction of the vapour will condense. From C to D the mixture is

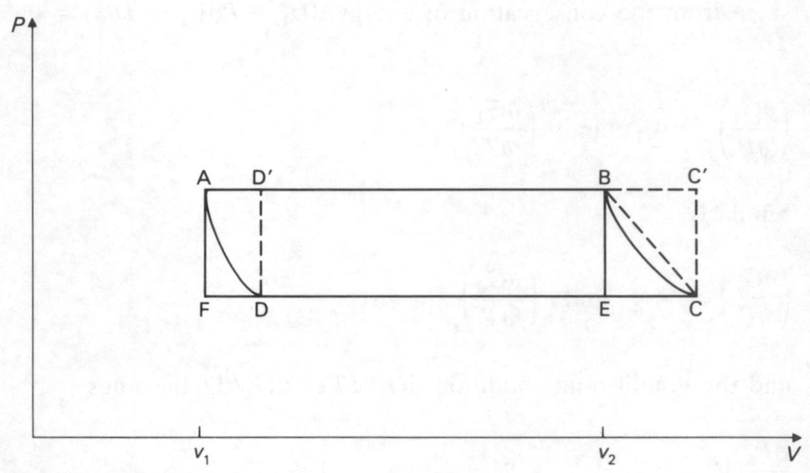

Fig. 7.1 Carnot's cycle for a vapour.

compressed isothermally and most of the liquid condenses. Finally the mixture is compressed adiabatically until all the vapour has condensed. The point D is chosen so that the latent heat of condensation of the remaining vapour just raises the temperature of the liquid to T thus completing the cycle. Carnot's theorem (Fig. 5.2) states that $Q_1/T_1 + Q_2/T_2 = 0$. This can be rewritten as

$$\frac{T_1 - T_2}{T_1} = \frac{Q_1 + Q_2}{Q_1} = \frac{\text{work done by cycle}}{\text{heat taken in}}$$

Since $Q_1 + Q_2$ is the total heat energy taken in on traversing the cycle this must all flow out again as work to conserve energy. The work done is the area ABCD so that by Carnot's theorem

$$\frac{\text{work done}}{\text{heat taken in}} = \frac{\text{area ABCD}}{\lambda} = \frac{T - T + \Delta T}{T} = \frac{\Delta T}{T}.$$

Area ABCD = area of the rectangle ABEF + area BEC − area AFD. Now the latter areas are less than the area BC′CE − AFDD′.
Thus

$$\frac{\Delta T}{T} = \frac{\text{Area AEDF}}{\lambda} + \mathrm{K}(\Delta p)^2$$

in which K is some finite constant representing the mean slope of the adiabatic curves BC and AD. The area ABEF is $\Delta p\,(v_2 - v_1)$ so that

$$\frac{\Delta T}{\Delta p} = \frac{(v_2 - v_1)T}{\lambda} + \text{terms of the order } \Delta p.$$

Thus in the limit of $\Delta p \to$ zero

$$\frac{\mathrm{d}T}{\mathrm{d}p} = \frac{(v_2 - v_1)T}{\lambda}$$

which is the result previously obtained. There is nothing approximate about the above derivation. It simply applies directly in a special case the arguments on which the analytical formulae are based, while Carnot's theorem is another way of expressing that entropy is a function of state. Experimentally the Clausius–Clapeyron equation is the best verified of all the thermodynamic relations. The classical investigation of James and William Thomson (1849, 1850) on the lowering of the melting point of ice with pressure are a landmark in the development of physics. The numerical agreement (experiment 7.2×10^{-2} theory 7.1×10^{-2} K/Pa) is as close as the date justify. The correct prediction of the anomalous sign due to the anomalously low density of ice is equally important. Not only was it unknown at that time that the melting point of ice depended upon the external pressure but it drew attention to the important part played by the ice/water equilibrium in the economy of nature.

The vapour pressure equation cannot be immediately integrated since λ is an unknown function of the temperature. To make progress one usually assumes that the vapour behaves like a perfect gas, which is a plausible assumption when the vapour pressure is low and the volume of the vapour is so enormous that the volume of the condensed phase can be neglected. Under these conditions

$$\frac{1}{p}\frac{\mathrm{d}p}{\mathrm{d}T} = \frac{\lambda}{RT^2}$$

which for constant $\lambda = \lambda_0$ integrates to

$$p = p_0 \mathrm{e}^{-\lambda/RT}$$

On the more realistic assumption that $\lambda = \lambda - aTp$, $p = p_0 T^{-a/R} \mathrm{e}^{-\lambda_0/RT}$ (Kirchoff's formula). The constant p_0 is often referred to as the internal pressure of the liquid. The Clausius–Clapeyron equation can be integrated in the temperature range within which the vapour can be treated as a perfect gas. The object of the integration is to obtain a theoretical vapour pressure equation which is to be compared with experimentally measured values. Since vapour pressures can be measured quite accurately down to 10^{-4} Pa, this restriction leaves an ample range in which the integrated vapour pressure equation can be compared with experiment. The first step in the integration is to replace the unknown variation in latent heat $\mathrm{d}\lambda/\mathrm{d}T$ by an expression depending upon the heat capacity of the condensed phase which can be measured relatively easily. Consider the equilibrium between a solid and its vapour. The general heat capacity equation (section 7.4)

$$C = C_p - T\left(\frac{\partial v}{\partial T}\right)_p \frac{\mathrm{d}p}{\mathrm{d}T}$$

can be applied to either phase to give

$$C_s^g - C_s^s = \frac{d\lambda}{dT} - \frac{\lambda}{T} = C_p^g - C_p^s - T\left\{\left(\frac{\partial v^g}{\partial T}\right)_p - \left(\frac{\partial v^s}{\partial T}\right)_p\right\}\frac{dp}{dT}$$

in which C_s^g, C_s^s are the molar heat capacities of the gaseous and solid phases taken along the saturation vapour pressure curve and dp/dT is the slope of the vapour pressure curve at T. At the low pressures being considered the second term in the bracket is negligible and the bracket reduces to $(\partial v^g/\partial T)_p = R/p$ per mole. Replacing dp/dT by $p\lambda/RT^2$ gives at once that

$$\frac{d\lambda}{dT} - \frac{\lambda}{T} = C_p^g - C_p^s - \frac{\lambda}{T}$$

so that

$$\frac{d\lambda}{dT} = C_p^g - C_p^s = \Delta C_p$$

$$\text{and } \lambda = \lambda_0 + \int_0^T \Delta C_p dT$$

in which λ_0 is latent heat of evaporation per mole at absolute zero and C_p^g, C_p^s are the conventional molar heat capacities of the two phases. Substituting in the simplified vapour equation which holds at low temperatures gives at once that

$$\log p = \int \frac{\lambda dT}{RT^2} = -\frac{\lambda_0}{RT} + \int \frac{1}{RT^2}\int \Delta C_p dT \cdot dT + C$$

The second integral can be simplified by integrating

$\int \frac{\Delta C_p dT}{RT}$ by parts to give

$$\log p = -\frac{\lambda_0}{RT} + \int \frac{\Delta C_p dT}{RT} - \frac{1}{RT}\int \Delta C_p dT + C$$

If the vapour is a monatomic gas $C_p = \frac{5}{2}R$ and for a substance such as sodium or Mercury

$$\log p = -\frac{\lambda_0}{RT} + \tfrac{5}{2}\log T - \frac{1}{R}\int_0^T \frac{C_p^s}{T}\,\mathrm{d}T + \frac{1}{RT}\int_0^T C_p^s\,\mathrm{d}T + i - \tfrac{5}{2}$$

in which i is an integration constant. The integrals are all finite and the lower limit can be set to zero because C_p^s falls off as T^3 at low temperatures. The constant i is arbitrary and the physical requirement that $p \to 0$ as $T \to 0$ is met whatever the value of i. This is because both $\log p$ and $-\lambda_0/RT$ both $\to -\infty$ as $T \to 0$. For finite temperatures the vapour pressure has a definite value so that i can always be found experimentally if the experimental values of the molar heat capacity are known. This adjustment is unique. If the vapour pressure is correctly given at any one temperature the classical Clausius–Clapeyron equation ensures that it retains its correct value at all temperatures low enough for the vapour to be a perfect gas. The theoretical ambiguity in the value of i can be removed by the requirement that the entropy of the vapour must agree with the value given in section 6.10 for the entropy of a perfect monatomic gas. This important comparison between the theoretical and experimental values of i is further discussed in section 8.6.

Worked example

Desalination is an important technical process. On the simplifying assumption that water vapour obeys Boyle's law find the work which must be done per mole of separated water if the desalination is carried out by the ideal process of reversible isothermal distillation. Show very approximately that if ΔT is the difference between the freezing (or boiling) points of water and sea water that reversible evaporation produces $T/\Delta T$ more pure water than would be obtained by converting the work into heat and using it for conventional irreversible evaporation at the working temperature T.

Let p_1 and p_2 be the saturated vapour pressures of water and sea water respectively. Evaporate reversibly one mole of sea water doing work = pressure × (increase in volume) = $p_2 v_2 = RT$ if Boyle's law is obeyed. Compress the vapour isothermally to the vapour pressure p_1 of pure water absorbing work = $\int p\,\mathrm{d}v = RT \log p_1/p_2$. This vapour can now be placed in contact with pure water and compressed until it has all condensed absorbing work $p_1 v_1 = RT$. The net work absorbed in the process is therefore $-RT + RT\log p_1/p_2 + RT = RT \log p_1/p_2$. The value of p_1/p_2 can be roughly evaluated in terms of ΔT. For an ideal liquid $p = p_0\,\mathrm{e}^{-\lambda/RT}$. For dilute solutions the difference in the vapour pressures

$p_1 - p_2 = \Delta p = dp/dT\,\Delta T$ and $RT \log p_1/p_2 = RT \log(p + \Delta p)/p = RT\,\Delta p/p = RT\,1/p\,dp/dT\,\Delta T = \lambda\Delta T/T$. Thus the work required to produce 1 mole of water by reversible distillation is $\lambda\Delta T/T$. If the same amount of work were converted into heat it would evaporate $\lambda\Delta T/T\lambda = \Delta T/T$ moles of water and direct distillation is an approximately $T/\Delta T$ less effective way of using power for desalination than reversible distillation. At 0 °C the factor is about 142.

7.6 Fluid flow

The way in which the application of thermodynamic principles assists in understanding the flow of fluids is illustrated by the following example.

Throttling

Consider the slow adiabatic flow of a gas controlled by means of a valve in a pipe of otherwise uniform cross-section. The flow of the gas is so small that the kinetic energy of the gas due to its mass motion can be neglected. The gas may conveniently be divided into three sections as in Fig. 7.2: the initial high pressure gas flowing towards the valve at a pressure p_1; a section which includes the valve or throttle in which the flow can be turbulent, is irreversible and the k.e. of the gas is not negligible (all that is postulated about this section is that no heat flows into the gas); and the final section in which the gas again flows slowly and uniformly at a new and lower

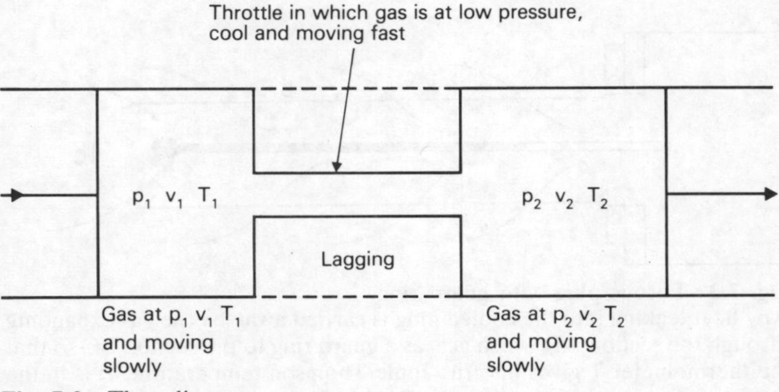

Fig. 7.2 Throttling.

pressure p_2. The expansion of a gas under these conditions is usually referred to as a 'throttling' process. It is also referred to as a Joule–Thomson, or Joule–Kelvin expansion after the two men who first investigated it. It is approximated to in practice wherever the flow of a fluid is controlled by a tap. An actual expansion carried out in the laboratory differs from an ideal throttling process chiefly because it is difficult to fulfil the adiabatic condition. In the central section which includes the tap or throttle, the cross-section of the pipe is severely restricted and in it the gas is moving at high speed at a low pressure. The corresponding expansion cools the gas and the adiabatic condition is only maintained if the tap is made of completely non-conducting material. The simplest way of achieving this ideal in practice is to replace the tap by a porous plug divided by a guard ring as shown in Fig. 7.3. For a strict throttling process one has by conservation of energy that the increase of internal energy $U_2 - U_1$ is equal to the work done on the gas. This is p_1v_1 done on the gas by the compressor, less p_2v_2 the work done by the gas against the final pressure p_2, so that

$$U_2 - U_1 = p_1v_1 - p_2v_2$$

or $U_1 + p_1v_1 = U_2 + p_2v_2$ which can be written as $H_1 = H_2$

in which H the enthalpy of the gas per mole is a function of state of the gas which has already been introduced in section 7.3. The gas in the throttle is not changing through a series of equilibrium states which can be represented by a line in the p, v, T surface. However, after the throttling process it reaches a state which can be represented by a single point on the surface and has a definite value not only for p, v, T the observable parameters of state, but also for U

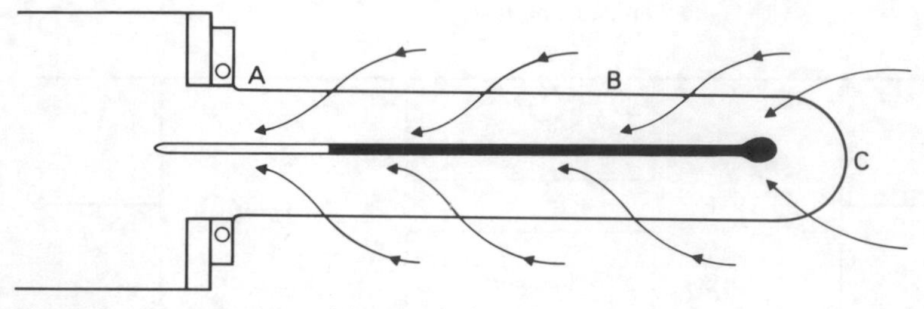

Fig. 7.3 Porous plug with guard ring.
Any heat leaking into the cooled plug is carried away by the gas expanding through the section AB which acts as a guard ring to the section BC so that the thermometer T gives the true Joule-Thomson temperature. It is in the pores of the porous plug that the flow is fast, turbulent and cold.

and S the concealed ones and for G and H the artificially constructed ones. Thus, through every point in a pv diagram (Fig. 7.4) one can draw a curve of constant H which will pass through the successive initial and final states of a throttling process. For a curve of constant H lying in the surface determined by the equation of state

$$dH = d(U + pv) = 0 = dU + p dv + v dp = T dS + v dp = 0$$

and $T\left(\frac{\partial S}{\partial p}\right)_H = -v.$

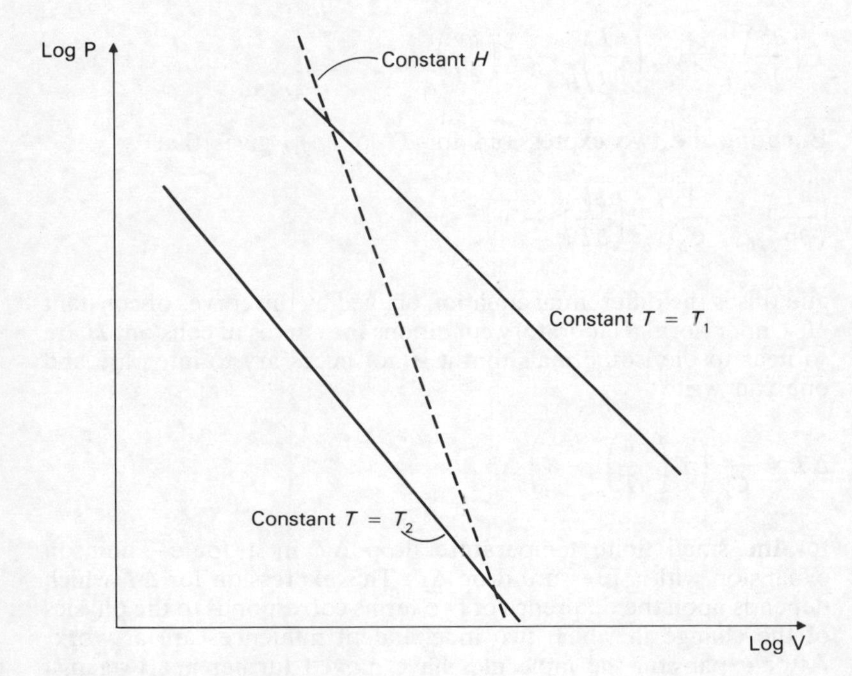

Fig. 7.4 Joule-Thomson cooling.
The full lines represents the PV curves for a real gas. They are not parallel straight lines because of the deviations from the perfect gas laws which have been exaggerated for clarity of illustration. The interrupted curve --- shows the PV relationship at constant enthalpy H. For a perfect gas $H = U + PV = 3RT/2 + RT = 5RT/2 = 5PV/2$ so that constant H and constant T curves coincide and there is no change of temperature in a throttling process. For a real gas the constant T and constant H curves do not coincide and the change in temperature can be read off from the corresponding values of P and V. Data for constructing these curves for H_2, A and CO_2 are given in the American Institute of Physics Handbook 1957.

If the equation of state is given in the form $S = S(p, T)$

$$\mathrm{d}S = \left(\frac{\partial S}{\partial p}\right)_T \mathrm{d}p + \left(\frac{\partial S}{\partial T}\right)_p \mathrm{d}T = \left(\frac{\partial S}{\partial p}\right)_T \mathrm{d}p + \frac{C_p}{T}\mathrm{d}T$$

Maxwell's relationship $(\partial S/\partial p)_T = -(\mathrm{d}v/\mathrm{d}T)_p$ transforms this to

$$\frac{\mathrm{d}S}{\mathrm{d}p} = \frac{C_p}{T}\frac{\mathrm{d}T}{\mathrm{d}p} - \left(\frac{\partial v}{\partial T}\right)_p$$

and

$$T\left(\frac{\partial S}{\partial p}\right)_H = C_p\left(\frac{\partial T}{\partial p}\right)_H - T\left(\frac{\partial v}{\partial T}\right)_p$$

Equating the two expressions for $T(\partial S/\partial p)_H$ gives that

$$\left(\frac{\partial T}{\partial p}\right)_H = \frac{1}{C_p}\left(T\left(\frac{\partial v}{\partial T}\right)_p - v\right)$$

and this is the differential equation obeyed by the curves of constant H. Under normal laboratory conditions the curves of constant H are so near to the isothermals that it is not necessary to integrate and one can write

$$\Delta T = \frac{1}{C_p}\left(T\left(\frac{\partial v}{\partial T}\right)_p - v\right)\Delta p$$

for the small finite temperature drop ΔT in a Joule–Thomson expansion with a pressure drop Δp. This expression for ΔT which depends upon the difference of two terms corresponds to the physics of the change in which two independent influences are at work. After expansion the molecules have moved further apart against the attractive forces and are thus moving more slowly, and the direct effect of the molecular forces is always to produce a cooling on expansion. At the same time work has been done on the gas $= p_1v_1 - p_2v_2$. This is the difference of two large terms, and its value depends upon the deviations from the perfect gas laws. It is a term which can have either sign, according to the particular gas and the pressures at which the expansion is carried out. Thus a general argument leads only to a general conclusion that the gas will heat up or cool down according to circumstances. The expression for ΔT is an important one in experimental physics. It is the fundamental equation governing the liquefaction of gases by the Linde–Hampson process, in which the gas is cooled by a series of Joule–Thomson expansions. It is also the basis of one of the better methods of establishing the absolute scale of temperature. The experimental value of $\Delta T/\Delta p \rightarrow (\partial T/\partial p)_H$ is referred to in the literature as the Joule–Thomson coefficient μ.

7.7 Liquefaction of gases

For a preliminary survey of the liquefaction of gases by Linde's method a discussion in terms of van der Waals' equation is adequate.

Simple differentiation immediately yields

$$\left(\frac{\partial v}{\partial T}\right)_p = \frac{R}{p - a/v^2 + 2ab/v^3}.$$

substituting this value in the expression for ΔT yields at once

$$C_p \frac{\Delta T}{\Delta p} = \frac{RT - pv + a/v - 2ab/v^2}{p - a/v^2 + 2ab/v^3}$$

$$= \frac{2a/v - 3ab/v^2 - pb}{p - a/v^2 + 2ab/v^3}$$

The denominator is always positive so the sign of ΔT is determined by the numerator and there is heating or cooling according as $pv >$ or $< 2a/b - 3a/v$. The condition for no change of temperature can be simplified by using the perfect gas value of $pv = RT$ and neglecting the very small term $3ab/v^2$ to give $T_{\text{inv}} = 2a/Rb$. Unless the temperature is below the inversion temperature T_{inv} there is no cooling, whatever the working pressure. Using the standard van der Waals' expression for the critical temperature, $T_c = 8a/27Rb$ one finds that $T_{\text{inv}} = 27/4T_c$ so that a permanent gas which cannot be liquefied by compression at room temperature can still be cooled, and so liquefied by the Linde Process. If the equation defining the line between heating and cooling is written in terms of the parameters $pv = y$, $p = x$ used by Amagat it becomes

$$y^2 - \frac{2ay}{b} + 3ax = 0.$$

This is the equation of a parabola with an axis parallel to the x axis.

If the gas is to cool, the initial state must lie inside the parabola and $p < a/3b^2$. If the pressure exceeds this value it will always heat. (Fig. 7.5). These considerations are sufficient for the preliminary choice of the conditions under which a liquefier must operate. Further work must depend upon experimentally determined quantities, since near the liquifying point van der Waals' equation is inadequate. For the permanent gases at room temperature the Joule–Thomson cooling is not nearly enough to liquify the gas. To liquify a gas like nitrogen (critical temperature 126 K, inversion temperature ~850 K) the cooling must be increased by repeated passage through the throttle. The temperature changes which occur at each passage are accumulated by means of a heat exchanger (Fig. 7.6). It is obvious on general grounds that the efficiency of the liquifier depends upon how effective the heat exchanger is in reducing the temperature difference between the incoming gas and the low pressure gas leaving the exchanger after having been cooled by passage through the throttle. The need for an efficient heat exchanger can be expressed quantitively by looking upon the thermally isolated heat exchanger and the expansion valve as the 'porous plug' of a Joule–Thomson experiment.

The gas enters at a pressure p_1 with an enthalpy h_1 per unit mass and leaves at a pressure p_2 with an enthalpy h_2, while a fraction f is liquified and can be drawn of with an enthalpy h_3.

Since there is no change of enthalpy in a throttling process

$$h_1 = (1 - f)h_2 + fh_3 \quad \text{and} \quad f = \frac{h_2 - h_1}{h_2 - h_3}$$

In this expression the denominator is effectively the latent heat of evaporation and is not much affected by the efficiency of the heat exchanger. The numerator can be evaluated approximately on the assumption that the heat capacity c_p is constant. If the gas is a perfect gas $h_1 = h_2$ and there is no liquefaction. If a real gas is passed once through the throttle its temperature falls by ΔT to $T - \Delta T$ but its enthalpy remains unchanged at h_1. If it is returned to its original temperature at pressure p_2 its enthalpy increases by $c_p\Delta T$ and $h_2 - h_1 = h_1 + \Delta T c_p - h_1 = \Delta T c_p$. If the heat exchanger is inefficient, so that the gas leaves the system at $T - \Delta T'$, and the temperature will only have to be raised to $T - \Delta T'$ the corresponding value of $h_2 - h_1$ is $c_p(\Delta T - \Delta T') = c_p\Delta T(1 - (\Delta T'/\Delta T))$.

Accordingly,

$$\frac{\text{f with an imperfect heat exchanger}}{\text{f with a perfect heat exchanger}} = 1 - \frac{\Delta T'}{\Delta T}$$

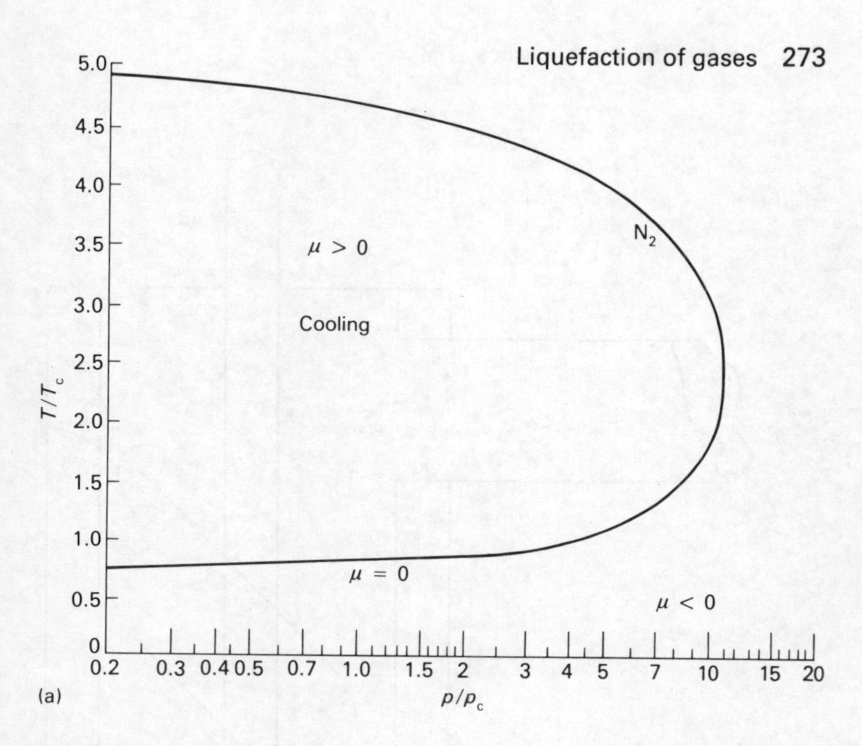

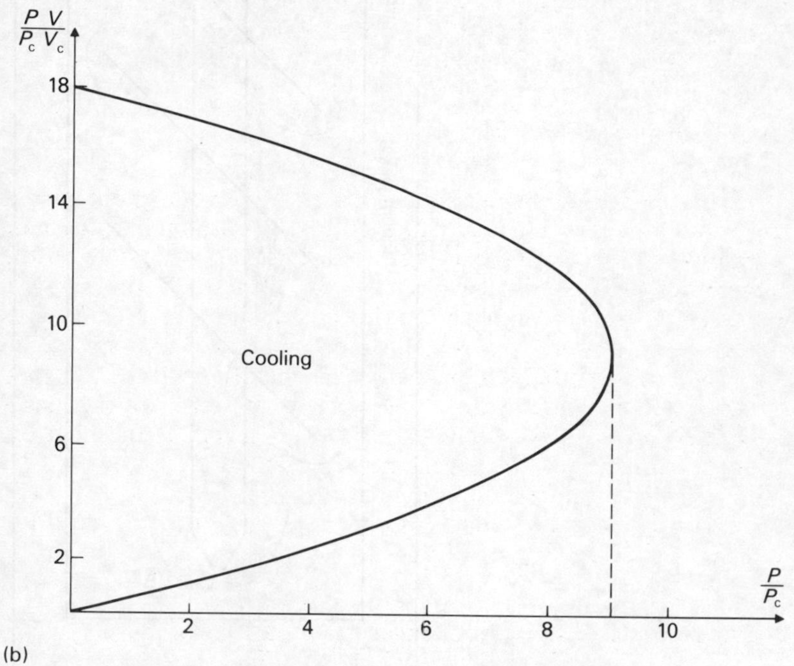

Fig. 7.5 Joule–Thomson inversion curve for nitrogen,
(a) experimental
(b) calculated from van der Waals' equation.

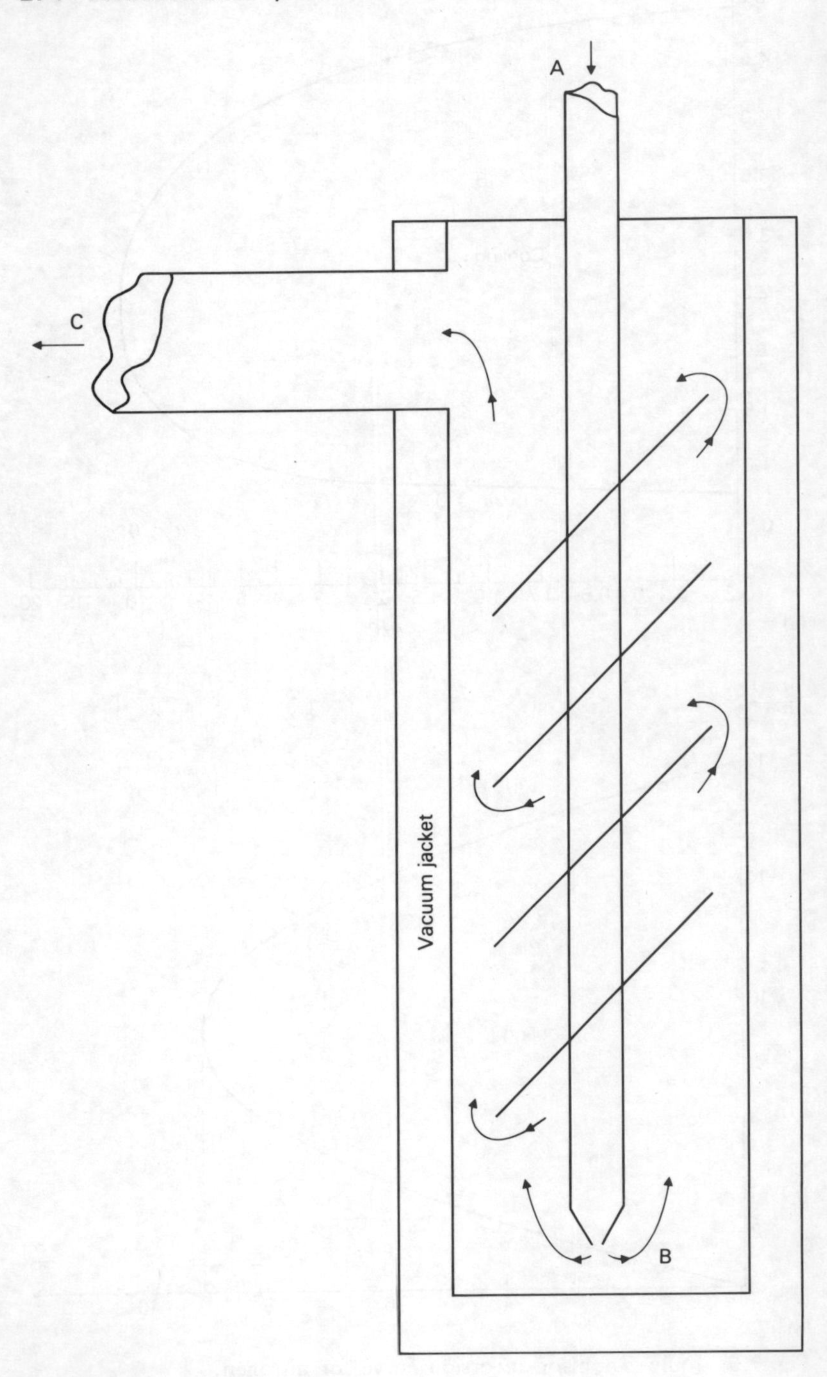
A
C
Vacuum jacket
B

The fraction $1 - \Delta T'/\Delta T$ is sensitive to the efficiency of the heat exchanger because ΔT is the small temperature change due to a single passage through the plug while $\Delta T'$ applies to a heat exchanger acting over the larger temperature range between room temperature (T) and the temperature at which the gas is liquifying. If $\Delta T' = \Delta T$ there is no liquification, the heat exchanger is not operating and the conditions are those of a simple Joule–Thomson expansion.

7.8 Absolute temperature

Since T in the theoretical formula for the Joule–Thomson cooling is by definition the absolute Kelvin temperature it would seem at first sight that the formula could not be checked experimentally unless one had access to an absolute thermometer. This difficulty can be avoided by making a series of measurements using an arbitrary thermometer whose readings T^* are related to the absolute temperature by the relationship $T^* = F(T)$. The unknown function $F(T)$ is chosen so that T^* changes from 0 to 100 in going from the ice point to the boiling point of water while T changes from T_0 the unknown absolute temperature of the ice point to $T_0 + 100$. This arbitrary and unknown relationship is used to relate the quantities measured on the arbitrary scale with the corresponding quantities measured on the absolute scale.

If two temperature scales T and T^* are related so that $T^* = F(T)$ then $\mathrm{d}T^*/\mathrm{d}T = \mathrm{d}F(T)/\mathrm{d}T = F'(T)$ and $\mathrm{d}T^* = F'(T)\mathrm{d}T$. An example would be a platinum resistance thermometer based upon the law $R = R_0(1 + \alpha T + \beta T^2)$ so that

$$T^* = \frac{R - R_0}{R_0} + T_0 = T_0 + \alpha T + \beta T^2 \text{ and}$$

$$\mathrm{d}T^* = (\alpha + 2\beta T)\mathrm{d}T.$$

If a quantity of heat energy $\mathrm{d}Q$ joules measured mechanically is added to a mole of gas at constant pressure it will produce a rise of temperature given by $\mathrm{d}Q = C_p\mathrm{d}T$. An observer using a T^* thermometer will observe a rise of temperature $\mathrm{d}T^*$ and report a heat capacity C_p^* such that $C_p^*\,\mathrm{d}T^* = \mathrm{d}Q$. Equating the two measures of $\mathrm{d}Q$ gives $C_p = C_p^*\,\mathrm{d}T^*/\mathrm{d}T = C_p^*\,F'(T)$.

Fig. 7.6 (opposite) Primitive heat exchanger.
The high pressure gas enters at A to expand through the jet at B. The unliquified gas at low pressure returns to the circulating pump at C. Heat exchange is promoted by the spiral fin attached to the central tube and by the maintenance of turbulent flow.

Thus $(\partial v/\partial T)_p = (\mathrm{d}v/\mathrm{d}T^*)\,F'(T)$ and μ^* the measured value of the Joule–Thomson coefficient becomes $\mu F'(T)$. Making these substitutions one readily finds that

$$\mu^* = \frac{1}{C_p^*}\left[T\left(\frac{\partial v}{\partial T^*}\right)_p F'(T) - v\right]$$

Collecting the terms in T one obtains

$$\frac{1}{TF'(T)} = \frac{\left(\dfrac{\partial v}{\partial T^*}\right)_p}{C_p^*\mu^* + v}$$

The unknown function $F'(T) = \mathrm{d}T^*/\mathrm{d}T$ can be eliminated by integrating to give

$$\int_{T_0}^{T_0+100}\frac{\mathrm{d}T}{T} = \int_0^{100}\frac{\left(\dfrac{\partial v}{\partial T^*}\right)_p \mathrm{d}T^*}{\mu^* C_p^* + v} = \mathrm{A}$$

The second integral can be evaluated experimentally since all the quantities are either absolute like v or like μ^* are measured with a real arbitrary thermometer calibrated so that there are 100 arbitrary degrees between the ice and steam points. The resulting relationship $\log (T_0 + 100)/T_0 = \mathrm{A}$ thus determines T_0 the absolute temperature of the ice point. The theoretical expression for the Joule–Thomson cooling can thus be checked by verifying that different measurements using different gases and different mean pressures give the same value for T_0.

7.9 The isothermal Joule–Thomson effect

Instead of measuring the cooling in a throttling effect one can incorporate a heating coil in the throttle and measure the rate at which heat must be added to ensure that the Joule–Thomson cooling is just cancelled. Conservation of energy gives that

$$u_2 - u_1 = p_1v_1 - p_2v_2 + Q$$

so that for a small pressure drop $\mathrm{d}p$

$$\mathrm{d}u = \mathrm{d}Q - \mathrm{d}(pv)$$

and $dQ = du + pdv + vdp = TdS + vdp$. So

$$\left(\frac{\partial Q}{\partial p}\right)_T = T\left(\frac{\partial S}{\partial p}\right)_T + v = -T\left(\frac{\partial v}{\partial T}\right)_p + v$$

in which $\left(\frac{\partial Q}{\partial p}\right)_T$ is the measured isothermal Joule-Thomson coefficient. If the equation of state is expressed in terms of virial coefficients (section 2.6) so that $pv = RT(1 + Ap + Bp^2)$,

$$\left(\frac{\partial Q}{\partial p}\right)_T = -RT^2\left[\left(\frac{\partial A}{\partial T}\right)_p + p\left(\frac{\partial B}{\partial T}\right)_p\right]$$

This is a useful result since the temperature dependence of two virial coefficients are separately determined by a single relatively simple measurement of the change of the isothermal Joule–Thomson coefficient with pressure.

7.10 Free adiabatic expansion

In a wide tube there is no throttling effect and the gas can expand freely, adiabatically and, if certain precautions are taken, the flow remains laminar or streamlined. In laminar or streamlined flow each element of gas follows a definite path and does not mix with gas from a neighbouring element. The aggregate of these paths for a finite body of gas defines a tube of flow across whose longitudinal surface there is no gas flow except by the slow process of diffusion. This can usually be neglected and with it the viscous forces which the molecular exchange creates. The mass of gas traversing a tube of flow does not change along the length of the tube so that its cross-sectional area A, the velocity of the gas v and the density ρ are related so that $\rho v A = \mathrm{K}$.

If ρ_1 and v_1 are the density and velocity of the gas when the cross-section of the tube is A_1 and ρ_2 and v_2 are the corresponding quantities when the cross-section has changed to A_2 conservation of energy requires that

$$A_1 p_1 v_1 - A_2 p_2 v_2 = \mathrm{K}(u_2 - u_1) + \frac{1}{2}\mathrm{K}(v_2^2 - v_1^2)$$

rate of doing work on the gas	rate of increase in the internal energy of the gas at rest	rate of increase in k.e.

in which u_1 and u_2 are the internal energies of unit mass of the gas at rest. Since the volume per second of the gas passing any section A_1 of the tube is A_1v_1, this can be rewritten in terms of the enthalpy h of unit mass as

$$h_1 - h_2 = \frac{1}{2}(v_2^2 - v_1^2)$$

and since no heat is flowing into the gas

$$\left(\frac{\partial h}{\partial p}\right)_S = -\frac{1}{2}\left(\frac{\partial v^2}{\partial p}\right)_S$$

at any point in the tube of flow.
From the definition of H, $\mathrm{d}H = T\mathrm{d}S + V\mathrm{d}P$ and

$$\left(\frac{\partial H}{\partial p}\right)_S = V \quad \text{and} \quad \left(\frac{\partial h}{\partial p}\right)_S = \frac{1}{\rho}$$

so that the state of the gas at any point in the flow can be calculated if the equation of state is known in the form $p = f(\rho, S)$. Throughout the flow $A = \dfrac{K}{\rho v}$. These are the relationships needed in designing a nozzle. A nozzle is a tube of varying cross section A in which a fluid at a high pressure and nearly at rest is converted by the pressure gradients into a stream of fluid with a large momentum at a low pressure. If the fluid is water whose density is effectively constant, a nozzle is relatively simple to design. Whether it is a small nozzle for watering the garden or a more ambitious affair for driving a 100 hp Pelton wheel, the relationship $A = \mathrm{K}/\rho v$ ensures that A must always decrease in the direction of the flow and the nozzle is everywhere convergent. If the fluid is a gas with a variable density the problem is less simple. In fact, a nozzle in which the gas is everywhere accelerated converges to a minimum cross-section and then becomes a divergent nozzle. This is not surprising since A is clearly very large at the beginning of the expansion when $v \approx 0$ and must again be large at the end of the expansion when ρ is very small. At the minimum cross-section ρv and so $(\rho v)^2$ is a maximum. Setting $(\partial(\rho v)^2/\partial p)_S = 0$ for a maximum gives

$$\rho^2\left(\frac{\partial v^2}{\partial p}\right)_S + 2\rho v^2\left(\frac{\partial \rho}{\partial p}\right)_S = 0$$

but $$\left(\frac{\partial v^2}{\partial p}\right)_S = -2\left(\frac{\partial h}{\partial p}\right)_S = -\frac{2}{\rho}$$

so that $v^2 = (\partial p/\partial \rho)_S = c^2$, the velocity of sound at the minimum in the cross-section. After the gas has passed through the minimum its flow is supersonic and no mechanical signal from the low pressure side of the jet can pass through the throat of the nozzle and effect the flow of the gas. It should be noted that the degree of convergence of the jet is not determined in this simple calculation and independent steps must be taken to ensure that the flow is streamlined and not turbulent. This simplified theory of gas flow, which does not depend upon the gas being a perfect gas, is generally used in designing nozzles used in vacuum pumps (section 3.12) and turbines. For steam turbines the walls of the nozzle are usually prolonged beyond the minimum in the throat but for gas turbines in which thermodynamic efficiency is not usually the prime consideration the expansion after the throat is only controlled on one side as shown in Fig. 7.7.

Worked example

Find the velocity of a perfect gas expanding adiabatically from p_1 to p_2 by integrating the expression $(\partial h/\partial p)_S = 1/\rho$ of section 7.10. Show that the velocity will not be supersonic unless

$$\frac{p_1}{p_2} > \left(\frac{2}{\gamma + 1}\right)^{\gamma/\gamma-1}$$

and will exceed the velocity of sound in the gas at rest before expansion if

$$\frac{p_1}{p_2} \geqslant \left(\frac{2}{3 - \gamma}\right)^{\gamma/\gamma-1}$$

Since $(\partial h/\partial p)_S = -\frac{1}{2}(\partial v^2/\partial p)_S = 1/\rho$ one can integrate directly to get

$$\int \left(\frac{\partial v^2}{\partial p}\right)_S \mathrm{d}p = -2\int \left(\frac{1}{\rho_S}\right) \mathrm{d}p.$$

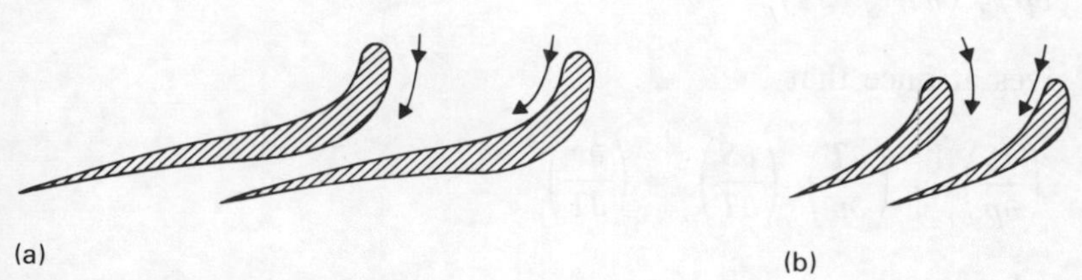

Fig. 7.7 Turbine nozzles.
(a) Convergent – divergent nozzle operating at a pressure ratio greater than the critical value.
(b) Simple convergent nozzle (after Cohen 1972).

For a perfect gas expanding adiabatically at constant S, $pv^{\gamma} = k = p(m/\rho)^{\gamma}$ and $1/\rho = k^{1/\gamma}/mp^{1/\gamma}$ so that direct integration gives

$$v_2^2 - v_1^2 = \frac{-2\gamma k^{1/\gamma}}{(\gamma - 1)m}\left[p_2^{(\gamma-1)/\gamma} - p_1^{(\gamma-1)/\gamma}\right].$$

Putting the initial velocity $v_1 = 0$ and using Newton's relationship $c^2 = \gamma p/\rho$ for the velocity of sound gives after a little reduction that

$$v_2^2 = \frac{-2c_2^2}{\gamma - 1}\left[1 - \left(\frac{p_1}{p_2}\right)^{(\gamma-1)/\gamma}\right] = \frac{-2c_1^2}{\gamma - 1}\left[\left(\frac{p_1}{p_2}\right)^{(1-\gamma)/\gamma} - 1\right].$$

From the first relationship $v_2 > c_2$, the local velocity of sound, if

$$\frac{p_1}{p_2} > \left(\frac{\gamma + 1}{2}\right)^{\gamma/\gamma-1} = \left(\frac{2}{\gamma + 1}\right)^{\gamma/1-\gamma}.$$

From the second relationship $v_2 > c_1$, the velocity of sound in the unexpanded gas, if

$$1 - \gamma = 2\left[\left(\frac{p_1}{p_2}\right)^{1-\gamma/\gamma} - 1\right] \quad \text{and} \quad \frac{p_1}{p_2} > \left(\frac{2}{3 - \gamma}\right)^{\gamma/\gamma-1}$$

7.11 Adiabatic compression

The change of temperature produced by a sudden (and so nearly adiabatic) compression can be calculated from Maxwell's third relationship.
The chain relationship

$$\left(\frac{\partial S}{\partial p}\right)_T \left(\frac{\partial p}{\partial T}\right)_S \left(\frac{\partial T}{\partial S}\right)_p = -1$$

gives at once that

$$-\left(\frac{\partial S}{\partial p}\right)_T = \left(\frac{\partial T}{\partial p}\right)_S \left(\frac{\partial S}{\partial T}\right)_p = \left(\frac{\partial v}{\partial T}\right)_p$$

and

$$\left(\frac{\partial T}{\partial p}\right)_S = \left(\frac{\partial T}{\partial S}\right)_p \left(\frac{\partial v}{\partial T}\right)_p = \frac{T}{c_p \rho v}\left(\frac{\partial v}{\partial T}\right)_p$$

For a substance with a volume expansion coefficient

$$\beta = \frac{1}{v}\left(\frac{\partial v}{\partial T}\right)_p, \left(\frac{\partial T}{\partial p}\right)_S = \frac{\beta T}{\rho c_p}$$

and the adiabatic rise in temperature ΔT, produced by a sudden compression Δp, is given by

$$\Delta T = \frac{\beta T}{\rho c_p}\Delta p.$$

For a solid this rise in temperature is quite small. $\sim 10^{-4}$ K/atmosphere. The corresponding cooling of a suddenly stretched wire is conveniently calculated by using a modified Gibbs' function as explained in section 7.3. Suppose

$$G' = u - TS + pv - lF$$

in which F is the tension applied to a wire of length l (initial length l_0) cross section A, Young's modulus E and expansion coefficient α. Then

$$\mathrm{d}G' = \mathrm{d}u - T\mathrm{d}S - S\mathrm{d}T + p\mathrm{d}v + v\mathrm{d}p - l\mathrm{d}F - F\mathrm{d}l$$

Conservation of energy requires that

$$\mathrm{d}Q = T\mathrm{d}S = \mathrm{d}u + \text{external work} = \mathrm{d}u + p\mathrm{d}v - F\mathrm{d}l$$

and

$$\mathrm{d}G' = -S\mathrm{d}T + v\mathrm{d}p - l\mathrm{d}F.$$

Since the extension is carried out at constant (atmospheric) pressure $\mathrm{d}p = 0$ and

$$\left(\frac{\partial G'}{\partial T}\right)_{pF} = -S, \left(\frac{\partial G'}{\partial F}\right)_{pT} = -l$$

Since

$$\frac{\partial^2 G'}{\partial F \partial T} = \frac{\partial^2 G'}{\partial T \partial F}$$

one has (dropping the subscript p since the whole process is understood to be at constant atmospheric pressure) that

$$\left(\frac{\partial S}{\partial F}\right)_T = \left(\frac{\partial l}{\partial T}\right)_F$$

Transforming $(\partial S/\partial F)_T$ as before one has that

$$\left(\frac{\partial T}{\partial F}\right)_S = -\left(\frac{\partial T}{\partial S}\right)_F \left(\frac{\partial l}{\partial T}\right)_F = -\frac{T}{c_p \rho A l_0}\left(\frac{\partial l}{\partial T}\right)_F$$

The sign of the temperature change is thus determined by the sign of $(\partial l/\partial T)_F$. If a fixed weight ($F$) is hung from the wire the load usually descends when the wire is heated, $(\partial l/\partial T)_F$ is positive and the wire cools on stretching. For exceptional substances the weight rises, $(\partial l/\partial T)_F$ is negative and the wire heats on stretching. An example of this behaviour is the well-known lecture experiment which purports to show the contraction of rubber on heating. The temperature changes which have been discussed are all reversible changes and the substance returns to its original temperature when the strain is removed. There are of course irreversible changes of temperature due to internal friction or mechanical hysteresis which are always positive. Formal thermodynamics has nothing to say about irreversible heating and it can only be calculated from detailed study of an appropriate molecular model.

7.12 Magnetic effects

Magnetic cooling

Reversible temperature changes occur when non-conducting paramagnetic salts (such as chrome alum and gadolinium sulphate) are magnetised by means of an external field. The temperature change can be calculated thermodynamically by setting up a modified Gibbs' function. To do this one must have a clear idea of the flow of energy between magnetised matter and the surrounding magnetic field.

An unmagnetised paramagnetic salt contains ions with constant magnetic moments μ orientated so that the average value of their component of magnetic moment in any direction is zero.

In the presence of a magnetic field of flux density B, the total magnetic moment per unit volume $M = \mu\Sigma_i n_i \cos\theta_i$ assumes a finite value: n_i = the number of ions whose magnetic moment makes an angle θ_i with B, Σn_i the number of paramagnetic ions per unit volume.

If M changes to $M + \mathrm{d}M$ magnetic ions have rotated in the magnetic field which does work $B\mathrm{d}M$ upon them. This work is converted reversibly into heat energy by molecular collisions. Thus, the general statement:

energy flowing into a system = increase in internal energy + external work done

becomes, per unit volume in the presence of a magnetic field,

$\mathrm{d}Q$	$+ B\mathrm{d}M$	$= \mathrm{d}U$	$+ P\mathrm{d}V$
Heat energy added = $T\mathrm{d}S$	magnetic energy flowing in from the field	increase in internal energy	work done against external pressure

Bearing this in mind one modifies the Gibb's function so that

$$G' = U - TS + PV - BM$$

and taking into account the conservation of energy to eliminate $T\mathrm{d}S$

$$\mathrm{d}G' = -S\mathrm{d}T - M\mathrm{d}B + V\mathrm{d}P.$$

The calculation then proceeds as in the previous example of the stretched wire to give

$$\left(\frac{\partial T}{\partial B}\right)_S = -\left(\frac{\partial T}{\partial S}\right)_B \left(\frac{\partial M}{\partial T}\right)_B = \frac{-T}{\rho B c_{pB}} \left(\frac{\partial M}{\partial T}\right)_B$$

Since $(\partial M/\partial T)_B$ is always negative a paramagnetic substance always heats on magnetisation and cools when demagnetised. If the substance obeys Curie's law $M = H\chi = HC/T = BC/\mu\mu_0 T$ one has at once that

$$\mathrm{d}T = \frac{\mathrm{C}}{2T\mu\mu_0 c_{pB}\rho}\,\mathrm{d}B^2$$

in which B is the magnetic field.

The cooling which can be obtained by demagnetising at room temperature is negligible. At 1.4 K, the lowest temperatures which can conveniently be reached by evaporating helium, it becomes important because of the sharp fall in the heat capacity. It can be used to reach the temperature at which c_{pB} starts to rise again as Curie's law breaks down, owing to the interaction between magnetic moments of the individual molecules. This is illustrated by the first experiments in this field by Giauque and MacDougall in 1933. On demagnetising gadolinium sulphate from 3.4 K and 0.8 T the temperature fell to 0.53 K; starting from 1.5 K the temperature only fell to 0.25 K. To some extent the final temperature is under the control of the experimenter if he uses isomorphic double salts (such as chrome alum) in which the mean distance between the magnetic ions can be increased by dilution with a magnetically inactive salt (potassium alum). Thermodynamic theory is not dependent on the origin of the magnetic moments and the whole argument applies equally well to substances in which the magnetic moment M is due to the alignment of nuclear magnetic dipoles. That this alignment actually occurs has been shown by detecting and measuring the anisotropy of the β and γ-rays emitted from radioactive nuclei. These experiments in their turn have led to a complete change in our view about the mechanism of radioactive change and the part played in them by the neutrino.

Super conductivity

Thermodynamics cannot of course predict the existence of superconductivity or explain the phenomenon when it has been discovered. Nevertheless thermodynamic theory has played a considerable part in suggesting fruitful experiments. The temperature at which a superconductor loses its resistance depends upon the magnetic field in which it is placed and falls as the magnetic field is increased according to an approximately parabolic law (Rutger's 1933) as shown in Fig. 7.8. A moderate field (10^{-1} T) suppresses superconductivity in all pure metals down to temperatures as near to the absolute zero as have been reached. These superconductors

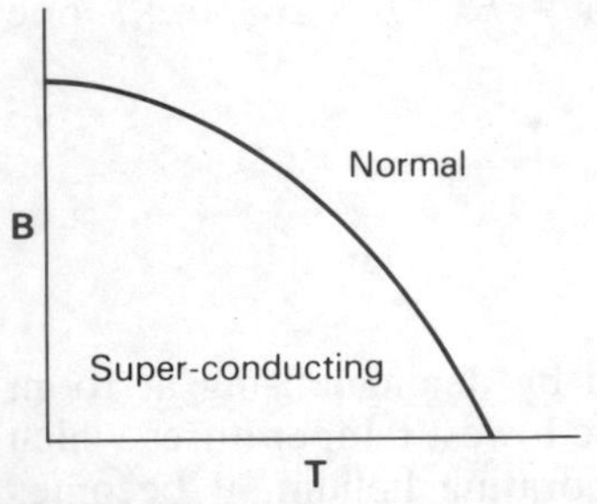

Fig. 7.8 Superconducting phase change in a magnetic field.

are known as type I or soft superconductors. The superconductors used in 'superconducting magnets' obey less simple laws and are known as hard or type II superconductors. The phenomena in type I superconductors are reversible and, as discovered by Meissner and Ochsenfeld, the magnetic field is ejected as soon as the specimen becomes superconducting. From the point of view of formal thermodynamics, which takes no cognisance of the immediate cause (as represented by a model) a superconductor can be represented by a region into which a magnetic field B cannot penetrate, and thus has zero permeability. The relationship $B = \mu\mu_0 H$ ensures that whatever the value of H, due for example to an outside solenoid, the magnetic flux B remains zero. Since $B = \mu_0 H(1 + \chi)$ the susceptibility $\chi = -1$ and the magnetic moment per unit volume $M = H\chi = -H$ thus ensuring that $B = \mu_0(H + M)$ is always zero. This artificial selection of the magnetic constants enables simple thermodynamic arguments to be applied to a superconductor and allows a discussion of the relationship between the critical field B_c at which superconductivity ceases and the thermal properties of the system. The curve Fig. (7.8) relating B_c and T is looked upon as a transition curve between the phases of the same simple substance, namely the normal phase and the superconducting phase of the metal. Invent as before a modified Gibb's function

$$G' = U - TS + PV - BM = U - TS + PV - \mu_0 HM$$

The term $-BM$ is the magnetic energy of a dipole M in a field B and can be included in the total internal energy to give $G' = U' - TS + PV$ with $U' = U - BM$. The simplified argument of section 5.6 can now be repeated for two phases (1 and 2) in a magnetic field so that

$$S_2 - S_1 = \frac{\lambda}{T} = \frac{U_2' - U_1' + P(V_2 - V_1)}{T},$$

a relationship which gives at once that for equilibrium $G_1' = G_2'$ per mole for the two phases. Thus for the normal phase which is unpolarised ($M = 0$)

$$G_n' = U_n - TS_n + PV_n$$

and for the superconducting phase

$$G_s' = U_n - TS_s + PV_s - \mu_0 HM.$$

Differentiating and eliminating TdS through the relationship $TdS = dQ = dU + PdV - \mu_0 HdM$ gives that at constant pressure ($dP = 0$) the two relationships $dG'_n = -S_n dT$ and

$$dG'_s = -S_s dT - \mu_0 M dH = -S_s dT + \mu_0 H dH$$

since $M = \chi H = -H$ for a superconductor. Thus since to maintain $G'_n = G'_s$, $dG'_n = dG'_s$ along the equilibrium curve

$$S_s - S_n = \mu_0 H_c \left(\frac{\partial H_c}{\partial T}\right)_{\text{trans}} = \frac{1}{2}\mu_0 \frac{\partial}{\partial T}(H_c^2)$$

in which H_c is the critical field and the differentiation is along the transition curve. Since $\partial H_c/\partial T$ is always negative the entropy of the normal phase is always greater than that of the superconducting phase except at the threshold temperature when $B_c = \mu_0 H_c = 0$. This difference of entropy corresponds to a latent heat of transition absorbed when the superconducting phase is transformed into the normal phase at constant temperature given by

$$\lambda = -T\mu_0 H_c \left(\frac{\partial H_c}{\partial T}\right)$$

There is also a difference in the heat capacity $c_s - c_n$ of the two phases, for

$$T\left(\frac{\partial S_s}{\partial T}\right) - T\left(\frac{dS_n}{dT}\right) = c_s - c_n = \frac{\mu_0 T_c}{\rho}\left[\left(\frac{\partial H_c}{\partial T}\right)^2 + H_c\left(\frac{\partial^2 H_c}{\partial T^2}\right)\right]$$

The discontinuity of the heat capacity persists up to the threshold temperature since $(\partial H_c/\partial T)^2$ is positive even when $H_c = 0$. It was this aspect of the problem which drew attention to the question (Rutger 1933) of the validity of applying thermodynamic arguments to superconducting substances. Both the specific heat discontinuity and the latent heat of transition have been observed experimentally and agree well with the theory (Kittel 1957). If instead of working at constant pressure one works at constant temperature and puts $\Delta T = 0$ in the relationship $dG'_n = dG'_s$ one has that

$$V_n - V_s = -\mu_0 M \left(\frac{\partial H_c}{\partial p}\right)_T = \mu_0 H_c \left(\frac{\partial H_c}{\partial p}\right)_T.$$

This small change of volume ($\Delta V/V \sim 10^{-8}$) has also been observed (Lasarew and Sudovstov, 1949) and is zero at the superconducting threshold as it should be ($H_c = 0$).

7.13 Electrical systems

Since electricity consists of particles (electrons) one can suppose that the conclusions drawn from molecular theory will be valid for electrical systems. This supposition is not rendered less probable by the consideration that all atoms and molecules are electrical systems whose nature is concealed from us because they happen to be neutral. From this point of view a metal is a more complicated system than a simple solid like camphor or iodine made up from one type of molecule only, since it consists of two parts; the free electrons which make it a metal and the crystalline lattice formed by the positive ions which makes it a solid. The formal thermodynamics of substances (phases) containing more than one type of molecule (components) is far from simple and was given by Gibbs in a famous memoire on Heterogeneous Equilibrium. It is too difficult for a simple treatment. However, metals can be treated as simple substances by assuming that the positive ion lattices play no part in the equilibrium and can be ignored except for the part they play in defining the region in which the electrons can move freely as in a gas. On this hypothesis the metal is looked upon merely as a source of electrons which exerts a definite vapour pressure. This electron vapour pressure is usually too small to measure (like for example the vapour pressure of iron at room temperature) but as it obeys the Clausius–Clapeyron equation it becomes sensible at high temperatures when it is the basis of the well-known thermionic current emitted by hot filaments into a good vacuum. In this simplified model the difference between one metal and another is expressed in terms of a single parameter λ_0, the latent heat of evaporation of the electrons from the metal at absolute zero. If ϕ is the work function of the metal in the photoelectric equation

$$h\nu = e\phi + E$$

$$\lambda_0 \approx e\phi$$

(Strictly speaking the photoelectric threshold should be measured at the absolute zero, but the difference between ϕ_T and ϕ_0 is usually smaller than the experimental error of a difficult measurement). If the electrons in a composite bimetallic wire made by welding the ends of say an iron and a nickel wire together are to be in equilibrium, the vapour pressures of the electrons must be the same for both metals. This is achieved by the migration of electrons from the electropositive iron to the electronegative nickel. This migration continues until the repulsion due to the negative charge acquired by the nickel prevents the arrival of more electrons from the iron. The space charge of the electrons in the nickel is now greater than it was, ϕ_{nickel} is reduced because the electrons have a negative charge, and more electrons have sufficient kinetic energy (k.e. $> e\phi = \lambda_0$) to leave the nickel. The vapour pressure is accordingly increased. Correspondingly the space charge of the electrons in the iron is decreased and ϕ_{iron} is increased so that fewer electrons have sufficient kinetic energy (k.e. $> e\phi_{\text{iron}}$) to leave the metal. The vapour pressure of the electrons is accordingly decreased. The process of electron transfer continues until the electron vapour pressures of the two metals are the same. The bimetallic wire is then in equilibrium with a common electron density in an evacuated space. This ability to adjust the vapour pressure by electron transfer is characteristic of metals. It provides a valid language in which to describe the equilibrium because of the minute fraction (10^{-14} per cent) of the available electrons which are transferred. The other characteristics of the metal are virtually unchanged and the metals before and after contact are reasonably described as iron and nickel. Since metals are electrical conductors there can be no electric field in the bulk of the metal and the potential jump between the two metals is confined to a thin layer a few atoms thick in the vicinity of the metallic junction. The potential difference between the two metals is referred to as the contact potential; it is ~ 1 V. It is in practice difficult to obtain metals so pure and so homogeneous that there are not internal contact potentials between different parts of the metal $\sim 10^{-4}$ V. These small differences play a part in metallic corrosion. That the transfer of electricity really takes place can be shown directly by the experimental methods illustrated in Fig. 7.9. This is a difficult branch of experimental physics because very small amounts of foreign material (e.g. oxygen) collecting in the boundary layer change the system being investigated. It can however be shown experimentally in selected examples that the contact potential is the same as the difference in the latent heats of evaporation and the photoelectric thresholds, as it should be.

If the free ends of the iron and nickel wires are joined together to form a bimetallic circuit, Fig. 7.10 the two metals are already in equilibrium and no further electron migration takes place. If however the circuit is completed by means of an ionic conductor (such as dilute sulphuric acid) whose conductivity is not due to an electron gas the situation is quite different; the contact potentials do not balance out and an electric current flows round the circuit. This was the fundamental discovery made by Volta (1799) in following up Galvani's famous observation (1786) of the twitching frogs legs, skewered on iron and copper wires, when the wires touched. The electrical equilibrium in a bimetallic circuit can be disturbed in a physically much simpler way by maintaining the junctions at dif-

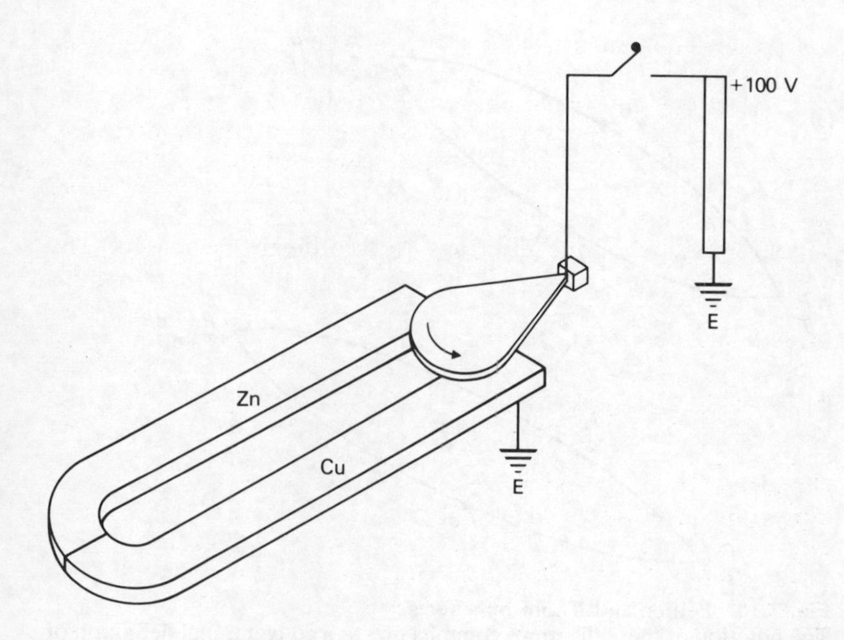

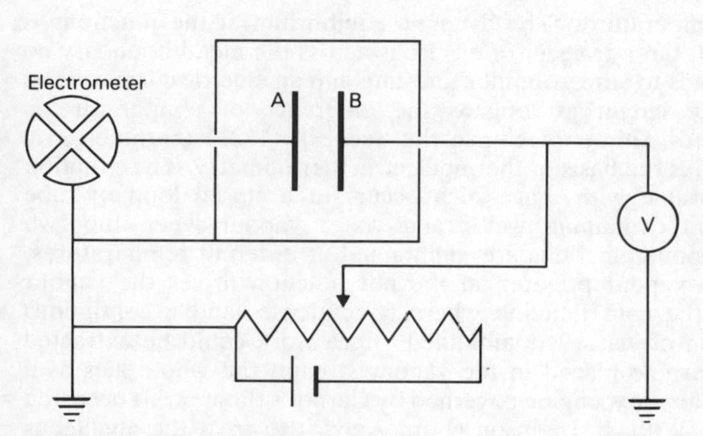

Fig. 7.9 Contact potential.

(a) On closing the switch the positively charged paddle is repelled by the positively charged zinc and moves over the copper strip.

(b) The contact potential between the fixed copper plate A and the moveable zinc plate B is found by adjusting the potentiometer slider until moving B produces no deflection of the electrometer. The plates A and B are then at the same potential, there is no field between them and the contact potential is V the reading of the voltmeter.

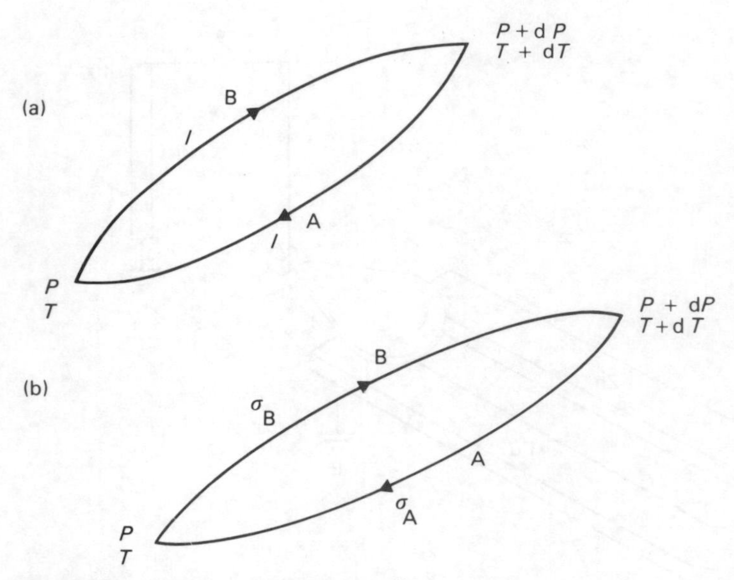

Fig. 7.10 Peltier and Thomson effects.
The notation of these diagrams commits one to a conventional definition of the sign of the thermal e.m.f. between A and B.

ferent temperatures. The dynamic equilibrium at the junctions is disturbed, but a transfer of electrons across the metal boundary no longer leads to a new equilibrium state and an electric current flows round the circuit as long as the difference of temperature is maintained. This is of course the Seebeck (1821) thermoelectric effect and is the basis of thermoelectric thermometry. The situation is comparable with that which occurs in a closed loop of tube (Fig. 7.11) containing water and water vapour when the two liquid/vapour junctions are maintained at different temperatures. The high vapour pressure at the hot junction drives the vapour towards the cold junction where it condenses and a continuous circulation of water is maintained. Since work could be extracted from a turbine placed in the vapour stream, the whole acts as a rudimentary heat engine governed by Carnot's theorem. It occurred (1854) to William Thomson (Lord Kelvin) to treat the analagous electric circuit by the methods of formal thermodynamics. Since the circulating thermoelectric current can be made to do external work by including an electric motor in the circuit, conservation of energy requires that energy must be flowing into the circuit from the heat sources which maintain the temperature difference between the junctions. This flow of energy takes place through the Peltier effect (1834). When a current I crosses the junction between metal A and

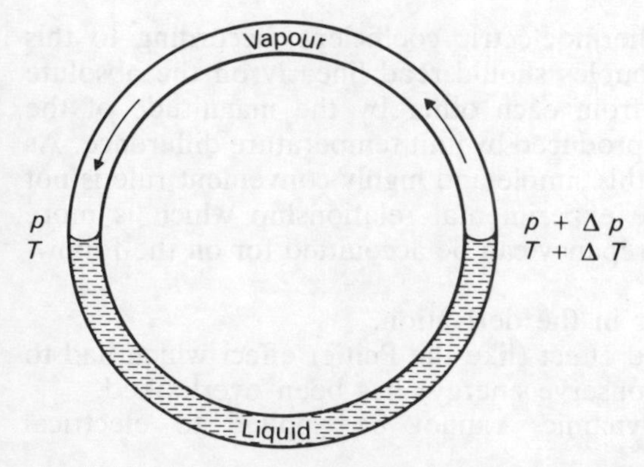

Fig. 7.11 Mechanical model of a thermoelectric circuit.

metal B heat energy is absorbed at the rate of IP_{AB} units per second. Whether heat is absorbed or liberated depends upon the sign of I with respect to the direction A → B. The Peltier coefficients P_{AB} are quite large ($\sim 10^{-2}$ J/C) and there is no difficulty in demonstrating the effect experimentally. The absorption of heat at a junction is used in the construction of compact refrigerators without moving parts when this type of simplicity is imperative. In the circuit of Fig. 7.10 equating the rate of absorption of heat to the rate of doing electrical work $I\Delta V$ gives at once that

$$\frac{dV}{dT} = \frac{dP}{dT} \text{ (conservation of energy)}$$

in which the Peltier coefficient $P = P_{BA}$.
The circuit as a whole is not changing with time. Although there is a net flow of heat energy $I(dP/dT)$ into the circuit the temperature of the system does not rise because equivalent work $I(dV/dT)$ is flowing out.

Similarly the entropy is not changing, the inward flow $I(P + \Delta P)/(T + \Delta T)$ being balanced by the outward flow $I(P/T)$ so that

$$\frac{P}{T} - \frac{P + \Delta P}{T + \Delta T} = 0 \quad \text{and} \quad \frac{dP}{dT} = \frac{P}{T}.$$

It follows that

$$P = AT \quad \text{and} \quad \frac{dV}{dT} = A = \text{constant}.$$

Since $\mathrm{d}V/\mathrm{d}T$ is the thermoelectric coefficient, according to this reasoning all thermocouples should read linearly on the absolute scale and only differ from each other by the magnitude of the constant A, the e.m.f. produced by unit temperature difference. As Thomson pointed out this simple and highly convenient rule is not in agreement with the experimental relationship which is more complicated. This discrepancy can be accounted for on the following hypotheses:

(a) There is an error in the derivation.
(b) Some unsuspected effect (like the Peltier effect which had to be called in to conserve energy) has been overlooked.
(c) Formal thermodynamics cannot be applied to electrical circuits.

Of these objections (a) is strictly correct but it is usual to accept arguments against it in view of the experimental justification of (b). In these circumstances it is unnecessary to fall back on (c). The assumption that the change of entropy of a system can be calculated from $\Sigma_i Q_i/T_i$ is only true if all the Q_i's refer to reversible processes. The Joule heat I^2R liberated throughout the circuit is not reversible and has been neglected. This neglect can be made plausible by saying that the Joule heat and the Peltier heat are quite independent processes and can be treated separately. An argument in favour of this can be made by pointing out that the ratio

$$\frac{\text{Joule heat}}{\text{Peltier heat}} = \frac{I^2R}{PI} = \frac{IR}{P}$$

can be made as small as one likes by making I small enough and therefore the Joule heat can be neglected. This is a reasonable argument but it is more difficult to justify the neglect of heat transfer by conduction. If one accepts the argument (as Thomson did) a simple picture of the electric current as the flow of a real fluid readily suggests the form of the missing effect postulated in (b). The electric current entering the cold junction at a temperature T and leaving the hot junction at a temperature $T + \Delta T$ has had its temperature raised by ΔT and may therefore be supposed to be absorbing heat at the rate of $\sigma_A I \Delta T$ J/s, in which σ_A is in some sense the specific heat of electricity in metal A. Similarly σ_B is the specific heat of electricity in metal B. (Fig. 7.10b). With this assumption the previous equations are modified so that the energy equation becomes

$$\frac{\mathrm{d}P}{\mathrm{d}T} - \sigma = \frac{\mathrm{d}V}{\mathrm{d}T} \text{ (conservation of energy)}$$

in which $\sigma = \sigma_A - \sigma_B$.

Similarly the entropy equation becomes $dP/dT - \sigma = P/T$. Combining the two equations gives

$$P = T\frac{dV}{dT}$$

$$\sigma = T\frac{d^2V}{dT^2}.$$

These equations are now compatible with the experimental results. They can also be tested since P, σ_A and σ_B can all be measured experimentally. A comparison between the directly measured quantities and their values calculated from thermoelectric data is shown in Tables 7.1 and 7.2. The agreement is as good as can be expected when the difficulty of the measurements is considered. The values of σ_A are much less than would be expected from the classical theory of a perfect electron gas which should be $3R/2$ per mole (1.29×10^{-4}

Table 7.1

Temp. = 0°C

Metal pair	$\frac{dV}{dT} \times 10^6$	$T\frac{dV}{dT} \times 10^4$	$P \times 10^4$
Ag Cu	2.12	5.79	4.81
Fe Cu	11.28	30.8	36.7
Pt Cu	−1.40	−3.82	−3.72
Cd Cu	2.64	7.21	7.15
Ni Cu	−20.03	−54.7	−50.6

(Recalculated from Pidduck, 1925.)

Table 7.2

Temp. = 100°C

Metal	$\sigma_A \cdot$ J/C $\times 10^7$	**Couple**	$\sigma \cdot$ J/C $\times 10^7$	$\frac{Td^2V}{dT^2} \times 10^7$
Cu	19.6	Cu Pb	15.4	29.7
Fe	−121	Fe Pb	−12.5	−112
Hg	−23.0	Hg Pb	−27.2	−32.1
Pt	−9.19	Pt Pb	−96.1	−92.0
Zn	33.4	Zn Pb	29.3	54.3
Cd	100	Cd Pb	96.1	117
Pb	4.2			

(Recalculated from Pidduck, 1925.)

in the units quoted) and not depend upon the metal. This is due to the electrons in the metal being a degenerate gas.

The occurrence of negative values of σ_A shows that it is not a heat capacity in the ordinary sense and the calculation of σ_A from a simplified electron model is usually carried out from a different point of view. If a wire has a temperature gradient in it there will no longer be a uniform electron density and there will be potential gradient $\mu\, dT/dx$ in it. The heat emitted reversibly when a current passes through the wire is due to the work done $I\mu\, dT/dx$ against this potential gradient. The agreement with experiment which can be achieved in these calculations is illustrated in Table 7.3. One concludes that although formal thermodynamics is not a very good way of investigating the structure of metals it has at least had the merit of drawing attention to and predicting some of their properties at least eighty years before any theoretical explanation could be given of them at all.

Table 7.3

$\frac{\mu}{T}$ in μV	Li	Na	K	Rb	Cs
observed	0.4	−0.0282	−0.0275	−0.0689	−0.062
calculated	−0.016	−0.036	−0.036	−0.041	−0.048

(Taken from Wilson, 1936.)

7.14 The electric cell

In a voltaic cell, as well as the contact potential between the two metals there are potential jumps at the two metal electrolyte contacts and in some cells using two electrolytes a further small potential difference across the boundary between the two electrolytes. Many practically useful batteries are too complicated to be discussed in terms of a simplified theory but there are some batteries in which the assumptions of an elementary treatment are well fulfilled. The simple battery whose thermodynamics is to be discussed has the following characteristics:

1 The chemical change in it is well defined and capable of being carried out independently in a calorimeter so that its energetics can be determined.

2 The e.m.f. E is independent of the state of the battery. This implies that the electrolyte does not change its composition during discharge. This condition is usually met by making all the electrolytes saturated solutions. For example the Daniel cell uses saturated solutions of copper sulphate and zinc sulphate, and the Weston standard cell a single solution saturated with mercurous sulphate and cadmium sulphate. Bearing these limitations in mind a simple battery can be looked upon as a 'black box' which maintains a difference of potential E between its two terminals. External work Eq is done when q coulombs of electricity are transferred from the positive to the negative terminal. This action is reversible and work Eq is done on the box when the electric current is reversed. The source of the work done by the box is for the most part the change in the internal energy of the constituents of the battery and the order of magnitude of the e.m.f. (E = 0–3 V) is correctly given by equating the work done to the heat released when the chemical change in the cell takes place in a calorimeter without doing any external work. There are however other small reversible changes which take place when a current is drawn from a battery at constant temperature. Heat Q' per coulomb may be absorbed and there may be volume changes $V_2 - V_1$ (per coulomb) which will cause mechanical work to be done against the constant atmospheric pressure.

Conservation of energy requires that for the change from state 1 to state 2 caused by the passage of one coulomb that

$$Q' = U_2 - U_1 + P(V_2 - V_1) + E$$

Heat flowing in	increase in internal energy	mechanical work	electrical work

$$= U_2 + PV_2 - (U_1 + PV_1) + E$$

which is often written as

$$E = \Delta H + Q'$$

in which ΔH is the *decrease* in enthalpy of the materials of the battery and Q' the heat which flows in to maintain the temperature constant when one coulomb circulates (reversibly) through the cell.

As will now be shown the quantities Q' and E are not entirely independent of each other.

Let G_1 be the Gibbs' function of the cell in state 1 and G_2 the Gibbs' function of the cell after it has discharged one coulomb reversibly and isothermally doing external work E at constant atmospheric pressure. Then,

$$G_1 - G_2 = U_1 - U_2 - T(S_1 - S_2) + P(V_1 - V_2)$$
$$= \Delta H + Q' = E$$

In general for any system whose Gibb's function is G

$$\left(\frac{\partial G}{\partial T}\right)_p = -S \quad \text{and} \quad G = U + T\left(\frac{\partial G}{\partial T}\right)_p + PV.$$

Applying this to the electric cell gives

$$G_1 - G_2 = E = \Delta H + T\left(\frac{\partial}{\partial T}\right)(G_1 - G_2)_p$$

so that $Q' = T\left(\dfrac{\partial E}{\partial T}\right)_p$

and $E = \Delta H + T\left(\dfrac{\partial E}{\partial T}\right)_p$

This equation is usually known as the Gibbs–Helmholz equation after its independent discoverers. It can be accurately tested, and is well obeyed as shown in the last column of Table 7.4.

Problems

1 Show that $\left(\dfrac{\partial C_p}{\partial p}\right)_T = -T\left(\dfrac{\partial^2 v}{\partial T^2}\right)_p$

2 Show that the change of entropy when a van der Waals' gas expands isothermally from v_1 to v_2 is

$$R \log \frac{v_2 - b}{v_1 - b}$$

3 Show that in general

$$\left(\frac{\partial p}{\partial T}\right)_S = \gamma\left(\frac{\partial p}{\partial T}\right)_v$$

and verify that it is true for a perfect gas.

[Problems continue on p 298]

Table 7.4

Reaction	E	T°C	$\frac{dE}{dT}$	ΔH measured	ΔH calculated	Source
$Zn + 2AgI = 2Ag + ZnI_2$	0.398 72	25	9.4×10^{-5}	83.11	83.32	Webb, 1925
$HgO + H_2 = Hg + H_2O$	0.933 45	0	-3.05×10^{-4}	197.0	196.2	Fried, 1926
$PbO + H_2 = Pb + H_2O$	0.260 40	0	-3.653×10^{-4} $\pm 5.1 \times 10^{-5}$	68.6	69.5 ± 2.7	Fried, 1926

4 Show that the change in the internal energy for a van der Waal's gas expanding isothermally from V_1 to V_2 is $(a/V_1) - (a/V_2)$. Use this result to justify the assumption that the internal forces can be represented by an internal pressure $\Delta P = a/V^2$.

5 Show that a gas which obeys the law $p(v - b) = rT$ will obey the law $p(v - b)^\gamma = K$ on adiabatic expansion. Show also that its internal energy is a function of T only as for a perfect gas.

6 Show that the general expression $\Delta T = \beta T/\rho c_p \, \Delta p$ of section 7.11 for the change of temperature on a sudden compression agrees with the known result for a perfect gas.

7 Show that the change of temperature ΔT in a rod of density ρ heat capacity c_p, Young's modulus E and expansion coefficient α, when it is suddenly extended by a stress σ is approximately given by

$$\Delta T = -\frac{T\sigma}{\rho c_p}\left(\alpha - \frac{\sigma}{2E^2}\left(\frac{\partial E}{\partial T}\right)\right)$$

Verify each term separately by a cycle calculation. What is the temperature drop in a steel rod of diameter 20 mm if it suddenly has to support a 100 kg weight?

8 Plot the logarithm of the saturated vapour pressure of water against the reciprocal of the absolute temperature. Use your result to calculate the approximate value of the latent heat of evaporation of water at 50 °C.

9 If the molar latent heats of evaporation of a solid and the corresponding liquid are λ and λ', and the melting point is T_0, show that the greatest difference between the vapour pressures of the solid and the supercooled liquid occurs approximately at a temperature T given by

$$\frac{1}{T} = \frac{1}{T_0} + \frac{R \log \lambda/\lambda'}{\lambda - \lambda'}$$

The rate of formation of ice crystals in a cloud is a maximum at about −14 °C. How does this compare with the temperature at which there is the maximum difference of vapour pressure between the two phases?

10 Show by making a drawing on a molecular scale that the vapour pressure of a liquid with a convex surface (radius ~10 molecular diameters) is likely to be higher than that over a flat surface.

11 When saturated air is cooled by expansion a fog is formed by condensation on the dust particles in the air. If the air is dust free condensation does not take place until the expansion ratio is about 1.38 : 1.
Discuss this result on the basis of Thomson's formula for the vapour pressure p_r of a spherical drop of radius r, density ρ, surface tension S, and saturated vapour pressure p,

$$\log \frac{p_r}{p} = \frac{2mS}{kTr\rho}.$$

The mass of the water molecule in the vapour state is m and k is Boltzmann's constant. (For an elementary derivation of Thomson's formula see Poynting; J. H. and Thomson J. J. 1927 *Properties of Matter*, Griffin & Co.)

12 A copper cylinder is imbedded in a block of ice (at 0 °C) with its axis vertical. At what rate will it melt its way through the ice? (Make the simplifying assumption that all the heat flows through the copper parallel to the axis with a uniform temperature gradient and that there is no friction).

13 Discuss the view that skating is not to be explained in terms of the variation of melting point with pressure but in terms of the work done against friction. ($\mu = 0.019$ at -1.4 °C, 0.05 at -30 °C).
(You may with profit consult Bowden F. P. and Hughes, T. P. Proc. Roy. Soc. 1939, **172A**, 280).

14 Show that if the internal energy of a gas is given by

$$U = np(v - b) + A \qquad C_p = \left(\frac{\partial H}{\partial T}\right)_p = (n + 1)\, p \left(\frac{\partial v}{\partial T}\right)_p .$$

Show also that during an adiabatic expansion $T^n(v - b) = K$.

15 Make a correctly scaled drawing of Fig. 7.4 for argon or CO_2. Hence calculate the temperature drop when the gas at 400 K and 100 atm is expanded through a throttle to 1 atm pressure.

16 By a calculation similar to that made for paramagnetic salts obtain an expression for the reversible temperature change on polarising a dielectric. Liquid o-cresol (relative permittivity 11.5) leaks at 300 K from a faulty capacitator in which it is stressed to 10^5 V/m. Calculate the reversible change in temperature if $d\varepsilon_r/dT = 0.11$ and $c_p = 2 \times 10^3$ J/kg.

17 Justify the relationship given in section 7.9 between the isothermal Joule–Thomson coefficient and the virial coefficients. Calculate $(\partial A/\partial T)_p$ and $(\partial B/\partial T)_p$ from the following data for nitrogen at 167 K.

p(atm)	64.4	84.0	103.0	121.4
$\left(\frac{\partial Q}{\partial p}\right)_T$ (J/mole atm)	−28.4	−27.3	−25.6	−24.0

Data from L.B. Band 11/4 2432 p 769.

Further reading

Hill, T.L. (1960) *An Introduction to Statistical Thermodynamics*, Addison-Wesley, Reading Mass.

Kittel, C.K. (1957) *Introduction to Solid State Physics*, Chap. 16, Wiley and Sons, N.Y.

Llewellyn-Jones, F., (1957) *The Physics of Electrical Contacts*, C.P. Oxford.

Treloar, L.R.G. (1975) *The Physics of Rubber*, C.P. Oxford.

Wilks, J. (1961) *The Third Law of Thermodynamics*, O.U.P. London.

References

Courant, R. (1936) *Differential and Integral Calculus*, vol 2, chap 2, Blackie and Son, London.

Cohen, H. Rogers, C.F.C. and **Saravanamultoo, H.R.H.** Longmans, London.

Ewing, J.A. (1920) *Phil. Mag* **39**, 633.

Fried, F. (1926) *Zeit für Phys. Chem.*, *123*, 406.

Landolt-Bornstein Tabellen 6 Auf II Band 4 teil p. 762, 24313, Springer, Berlin.

Lazarew, B.G. and **Sudovstov, A.I.** (1949) *Dok. Akad. Nauk*, **69**, 345.

Meissner, W. and **Ochsenfeld, R.** (1933) *Naturwiss.*, **21**, 787.

Webb, T.J. (1925) *J. Phys. Chem.* **29**, 827.

CHAPTER 8

Experimental evidence

8.1 Importance of the perfect gas

It is the object of this chapter to present a brief account of the experimental results which show that the theoretical deductions about the properties of a collection of molecules which have been presented are actually true in practice. It will have been observed that in the course of the development of the kinetic theory which started with an elementary deduction of Boyle's law and the deviations from it, the interest returned (cf. Chapter 6) to the behaviour of perfect gases. Their study brings many scientific disciplines into a common framework in a way not even thought of when the early investigations started more than three hundred years ago. This renewed interest in perfect gases also occurs in the experimental work. It is the fundamental expression for entropy of a perfect gas which needs to be tested experimentally. This is the measurement which draws attention to the fact that a classical (i.e. based on Newtonian mechanics and the equipartition of energy) treatment of gases cannot yield results in agreement with experiment. The whole concept of entropy as a measurable quantity is tied to a quantum-mechanical treatment.

8.2 Avogadro's number L and the equipartition of energy

The molecular theory was not finally accepted until there was experimental evidence that collections of particles behaved like a gas. This evidence was provided by Perrin (1908) who observed the Brownian motion of small spheres made by pouring a solution of gamboge gum in alcohol into water; they were graded according to size by centrifuging. That small particles suspended in a fluid medium never came to rest but remained in a state of constant agitation was observed by Brown, (facile princeps Botanicorum) an English botanist, in 1827. The motion was not understood until it was predicted theoretically by Einstein (1905) and measured by Perrin. Two comparisons were made between the collection of gamboge spheres and the behaviour of a gas:

The vertical distribution under gravity

As is well known from the barometer formula (worked example of section 2.4) the density n_h of molecules at a height h is given by

$$n_h = n_0\, e^{-mgh/kT}.$$

Perrin observed that his gamboge spheres did not all settle at the bottom of the vessel but were distributed so that

$$n_h = n_0\, e^{-mgh(1-\rho/\rho')/kT}$$

in which ρ is the density of the fluid and ρ' the density of the particles. The height over which he could make the measurements was ~100 μm. He obtained a value of k (Boltzmann's constant) corresponding to a value of $L = (68 \pm 4) \times 10^{22}$. The measurements were on the limit of what was technically possible at the time. The particles could of course only be observed as diffraction rings and the whole measurement was only possible because ρ' was only ~1.2, so that the particles were nearly floating in the liquid. Present day techniques would have enabled Perrin to overcome some of his difficulties but subsequent methods of measuring L are all indirect and depend upon theoretical ideas drawn from many branches of atomic physics. The accepted value of L Avogadro's number is

$$L = 6.022 \times 10^{23}$$

based upon an atomic mass unit in which ${}^{12}_{6}C = 12$.

Direct measurements of the Brownian movement

A comparison of the mean squared distance $\overline{x^2}$ wandered by a particle in a time t with the value predicted by Einstein's theoretical formula yields a value for the mean kinetic energy of the particle. The value of $\overline{x^2}$ can be calculated by a simple approximate method similar to that used by Clausius in deriving the virial theorem (Langevin 1912). Consider the motion of a particle parallel to the x axis. Its motion is determined by two forces; X due to the molecular bombardment and $-\alpha\,\mathrm{d}x/\mathrm{d}t$ due to the viscous forces which would rapidly bring the particle to rest if it were not for the continual molecular bombardment. The equation of motion is thus

$$m\frac{\mathrm{d}^2x}{\mathrm{d}t^2} = X - \alpha\frac{\mathrm{d}x}{\mathrm{d}t}$$

Using the identity

$$\frac{1}{2}\frac{\mathrm{d}^2}{\mathrm{d}t^2}(x^2) = \left(\frac{\mathrm{d}x}{\mathrm{d}t}\right)^2 + x\frac{\mathrm{d}^2x}{\mathrm{d}t^2}$$

one readily finds that

$$\frac{m}{4}\frac{\mathrm{d}^2}{\mathrm{d}t^2}(x^2) = \frac{1}{2}m\left(\frac{\mathrm{d}x}{\mathrm{d}t}\right)^2 + \frac{1}{2}Xx - \frac{\alpha}{4}\frac{\mathrm{d}}{\mathrm{d}t}(x^2).$$

If this expression is averaged over a very large number of similar particles $\overline{Xx} = 0$, and

$$\frac{m}{4}\frac{\mathrm{d}^2}{\mathrm{d}t^2}(\overline{x^2}) = \frac{1}{2}m\overline{\left(\frac{\mathrm{d}x}{\mathrm{d}t}\right)^2} - \frac{\alpha}{4}\frac{\mathrm{d}}{\mathrm{d}t}(\overline{x^2})$$

a relationship between $\overline{x^2}$ and the mean value of the kinetic energy of the particles. According to the theorem of equipartition of energy the mean kinetic energy is $kT/2$ per particle per degree of freedom and

$$\frac{m}{4}\frac{\mathrm{d}^2}{\mathrm{d}t^2}(\overline{x^2}) + \frac{\alpha}{4}\frac{\mathrm{d}}{\mathrm{d}t}(\overline{x^2}) = \frac{kT}{2}$$

This is a first order differential equation in the variable $\mathrm{d}(\overline{x^2})/\mathrm{d}t$ which integrates directly to give

$$\frac{\mathrm{d}\overline{x^2}}{\mathrm{d}t} = \frac{2kT}{\alpha} + \mathrm{C}\,\mathrm{e}^{-\alpha t/m}$$

The second term represents the exponential decay of the unknown initial velocity of the particle. Consideration of the numerical value of the constants involved shows that this term can be neglected after a time $\sim 10^{-5}$ s so that for longer times of observations

$$\overline{x^2} = \frac{2kT}{\alpha} \cdot t$$

The linear variation of $\overline{x^2}$ with the time is confirmed by experiment. If the coefficient of viscous drag is put equal to $6\pi\eta r$, the Stokes' law value for a sphere of radius r, one has

$$\overline{x^2} = \frac{kT}{3\pi\eta r} \cdot t$$

the expression first given by Einstein.

This expression for the mean square distance which a small particle drifts gives very good agreement with the experimental values over a range of values of η, r and T. The measurement of $\overline{x^2}$ is not particularly difficult, especially in gases, for which η is much smaller than it is in liquids. The values of L, $(60.3 \pm 1.2) \times 10^{22}$ (Fletcher 1913) and $(60.5 \pm 0.3) \times 10^{22}$ (Westgren 1915) obtained from the observed Brownian motion in air and water are in excellent agreement with the accepted value.

Another system whose Brownian motion can be measured accurately is the suspended coil galvanometer. The particle whose Brownian motion is to be measured is no longer small but the whole suspended coil of the galvanometer. The oscillations are minute but can be measured accurately by using a divided photocell and electronic amplifier. If K is the torsion constant of the galvanometer, the mean potential energy of the fluctuations is $\frac{1}{2}K\,\overline{\theta^2}$, and according to the equipartition theorem this should be equal to $kT/2$. These fluctuations were observed fifty years ago by Moll and Burger in their study of the very stable and sensitive Moll galvanometer. Later measurements by Jones and McCombie (1952) show that there is good agreement between the observed and the theoretical value of $\overline{\theta^2}$; from a long series of measurements they found that $\overline{\theta^2}_{\text{observed}} = \overline{\theta^2}_{\text{theoretical}} \times 1.00 \pm 0.012$. For the very interesting experimental details the reader is referred to the original publication. This agreement is important as verifying the theorem of the equipartition of energy. Some authors had argued that the deflection should correspond to $\frac{1}{2}K\,\overline{\theta^2} = kT$; $kT/2$ for the molecular bombardment and $kT/2 = \frac{1}{2}L\,\overline{I^2}$ due to the fluctuating current in the inductance L of the circuit. This is not so. A shorted galvanometer suspended in a vacuum would show a Brownian motion corresponding to $\frac{1}{2}L\,\overline{I^2} = kT/2$. On admitting gas there would be no increase in $\overline{\theta^2}$, the increase due to the molecular bombardment being exactly compensated by the extra viscous damping produced by the gas.

Purely electrical effects due to the random motion of the electrons in a conductor can also be observed. This random motion disturbs the uniform distribution of the negatively charged electrons necessary to ensure that each part of a conductor is electrically neutral. The fluctuations in electric density produce fluctuating internal electric fields and there is always a fluctuating voltage across the ends of a resistor. The mean square value $\overline{V^2}$ of this voltage can be calculated indirectly by comparing the behaviour of a long loss-free coaxial line as looked upon by a physicist thinking in terms of the transmission of electromagnetic waves with velocity c, and the radio engineer thinking in terms of the impedance $Z = R + \mathrm{j}X$ presented by the line.

An isolated open-ended line of length l will support indefinitely a set of standing voltage waves $V = V_0 \cos \omega t \cos \pi n x/l$ formed from two equal sets of waves travelling in opposite directions with the velocity of light. Assume in accordance with Fourier's theorem that the random motion of the electrons in the loss-free line can be described in terms of these standing waves. Each possible standing wave determined by the whole number n has a characteristic wave length λ and frequency ν, related by the universal relationship $\lambda\nu = c =$ constant velocity of electromagnetic waves in free space. The wave length of the nth standing wave is clearly $2l/n$. The corresponding frequency ν is given by $\nu = nc/2l$, so that the number of possible standing waves n below an allowed frequency ν is given by $n = 2l\nu/c$. Correspondingly the number of possible standing waves Δn between two allowed frequencies ν and $\nu + \Delta\nu$ is

$$n + \Delta n - n = \Delta n = \frac{2l}{c}(\Delta\nu).$$

For the very large values of n which we shall be considering this can be written without appreciable error in the differential form

$$\mathrm{d}n = \frac{2l\,\mathrm{d}\nu}{c}.$$

On the electromagnetic theory each standing wave represents two degrees of freedom (the voltage wave and the accompanying current wave) so that according to the equipartition theorem the total energy of the waves due to thermal agitation is

$$\frac{kT}{2} \times 2 \times \frac{2l\,\mathrm{d}\nu}{c} = \frac{2lkT\,\mathrm{d}\nu}{c}.$$

If the line were suddenly terminated (i.e. shorted) by a resistance R at either end, chosen so that the waves were no longer reflected at the ends, power equal to $KT\mathrm{d}\nu$ watts would flow into each resistance in the frequency band of width $\mathrm{d}\nu$. One can also treat the problem in terms of the transmission of power from a generator down a long line. The two terminals of a long line behave as if there were a resistance R (calculated from the dimensions of the line) across them. Nyquist (1928) imagined that the standing wave system in the open-ended line was maintained by the power injected in to the doubly terminated line (Fig. 8.1) from the voltage fluctuations in the two, terminating resistances. If each resistor is looked upon as a passive resistance R in series with an alternator with an r.m.s. voltage $\overline{V^2}\,\mathrm{d}\nu$ between the frequencies ν and $\nu + \mathrm{d}\nu$ each generator is delivering $(v^2\mathrm{d}\nu/4R)$ W into the line in this frequency band.

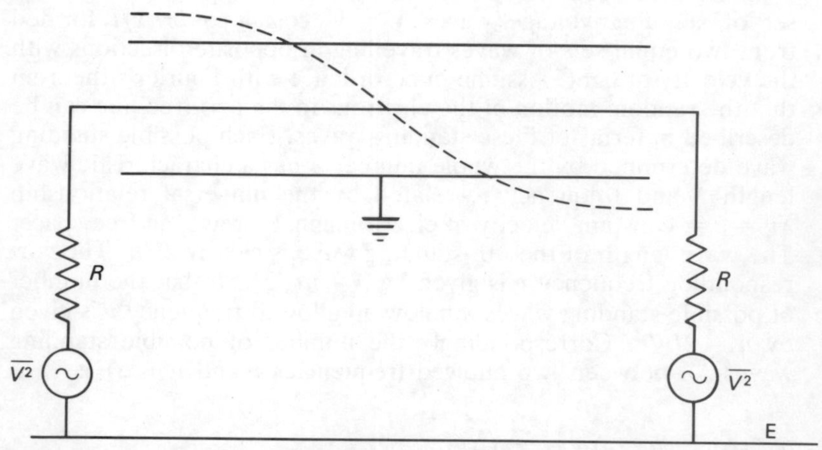

Fig. 8.1 Voltage fluctuations in a resistance.

Equating the power delivered to the power received from the standing wave system gives at once that $\overline{V^2}\mathrm{d}\nu = 4kTR\mathrm{d}\nu$, a value which agrees well with experiment. Since the phases of the generators are random, the values of $\overline{V^2}$ for different resistances add directly and no allowance has to be made for the phase difference between them.

Worked example

Use the circuit representation of a real resistor as an ideal resistance in series with a noise generator to find the mean square of the voltage fluctuations across a resistor of resistance R in parallel with a resistor of resistance r and comment on the result.

At a given time let the voltages be V and v as shown in Fig. 8.2. Then since there are no inductances or capacities in the circuit the voltage across the combined resistances can be calculated by a straightforward application of Ohm's law to be

$$\frac{Vr + vR}{R + r}$$

and the mean square voltage

$$\overline{\left(\frac{Vr + vR}{R + r}\right)^2} \quad \text{is} \quad \frac{1}{(R + r)^2}(\overline{V^2}r^2 + \overline{v^2}R^2 + 2\overline{Vv}Rr).$$

Since V and v are independent random voltages not related in phase, the last term averages to zero and the mean square voltage across the combined resistances in the frequency range $d\nu$ is

$$\frac{4kT(Rr^2 + R^2r)d\nu}{(R + r)^2} = \frac{4kT\,Rr}{R + r}\,d\nu = 4kT\,\mathcal{R}\,d\nu$$

in which $\mathcal{R}$ is the parallel resistance calculated from $1/\mathcal{R} = 1/R + 1/r$. This temperature dependent fluctuation is usually referred to as the 'Johnson noise', and is one of the phenomena which prevents weak signals being detected by electronic amplifiers with a sufficiently high voltage gain: the signals are said to be lost in the noise of the input circuit. The other source of unavoidable noise is 'shot noise' due to the finite electronic charge. An apparently steady current I is really a succession of I/e random

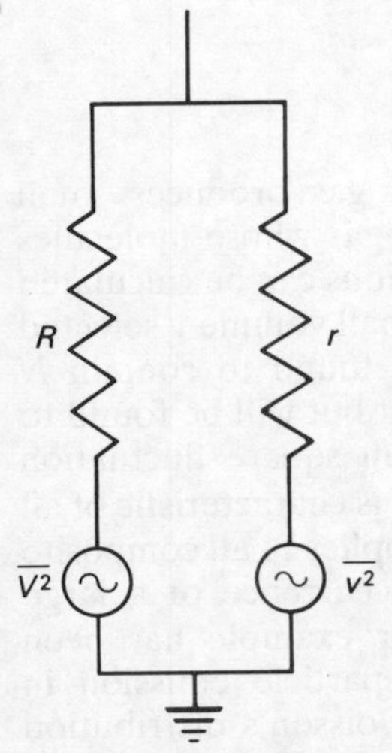

Fig. 8.2 Voltage fluctuations in resistances connected in parallel.

pulses per second and the fluctuations in I are analagous to the pattering noise heard when shot are poured onto a steel plate. Shot noise is temperature independent.

Worked example

Estimate the minimum signal power which can be detected when a signal from a signal generator of internal resistance R is connected to an amplifier of the same impedance and band width $\Delta\nu$. The only part of the signal which need be considered is that part S volts which lies in the frequency band $\Delta\nu$. The power P transferred to the amplifier in this frequency range is $S^2/4R$. The input circuit has two resistances R in parallel and so will generate a mean squared voltage

$$\overline{V^2} = \frac{4kTR}{2}\,\Delta\nu = \frac{4kT}{2}\cdot\frac{S^2}{4P}\,\Delta\nu = S^2kT\,\Delta\nu/2P.$$

Setting the limit of detection as occurring when $S^2 = \overline{V^2}$ gives the minimum detectable power as $\frac{1}{2}kT/2\;\Delta\nu = 2\times10^{-21}\;\Delta\nu$ W at room temperatures. The minimum band width for transmitting intelligible speech is 300 Hz and for the pulses of television transmission 10^7 Hz; the corresponding power levels are 6×10^{-21} and 2×10^{-11} W respectively.

The signal-to-noise ratio of radio amplifiers can be improved by cooling the input circuit down to about 3 K, the effective temperature of outer space.

8.3 Density fluctuations

The random motion of the molecules of a gas produces small random fluctuations in density. For a perfect gas whose molecules move independently of each other the fluctuations can be calculated from considerations of probability only. If a small volume v selected at random from a large volume V of gas is found to contain N molecules the value of N will not be a constant but will be found to fluctuate about a mean value $\bar{N}$ with a mean square fluctuation $\overline{(\Delta N)^2} = \overline{(N-\bar{N})^2} = \bar{N}$. This square root law is characteristic of all fluctuations of a purely statistical nature and applies to all composite events which are due to the simultaneous occurrence of a large number of independent events. A particular example has been worked out in detail for fluctuations in α particle emission in section 6.2; a more general proof based on Poisson's distribution function for biased coins is given by Born (1962). If the density of molecules in a gas is n per unit volume, the number of molecules in a small volume v should therefore be written as $nv \pm \sqrt{nv}$ or in terms of the densities

$$\frac{\overline{(\Delta\rho)^2}}{\rho^2} = \frac{1}{nv} = \gamma^2$$

in which ρ is the average density of the gas, $\overline{(\Delta\rho)^2}$ is the mean square deviation and γ^2 is a convenient abbreviation in printed formulae. Alternative expressions

$$\gamma^2 = \frac{-kT}{v^2\left(\dfrac{\partial p}{\partial v}\right)} = \frac{kT}{vK}$$

in which K is the bulk modulus are obtained by applying the perfect gas law $p = n\,kT$ and are useful in discussing the transition to real gases.

These fluctuations are too small for direct observation: a cubic mm of helium at s.t.p. contains 2.67×10^{16} molecules so that the number of molecules in a cubic millimetre isolated at random will fluctuate by about $\pm\sqrt{2.67 \times 10^{16}}$ and the density will be uncertain to about 6.12×10^{-7} per cent. If the volume v is about a cubic wave length of light ($\approx 10^{-19}$ m^3) the fluctuations ($\approx 6 \times 10^{-2}$ per cent) are large enough to scatter light passing through the gas as was first pointed out by Lord Rayleigh (1881). Rayleigh considered a beam of light of intensity I and wavelength λ passing through a medium of refractive index μ in which was embedded a small volume v ($\lambda^3 \gg v$) of refractive index $\mu + \Delta\mu$. By a purely optical calculation he was able to show that the intensity of light I_{rad} radiated per unit solid angle by the inhomogeneity in a direction at right angles to the initial beam was given by

$$I_{rad} = \frac{2\pi^2 I v^2}{\lambda^4}\left(\frac{\Delta\mu}{\mu}\right)^2.$$

(For a simple derivation see Wood (1923)). The small change in refractive index $\Delta\mu$ is related to the corresponding change in density by means of the experimentally and theoretically well-established Lorentz–Lorenz relationship

$$\frac{\mu^2 - 1}{\mu^2 + 2} = \rho \times \text{constant} = C\rho$$

equivalent to

$$\frac{\Delta\mu}{\mu} = \frac{(\mu^2 - 1)(\mu^2 + 2)}{6\mu^2}\frac{\Delta\rho}{\rho}$$

Worked example

The following data are available for oxygen.
Refractive index of the liquid = 1.221, density of the liquid = 1.12×10^3 kg/m^3, refractive index at s.t.p. $= 1 + 2.71 \times 10^{-4}$. How do these values agree with the Lorentz–Lorenz relationship?
From the data for the liquid one obtains immediately that

$$C\rho = \frac{\mu^2 - 1}{\mu^2 + 2} = 1.406 \times 10^{-1}$$

and $C = 1.255 \times 10^{-4}$. At s.t.p. the density of oxygen from the gas laws is 32/22.41 and $C\rho$ for the gas is

$$\frac{32 \times 1.255}{22.41} \times 10^{-4} = 1.792 \times 10^{-4}.$$

If μ for the gas $= 1 + \alpha$, $(\mu^2 - 1)/(\mu^2 + 2) = \frac{2}{3}\alpha +$ negligible terms in α^2 so that $\alpha = 2.69 \times 10^{-4}$ differing by less than 1 per cent from the experimental value.

These results enable the light scattered by a unit volume of a perfect gas to be estimated in terms of its measured refractive index. Divide the gas into $1/v$ small volumes v. When a beam of intensity I passes through the gas each volume v will act as a source of scattered radiation

$$I_{\text{rad}} = \frac{2\pi^2 I v^2}{\lambda^4} \cdot \frac{(\mu^2 - 1)^2(\mu^2 + 2)^2}{36\mu^4}\gamma^2$$

and the total intensity of the light radiated from unit volume of the gas at right angles to the incident beam is obtained by adding the intensities scattered by the $1/v$ volumes into which it has been divided, to give

$$I_{\text{scatt}} = \frac{\pi^2 I(\mu^2 - 1)^2(\mu^2 + 2)^2}{18n\lambda^4\mu^4}$$

It will be observed that this result does not depend upon the volume v, (which is only limited by the condition $\lambda^3 \gg v$) and the expression gives a value for the scattered light free from arbitrary assumptions. For air at low pressures such as occurs in the upper atmosphere the refractive index is so nearly unity that the expression simplifies to

$$I_{\text{scatt}} = \frac{\pi^2 I}{2n\lambda^4}(\mu^2 - 1)^2$$

which was used by Rayleigh to account for the blue colour of the sky. The agreement between theory and experiment is as good as can be hoped for in such an awkward calculation. The intensity of the scattered light is however rather small for a laboratory investigation as it requires several hundred km of air at s.t.p. to reduce the intensity of a beam of yellow light by a factor e.

For a real gas the optics of the problem is the same and the scattering depends on the fluctuations in γ^2 in volumes less than (wave length λ of the light)^3: these fluctuations can however no longer be calculated from purely statistical arguments. If there are forces between the molecules the probability of a molecule being in a small volume v is no longer independent of the presence of other molecules and one must allow for their interaction. Qualitatively one can see that molecular attraction is likely to increase the fluctuations (and so the scattering) owing to the tendency of the molecules to form clusters. A simple calculation of the fluctuations in density by elementary (or indeed any other) method is no longer feasible, but an indirect somewhat speculative method of attack initiated by Smoluchowski (1908) and Einstein (1910) is sufficient to allow a rational interpretation of the experimental data. Smoluchowski reversed Boltzmann's arguments. Instead of looking upon Ω as a theoretical quantity to be calculated from the molecular distribution function from which the entropy S could then be calculated from the relationship $S = k \log \Omega$, he looked upon S as an experimental quantity from which Ω and so the molecular distribution function could be inferred from the inverse relationship $\Omega = e^{S/k}$.

This can be interpreted to mean that if a gas (volume V) changes owing to molecular fluctuations from a state A with entropy S_A and a uniform density ρ to a state B with entropy S_B in which a small volume v ($v \ll V$) of the gas has a density $\rho + \Delta\rho$, then

$$\frac{\text{time spent in state B}}{\text{time spent in state A}} = \frac{\Omega_B}{\Omega_A} = e^{S_B - S_A/k} = e^{\Delta S/k}.$$

To calculate the change in entropy ΔS produced by a fluctuation Smoluchowski assumed that the fluctuation was a replica on a small scale of the change shown in Fig. 8.3 in which a small volume v of a gas expands to a volume $v + \Delta v$, at the same time compressing a gas whose volume is so large that its pressure p_0 is not appreciably changed by the small compression Δv. If this change is carried out isothermally and reversibly and without loss of energy, ΔS the difference in entropy between the initial and final states is (W the external work done)$/T$ and $e^{\Delta S/k} = e^{W/kT}$. In such an approximate calculation it is sufficient to work to the first order in Δv so that

$$W = \Delta v \text{ (average difference of pressure)}$$

$$= \frac{1}{2}\Delta v \left(\frac{\partial p}{\partial v}\right) \Delta v$$

$$= \frac{1}{2} v^2 \left(\frac{\Delta v}{v}\right)^2 \left(\frac{\partial p}{\partial v}\right)$$

$$= \frac{1}{2} v^2 \left(\frac{\partial p}{\partial v}\right) \left(\frac{\Delta \rho}{\rho}\right)^2$$

Adopting this approximate value for W gives at once that the probability of a fluctuation $\Delta\rho/\rho$ is proportional to

$$e^{v^2 \left(\frac{\partial p}{\partial v}\right) (\Delta\rho)^2/2kT\rho^2}.$$

Noting that $(\partial p/\partial v)$ is essentially negative one sees that the density fluctuations $\Delta\rho/\rho$ in a small volume v are distributed about the mean density ρ according to a Gaussian law with a mean square

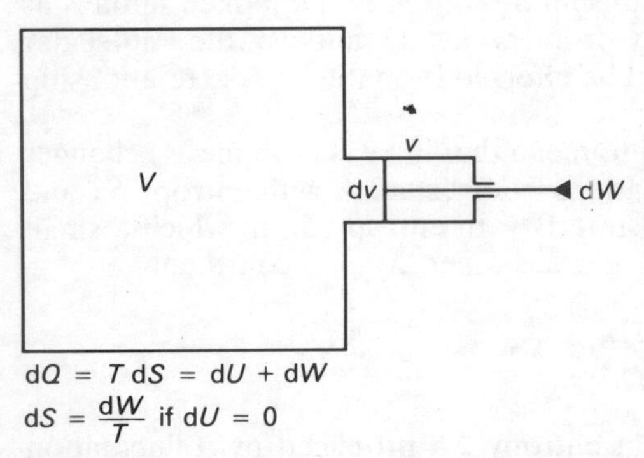

Fig. 8.3 V. Smoluchowski's model of a fluctuation.

deviation

$$\left(\frac{\overline{\Delta\rho}}{\rho}\right)^2 = \frac{-kT}{v^2\left(\frac{\partial p}{\partial v}\right)} = \frac{kT}{vK} = \gamma^2.$$

It is encouraging to observe that for a perfect gas γ^2 has the same value as was found by purely statistical arguments: for a real gas γ^2 retains its mathematical form but K is now the measured value of the bulk modulus which can be calculated from the equation of state. Accordingly for a real gas

$$I_{\text{scatt}} = \frac{\pi^2 I}{18\mu^4\lambda^4}(\mu^2 - 1)^2(\mu^2 + 2)^2\frac{kT}{K}$$

This formula, which contains no arbitrary constants, was originally used as an alternative method of measuring Boltzmann's constant k. The value obtained agreed with the known value within the experimental error (about 15 per cent) of a difficult measurement.

The most interesting feature of the expression for the intensity of the scattered light is that I_{scatt} becomes very large in the critical region since the bulk modulus K vanishes as $(\partial p/\partial v)_T \to 0$ at the critical point. This sudden increase of the scattered light was first observed by Avenarius (1874) and has been the subject of many investigations which have been summarised by the Bureau of Standards (1960).

Measurements on ethene (ethylene) by Keesom (1911) show good quantitative agreement in the temperature range 13.5–11.24 °C in a specimen whose critical temperature was 11.18 °C. Subsequent work has shown that below the critical temperature the dependence of the scattering on λ^{-4} changes to a λ^{-2} dependence as the critical point is approached. This indicates that the scattering centres are no longer small compared with λ^3 as postulated in Rayleigh's analysis but are clusters of molecules up to 1000 molecular diameters in size.

8.4 Maxwell's distribution function

The velocities, c_1, c_2, $c_3 \ldots c_N$ of the molecules of a gas may be represented graphically by taking a fixed point 0 and a series of points P_1, P_2, $P_3, \ldots$ (one for each molecule) chosen so that the vectors OP_1, OP_2, $OP_3 \dots OP_N$ represent the magnitude and direction of the velocity of the corresponding molecule. There will be so many points in this representation that they can without appreciable error be treated as forming a continuous fluid of density ρ in which ρ is the number of points per unit volume of velocity space. In a steady state this density is constant and can be described in mathematical terms by setting up a Cartesian coordinate system u, v, w, with origin at 0. The density ρ which has a definite value at every point in the representation will be a function of the coordinates so that one can write

$$dN = N\rho(u, v, w) \cdot d\tau$$

in which dN/N is the fraction of molecules whose representative points lie in the elementary volume $d\tau$ surrounding the point whose coordinates are u, v, w. A mathematical expression for ρ in terms of u, v and w is the required velocity distribution function of the gas.

One can see at once that $\rho(u, v, w)$ cannot be a completely arbitrary function since it is tied to the properties of the gas from which it is derived. For example, ρ must have spherical symmetry and have the same value for all points on the surface of a sphere centred on the origin 0. If this were not so the gas would not be isotropic, i.e. it would have different properties in different directions and not be the gas we are talking about. Similarly, ρ can never be negative and must come down to zero at large distances from the origin; otherwise some of the molecules would have infinite energies, which is not compatible with the finite energy of the gas.

By applying general arguments of this type Maxwell was able to obtain an expression for the distribution function of a gas long before an experimental approach was technically possible. This is done by giving mathematical form to the general statement that since ρ is a property of the gas its value at any point must be independent of the coordinate system in which the distribution of ρ is described. First choose a system of polar coordinates c, θ, ϕ with origin at 0, in which the radius vector c represents the numerical value of the velocity of the molecules in the gas. Then the fact that the gas is isotropic requires that the general form of $\rho = F(c, \theta, \phi)$ is restricted to the spherically symmetric form $\rho = \psi(c \text{ only})$.

Next, choose a cartesian system of coordinates u, v, w with origin 0, and consider the number of molecules dN' whose representative points lie between u and $u + du$. This number must be a function of u only and we can write

$$\frac{dN'}{N} = f(u)du.$$

Since the gas is isotropic there can be no physical difference between the cartesian axes so we can equally well write

$$\frac{dN'}{N} = f(v)dv \quad \text{and} \quad \frac{dN'}{N} = f(w)dw$$

for the number of molecules whose velocities lie between v and $v + dv$, and w and $w + dw$, respectively. We need to know the fraction of the molecules which lie in an element of volume $du\,dv\,dw$ surrounding the point u, v, w. To do this take the $dN' = Nf(u)du$ molecules whose velocities lie between u and $u + du$ and select from them the fraction $f(v)\,dv$ whose velocity lies between v and $v + dv$ to give $Nf(u)du\,f(v)dv$ molecules whose velocity lies between u and $u + du$ and v and $v + dv$. From these again select the fraction $f(w)dw$ whose velocity lies between w and $w + dw$ to give $dN = Nf(u)\,f(v)\,f(w)\,du\,dv\,dw = N\rho d\tau$ so that $\rho = f(u)\,f(v)\,f(w)$. The two expressions for ρ in cartesian and polar coordinates must give the same numerical value when they refer to the same point, so that

$$\psi(c) = f(u)\,f(v)\,f(w) \quad \text{if} \quad u^2 + v^2 + w^2 = c^2.$$

This functional equation (as is usually the case with functional equations obtained in arguments of this sort) is easily solved by inspection as is seen by writing

$$\begin{aligned} f(u) &= \mathrm{A}\,e^{-\mathrm{B}u^2}, f(v) = \mathrm{A}\,e^{-\mathrm{B}v^2}, f(w) = \mathrm{A}\,e^{-\mathrm{B}w^2}, \\ \psi(c) &= \mathrm{A}^3\,e^{-\mathrm{B}c^2} = \mathrm{A}^3\,e^{-\mathrm{B}(u^2+v^2+w^2)} \\ &= f(u)\,f(v)\,f(w), \end{aligned}$$

the alternative solution with a positive exponent having already been excluded on physical grounds. What is remarkable and extemely unusual is that this simple solution of the functional equation is unique. For, forming the partial derivative with respect to u

$$f'(u)\,f(v)\,f(w) = \psi'(c)\left(\frac{\partial c}{\partial u}\right)_{vw} = \psi'(c)\,\frac{u}{c}$$

since $(\partial c/\partial u)_{vw} = u/c$ as is readily seen by differentiating $c^2 = u^2 + v^2 + w^2$ with respect to u only.
Dividing by $\psi(c) = f(u)\,f(v)\,f(w)$ one gets

$$\frac{f'(u)}{u\,f(u)} = \frac{\psi'(c)}{c\psi(c)}$$

The l.h.s. of this equation is a function of u only while the r.h.s. is a function of the independent variables u, v, w out of which the variable c is constructed and so this relationship can only be true if both sides are equal to a constant = 2B. That $(f'(u)/uf(u)) = -2\mathrm{B}$ is solved by $f(u) = \mathrm{A}\,\mathrm{e}^{-Bu^2}$ is easily verified. Since, however, this differential equation is essentially the definition of the logarithmic function there are no essentially different real solutions, as is shown in standard mathematical texts. (Hardy 1955). The constants A and B are determined through the relations

$$\int_0^N \mathrm{d}N = N = \text{total number of molecules in the gas}$$

$$\int_0^N \frac{1}{2}mc^2\,\mathrm{d}N = \frac{1}{2}N\,mc^2 = \frac{3NkT}{2}$$

$$= \text{total kinetic energy of the gas.}$$

These integrals are all of the type

$$\int_0^\infty x^n\,\mathrm{e}^{-\alpha^2x^2}\,\mathrm{d}x = I_\mathrm{n}$$

Integration by parts shows that $I_{\mathrm{n}+2} = (\mathrm{n}+1)/2\alpha^2\,I_\mathrm{n}$ and the integrals fall into sets according whether n is odd or even.
For odd n there is no difficulty since the lowest member

$$\int_0^\infty x\mathrm{e}^{-\alpha^2x^2}\,\mathrm{d}x = \frac{1}{2\alpha^2} = I_1$$

For even n the lowest member (n = 0) is Euler's integral

$$\int_0^\infty \mathrm{e}^{-\alpha^2x^2}\,\mathrm{d}x = \frac{\sqrt{\pi}}{2\alpha} = I_0$$

Worked example

Evaluate Euler's integral.
Euler's integral can be evaluated by finding the volume of the solid obtained by rotating the bell shaped curve $z = \mathrm{e}^{-\alpha^2x^2}$ about the z axis. This is the same type of problem as finding the volume of a cone, and defining the cone as the solid obtained by rotating a line making an angle θ with the z axis about that axis.

In cartesian coordinates the element of volume is dx dy dz and the volume V is given by

$$V = \int_{-\infty}^{+\infty}\int_{-\infty}^{+\infty}\int_{0}^{z} \mathrm{d}x\, \mathrm{d}y\, \mathrm{d}z$$

$$= \int_{-\infty}^{+\infty}\int_{-\infty}^{+\infty} \mathrm{e}^{-\alpha^2(x^2+y^2)}\, \mathrm{d}x\, \mathrm{d}y$$

$$= \int_{-\infty}^{+\infty} \mathrm{e}^{-\alpha^2x^2}\, \mathrm{d}x \times \int_{-\infty}^{+\infty} \mathrm{e}^{-\alpha^2y^2}\, \mathrm{d}y = 4I_0^2.$$

The volume can be found explicitly in cylindrical coordinates by dividing it up into elementary tubes of height z, radius r, and thickness dr, to give

$$V = \int_0^\infty 2\pi r\, \mathrm{e}^{-\alpha^2r^2}\, \mathrm{d}r = \pi \int_0^\infty \mathrm{e}^{-\alpha^2 s}\, \mathrm{d}s = \frac{\pi}{\alpha^2}$$

Equating the two values of V gives $I_0 = \sqrt{\pi}/2\alpha$ as stated

The values of I_n calculated in this way are given in Table 8.1 in terms of α and for $\alpha^2 = m/2kT$ its value in Maxwell's distribution function. Reference to these Tables shortens the evaluation of the integrals which occur in kinetic theory calculations.

Worked example

Calculate the values of A and B in Maxwell's distribution function. The density ρ is given by $\psi(c) = \mathrm{A}^3\mathrm{e}^{-\mathrm{B}c^2}$, so the number of molecules in the spherical shell between c and c + dc is $4\pi c^2\psi(c)\mathrm{d}c$ and the total number of molecules N is given by

$$\int_0^\infty 4\pi \mathrm{A}^3 c^2\, \mathrm{e}^{-\mathrm{B}c^2}\, \mathrm{d}c = 4\mathrm{A}^3 \cdot \frac{\sqrt{\pi}}{4\mathrm{B}^{3/2}}$$

$$= \pi^{3/2}\left(\frac{\mathrm{A}}{\sqrt{\mathrm{B}}}\right)^3 = N.$$

Similarly, since each molecule has energy $\frac{1}{2}mc^2$ the total energy is

$$\int_0^\infty 4\pi\mathrm{A}^3 \cdot \frac{1}{2}mc^2 \cdot c^2\, \mathrm{e}^{-\mathrm{B}c^2}\mathrm{d}c = 2\pi\mathrm{A}^3\left(\frac{3\cdot\sqrt{\pi}}{8\mathrm{B}^{5/2}}\right) m$$

$$= \frac{3mN}{4\mathrm{B}} = \frac{3NkT}{2}.$$

Table 8.1 Values of $I_n = \int_0^\infty x^n e^{-\alpha^2 x^2} dx$

n	0	1	2	3	4	5	6	7	8
I_n	$\frac{\sqrt{\pi}}{2\alpha}$	$\frac{1}{2\alpha^2}$	$\frac{1}{4}\frac{\sqrt{\pi}}{\alpha^3}$	$\frac{1}{2\alpha^4}$	$\frac{3}{8}\frac{\sqrt{\pi}}{\alpha^5}$	$\frac{1}{\alpha^6}$	$\frac{15}{16}\frac{\sqrt{\pi}}{\alpha^7}$	$\frac{3}{\alpha^8}$	$\frac{105}{32}\frac{\sqrt{\pi}}{\alpha^9}$
I_n for $\alpha^2 = \frac{m}{2kT}$	$\sqrt{\frac{\pi kT}{2m}}$	$\frac{kT}{m}$	$\sqrt{\frac{\pi k^3T^3}{2m^3}}$	$2\left(\frac{kT}{m}\right)^2$	$3\sqrt{\frac{\pi k^5T^5}{2m^5}}$	$8\left(\frac{kT}{m}\right)^3$	$15\sqrt{\frac{\pi k^7T^7}{2m^7}}$	$48\left(\frac{kT}{m}\right)^4$	$105\sqrt{\frac{\pi k^9T^9}{2m^9}}$

So

$$B = \frac{m}{2kT}, \quad A^3 = N\left(\frac{B}{\pi}\right)^{3/2} = N\left(\frac{m}{2\pi kT}\right)^{3/2}$$

$$\text{and} \quad dN = 4\pi N\left(\frac{m}{2\pi kT}\right)^{3/2} c^2 \, e^{-mc^2/2kT} dc$$

Maxwell's discovery of the distribution function was very important at the time it was made. In the first place its rather splendid mathematical form gave a welcome air of respectability to what was at the time a very speculative subject, and greatly encouraged experimental and theoretical work on the molecular theory of gases. Secondly, it represented a completely new approach to physical problems. It was the first problem successfully solved by applying general mathematical reasoning to general views about the nature of matter. It was in a sense a return to the methods of Descartes. The next successful application of the method was made by Maxwell himself in his electromagnetic theory of light. Einstein's theories of relativity and gravitation and Schrödinger's development of wave mechanics are in the same tradition.

Many students on first reading Maxwell's arguments feel that they have been tricked. After all 'Nowt for nowt' is as true in the realm of mathematics as in Yorkshire. They are, of course, quite right. The most unsatisfactory part of the derivation of Maxwell's distribution function is that it has been deduced without any reference to the molecular collisions through which the molecular distribi tion is maintained. In a perfect gas in which the molecules do not interact with each other their velocities will never change and any distribution will be maintained indefinitely. This is born out by the derivation of Boyle's law given in section (2.1) and section (2.2) in which pv depends only on $\overline{c^2}$ and not at all on the distribution function. Scrutiny of the derivation of the distribution function shows that it is based upon a concealed assumption, of which Maxwell was, of course, well aware. The treatment assumes that the molecules selected to have velocities between u and $u + du$ will have the same distribution function $f(u, v, w)$ as molecules selected at random. This is a special and not very plausible assumption which needs to be justified. The reasonable suggestion that molecules with a high value of u are fast molecules which have just come from a particularly energetic collision and will therefore have a higher than average value of v and w certainly needs investigating before setting it aside. Without this assumption the demonstration collapses and no further progress can be made. A great deal of statistical mechanics and the mathematical development of kinetic theory have sprung from a search for a more satisfactory derivation. An account of this work is given in texts of statistical mechanics. Since the object is to avoid being tied to a particular molecular model the treatment is of necessity rather abstract and mathematical.

General confirmation that the Maxwellian function is obeyed in a gas is obtained by measuring the width of spectral lines. It is the only method which measures the actual velocity spectrum in the body of the gas. Even if the spectral lines emitted by a gas were strictly monochromatic they would be broadened by the motion of the atoms through the Doppler effect. This broadening, of the order $v/c = \Delta\nu/\nu$ (v = molecular velocity in the line of sight, c = velocity of light) is small but quite measurable. For accurate work it is best to work in absorption and measure the absorption coefficient k_ν of white light in the neighbourhood of an absorption line of frequency ν_0. Careful measurement (using interference spectroscopy) shows that the mean half width of the line $\Delta\nu$ agrees well with the theoretical value

$$\Delta\nu_D = \frac{2\nu_0}{c}\sqrt{\frac{2\log 2 \cdot kT}{m}}$$

calculated for a Maxwell distribution but that there is still an appreciable absorption in the skirts of the line not accounted for. This is due to the natural width of the line $\Delta\nu_N$ and is independent of the Doppler effect. Although small in magnitude,

$$\frac{\Delta\nu_D}{\Delta\nu_N} \approx 10^2 - 10^3,$$

natural absorption falls off from the central value, frequency ν_0 as

$$\frac{(\Delta\nu_N)^2}{\Delta\nu_N^2 + 4(\nu - \nu_0)^2}$$

a much less powerful law than Maxwell's, which spreads the absorption over a much larger range of frequency. It is the natural absorption which enables the Frauenhofer lines to be observed with a pocket spectroscope. If they were broadened only by Doppler broadening they would not be detectable in this simple way. In conclusion, it is not possible to say more than that the observed width of spectral lines is consistent with a Maxwellian distribution but that owing to the complexity of the problem the measurements can provide no more than supporting evidence.

Much experimental work has been devoted to verifying Maxwell's distribution function directly. The easiest experiments are made with electron gases derived from thermionic emission. They are also the most difficult to interpret owing to the effects of space charge. The current between a filament and a negatively charged plate depends upon the velocity spectrum of the emitted electrons. The theoretical formulae due to Schottky based upon a Maxwellian distribution are reasonably well confirmed by experiment; the temperature which gives the best fit being within 5 per cent of the measured temperature of the filament. Even a rough verification is important since the electrons in the metal form a degenerate gas with a distribution function very far from Maxwellian.

The velocity of light molecules ($\sim 10^3$ m/s) is rather high for direct measurement but for heavy atoms and molecules the velocity distribution in a collimated beam can be measured directly. The apparatus, which must of course be mounted in a highly evacuated space, consists of three parts; a furnace or oven, a velocity selector and a detector. Molecules from a narrow slit in the furnace wall are collimated into a molecular beam whose intensity is measured by the detector after the beam has passed through the velocity selector. The detector is usually a diode with a hot tungsten filament which ionises the incident molecules because it has a high work function. The resulting positive ion current reduces the space charge and the resultant change in anode current is then detected by a sensitive DC amplifier. The best form of velocity selector consists of a duralumin cylinder rotating with angular velocity ω about an axis parallel to the direction of the molecular beam. A series of helical slots of pitch ϕ are cut into its periphery. In general a molecule entering one of the slots will strike the wall and be lost to the beam. If however its velocity c is ω/ϕ, its position relative to the walls will not change and it will transverse the velocity selector and reach the detector. A molecule whose velocity is $c \pm \mathrm{d}c$ can get through the selector by entering near one wall of the radial slot and leaving near the other wall having changed its angular position relative to the cylinder by $\Delta\theta$. As reference to Fig. 8.4 will show, the molecule entering at A with velocity c will reach B′ in a time $t = l/c$ in which time the wall of the slot at B will have reached B′ if $\phi l = \omega t$; so the condition for transmission is $\phi l = \omega l/c$ and $c = \omega/\phi$.

The molecule entering at A′ with a velocity $c + \mathrm{d}c$ will reach B′ in a time $l/(c + \mathrm{d}c)$, in which time it will appear to drift across the slit from the trailing to the leading edge. That is to say that the condition for transmission is obtained by equating the time $\Delta\theta/\omega$ for which the channel is open to the difference in time

$$\mathrm{d}t = \frac{\mathrm{d}}{\mathrm{d}c}\left(\frac{l}{c}\right) = -\frac{l}{c^2}\,\mathrm{d}c$$

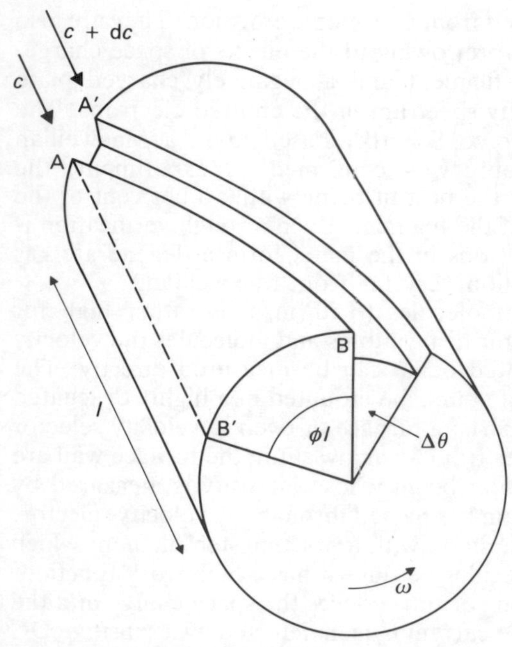

Fig. 8.4 Mechanical velocity selector.

which molecules of different velocities take to transverse the channel and $l/c^2\ \mathrm{d}c = \Delta\theta/\omega$; but $\omega = c\phi$ so that $\mathrm{d}c = \Delta\theta/l\phi \cdot c = \mathrm{K}c$ in which K is a constant of the rotor. The range of velocity $\pm\mathrm{d}c$ which is transmitted by the velocity selector is not constant but depends upon the velocity so that $\mathrm{d}c = \pm\mathrm{K}c$. This decrease in the resolving power of the velocity selector with increasing velocity must be allowed for in interpreting the experimental results. If the velocity distribution in the furnace is $\mathrm{d}N/N = f(c)\mathrm{d}c$, the velocity distribution of molecules leaving the slit depends upon $cf(c)\mathrm{d}c$ since the rate at which molecules reach the slit is proportional to their velocity (stationary molecules would not leave the furnace). The response I of the detector is therefore proportional to $cf(c)\mathrm{d}c = \mathrm{K}'c^2f(c)$ in which the constant K′ includes the constant in the distribution function, the solid angle the rotor subtends at the slit and the sensitivity of the detector.

In the experiments of Miller and Kusch (1955) the rotor was a Dural cylinder with 708 helical grooves cut into its circumference. The source was either the solid face of a crystal maintained at a fixed temperature in the oven or a slit (0.43 mm wide between 0.03–0.04 mm steel shims) through which the vapour in the oven escaped from the furnace by effusion. For a Maxwellian distribution $I = Ac^4 \, e^{-mc^2/2kT}$. This can be plotted as a universal curve

$$I' = \frac{I}{I_{max}} = V^4 \, e^{-2V^2+2}$$

The reduced velocity $V = c/c_{max}$ in which c_{max} is the velocity at which the response of the detector goes through its maximum value I_{max}. In this way Miller and Kusch were able to plot the results of their experiments at various temperatures on a single curve. The excellent agreement they obtained between experiment and theory is shown in Fig. 8.5. In particular it should be noted that there is good agreement at the low velocity end of the spectrum; the marked deficit of slow molecules found by earlier observers using inferior techniques is not confirmed. This deficit was usually attributed to scattering in the slit. Miller and Kusch were able to confirm this explanation by removing the very thin steel shims used to define the slit so that the effusion took place from a copper slit 0.08–0.12 mm thick. They observed a shift in the distribution function corresponding to an increase in the oven temperature of 3 per cent. The good agreement with theory is also shown in Table 8.2. The overall agreement can be roughly expressed by saying that while the oven temperature was controlled to $\pm\frac{1}{2}$ per cent the maximum discrepancy between theory and experiment correspond to an uncertainity in the temperature of $\pm 1\frac{1}{2}$ per cent. It will be noted that the measurements have only been made over a limited velocity range. For low velocities there is no experimental evidence (except for neutrons) and at higher velocities one has to rely on the chemical evidence, i.e. the measurement of chemical activation energies section 3.11b. This indicates that the number of molecules whose energy exceeds

$$E = f\left(\frac{E}{kT}\right) e^{-E/kT};$$

for large values of E the exponential term is so dominant that the effect of the $f(E/kT)$ factor is negligible. The absence of direct confirmation at high velocities is regrettable since it is the part of the energy spectrum which determines the calculated rate of liberation of nuclear energy in the sun. (The failure to find the expected neutrino emission is one of the outstanding discrepancies in the general picture of the universe). Indeed for both large and small energies (compared with kT) Maxwell's distribution function like the Gaussian distribution law is in danger of falling into that class of

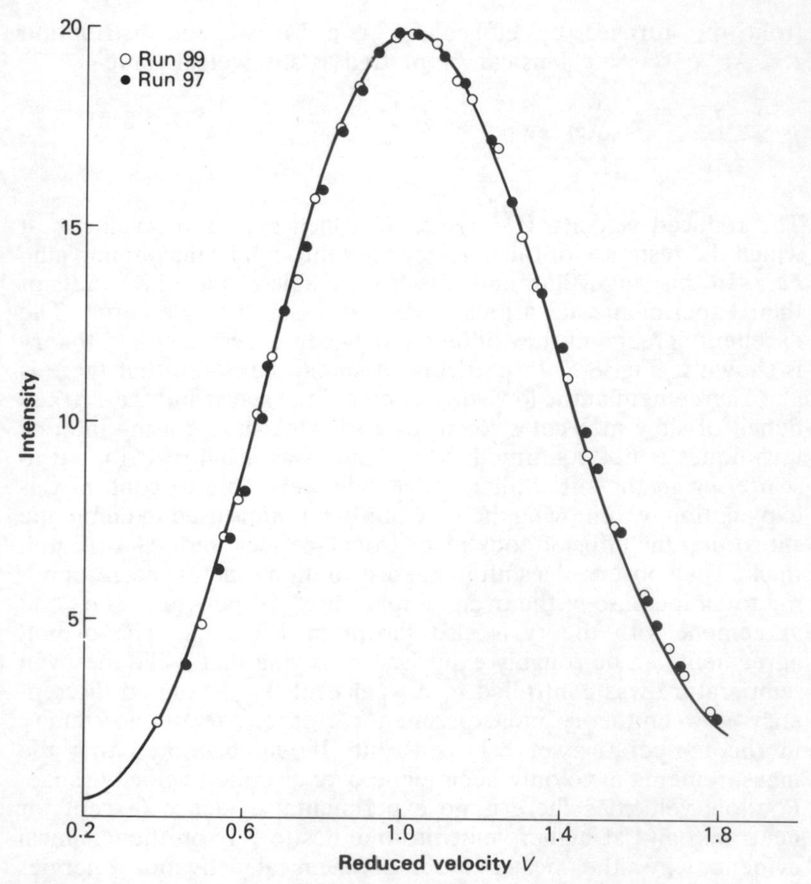

Fig. 8.5 Velocity distribution in thallium vapour.
The full line represents the reduced velocity calculated for a Maxwell distribution.
○ $T = 870 \pm 4K$
● $T = 944 \pm 5K$
Taken from Miller and Kusch 1955.

relationships of which Poincaré remarked 'Everybody firmly believes in it because mathematicians imagine that it is a fact of observation and observers that it is a theorem of mathematics'.

Worked example

Find an approximate expression for the fraction of molecules of a perfect monatomic gas whose energy exceeds E if $E \gg kT$.
If the energy exceeds E the velocity c exceeds $\sqrt{2E/m} = x$ and the required

Table 8.2

Beam	Temp. K	Oven pressure Pa	I_{max} m/s from oven temperature	Direct measurement
K	466 ± 2	6.0 10^{-1}	628 ± 2	630 ± 3
K	544 ± 3	1.6 10^{1}	678 ± 3	679 ± 3
K	489 ± 2	1.87	644 ± 2	682 ± 3
Tl	870 ± 4	4.27 10^{-1}	376 ± 1	376 ± 2
Tl	944 ± 5	2.8	392 ± 1	395 ± 2

(From Miller and Kusch, 1955.)

fraction is from Maxwell's distribution function

$$\frac{1}{N}\int_x^\infty \mathrm{d}N = \frac{4\alpha^3}{\sqrt{\pi}}\int_x^\infty c^2\,\mathrm{e}^{-\alpha^2c^2}\,\mathrm{d}c$$

in which $\alpha^2 = m/2kT$ and $\alpha^2x^2 = E/kT$ is a suitable dimensionless parameter. Integration by parts shows at once that

$$\int_x^\infty c^2\,\mathrm{e}^{-\alpha^2c^2}\,\mathrm{d}c = \frac{1}{2\alpha^2}\int_x^\infty \mathrm{e}^{-\alpha^2c^2}\,\mathrm{d}c + \frac{x}{2\alpha^2}\,\mathrm{e}^{-\alpha^2x^2}.$$

The integral on the r.h.s. whose approximate value is

$$\frac{\mathrm{e}^{-\alpha^2x^2}}{2\alpha^2x}\left(1 - \frac{1}{2\alpha^2x^2}\right)$$

is tabulated as $1 - erf\,(x)$ and has a negligible value for the large values of x being considered. Accordingly the required fraction

$$\frac{\mathrm{d}N}{N} = \frac{4\alpha^3}{\sqrt{\pi}}\cdot\frac{x}{2\alpha^2}\cdot\mathrm{e}^{-\alpha^2x^2}$$

$$= \frac{2\alpha x}{\sqrt{\pi}}\,\mathrm{e}^{-\alpha^2x^2}$$

$$= \frac{2}{\sqrt{\pi}}\sqrt{\frac{E}{kT}}\cdot\mathrm{e}^{-E/kT}.$$

8.5 Absolute entropy and Nernst's theorem

Nernst (Mendelssohn 1973) seems to have been the first to realise that classical thermodynamics only enabled changes in quantities like vapour pressure and chemical equilibrium constants to be calculated in terms of measurable thermal quantities and observed deviations from the perfect gas laws. It was quite incapable of giving any information at all about the absolute values of these quantities under any realisable conditions whatsoever. In short, the situation was always that presented by the Clausius–Clapeyron vapour pressure equation: the slope of the vapour pressure curve dp/dT could always be calculated but it could never be converted into an actual vapour pressure curve

$$p = p_0 \, e^{-\lambda/RT}$$

because there was no rational thermodynamic way of determining the integration constant p_0. Nevertheless, it is the actual pressure p which the engineer and technologist must know before they can start designing. Similarly physical chemists wished to be able to calculate equilibrium constants from purely calorimetric data which could be obtained without actually carrying out the reaction. To Nernst this was the goal of physical chemistry and he saw this way to reaching it as early as 1894, and considered he had reached it in 1906 when he correctly calculated the equilibrium constants of the gaseous reaction $H_2O + CO \rightleftharpoons H_2 + CO_2$ and the e.m.f. of the gaseous H_2/Cl_2 cell. He based his calculations on a new assumption to which he was led by considering Tamman's measurements of the aqueous vapour pressure of strong (up to 12N) solutions of sulphuric acid in water. These vapour pressures were so low that it was legitimate to treat the vapour as a perfect gas and so calculate the change in the Gibb's free energy G when sulphuric acid is diluted. He found that $G_1 - G_2 = \Delta H$. This was equivalent to saying that the term of the Gibbs–Helmholz equation

$$T\left(\frac{\partial(G_1 - G_2)}{T}\right)_p$$

was zero in this change (cf section 7.14). A similar result had been obtained by Kelvin and Helmholz in their early calculations of the e.m.f. of electric cells. They got good agreement with experiment because they happened to choose cells with very small temperature coefficients so that the Gibbs–Helmholz equation $E = \Delta H + T(\partial E/\partial T)_p$ reduced to $E = \Delta H$. Convinced that the small value of the temperature coefficient was no coincidence Nernst turned his attention to the behaviour (at that time quite unknown) of U and G near the absolute zero as shown in Fig. 8.6. He argued that

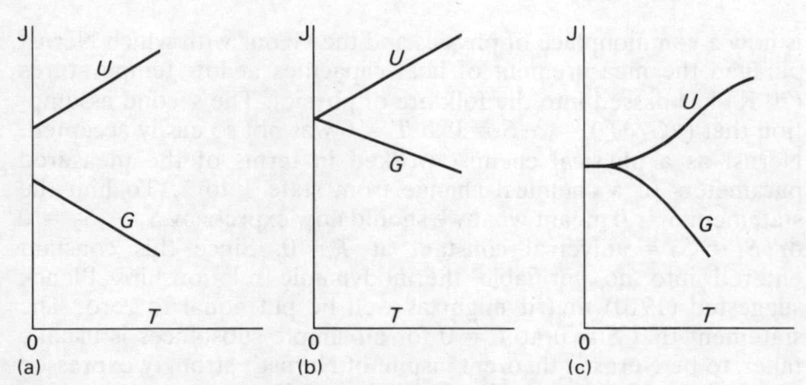

Fig. 8.6 Variation of U and G with temperature.
(a) not possible because $U \neq G$ at $T = 0$
(b) possible but unlikely to lead to the empirical result $U = G$ for finite T.
(c) $U = G$ for finite T.

Fig. 8.6c in which

$$\left.\begin{array}{l}\left(\dfrac{\partial G}{\partial p}\right)_T = -S \\ \left(\dfrac{\partial U}{\partial T}\right)_p\end{array}\right\} = 0 \text{ for } T = 0 \text{ was the correct universal form.}$$

Strong support for this assumption can (with hindsight) be obtained by integrating the Gibbs–Helmholz equation $G = U + T(\partial G/\partial T)_p$ to give

$$G = -T \int_0^T \frac{U}{T^2}\,\mathrm{d}T.$$

If G is to play the key role in determining chemical equilibria envisaged by Nernst the integral must be capable of being evaluated in finite terms. If one expands U in a power series $U = U_0 + \mathrm{a}T + \mathrm{b}T^2 + \mathrm{c}T^3$ one finds that G only has a finite value (no log T term) if a = 0 and

$$\left(\frac{\partial U}{\partial T}\right)_p = \left(\frac{\partial G}{\partial T}\right)_p = 0 \quad \text{for} \quad T = 0$$

in agreement with Nernst's views. That

$$\left(\frac{\partial U}{\partial T}\right)_p \approx C_p \to 0 \quad \text{as} \quad T \to 0$$

is now a commonplace of physics and the vigour with which Nernst pursued the measurement of heat capacities at low temperatures (20 K) has passed into the folklore of physics. The second assumption that $(\partial G/\partial T)_p = -S = 0$ at $T = 0$ was not so easily accepted. Nernst as a physical chemist worked in terms of the measured parameters of a chemical change from state 1 to 2. To him the statement $S = 0$ meant what we should now express by $S_1 - S_2 = 0$ or $S_1 = S_2 =$ universal constant at $T = 0$. Since this constant entered into no verifiable thermodynamic relationships Planck suggested (1910) that it might as well be put equal to zero. The statement that $S = 0$ at $T = 0$ for all simple substances is usually taken to be Nernst's theorem inspite of Nernst's strongly expressed protests that this is not what he had said. From the present theoretical point of view Nernst's theorem for pure simple substances is self-evident. By definition pure crystalline substances at the absolute zero have all their molecules in the lowest energy state so that there is no distribution function, $\Omega = 1$ and $S = k \log \Omega = 0$. Thus not only do all simple crystalline substances have the same constant entropy at absolute zero but Planck's more stringent formulation (which infuriated Nernst) that the constant is zero is also justified. The experimental verification was made in two stages. Early experiments were directed to showing that different crystalline forms of the same substance which demonstrably had different entropies at room temperatures had the same entropy at absolute zero (section 8.6). Later work (section 8.7) showed that the entropy at absolute zero was actually zero by comparing the measured and calculated value of the entropy of a perfect gas. The two results only agree if the entropy of the solid is chosen to be zero at $T = 0$. Since the entropy of a perfect gas depends upon m, the mass of the molecule, this is good evidence that all simple solids have zero entropy at absolute zero. The application of Nernst's theorem to the calculation of chemical equilibrium constants in simplified examples is touched upon in section 8.11. The application to real examples is not included since the very considerable difficulties have nothing to do with Nernst's theorem but lie in the calculation of the partition functions of real polyatomic gases at temperatures high enough for the equilibrium constant to be measured experimentally.

8.6 The entropy of crystalline solids

Nernst's theorem was first confirmed by measuring the difference in entropy between simple allotropes. The experimental demonstration depends upon the very slow rate at which different crystalline modifications of the same substance change into each other. Typical examples are sulphur, tin and diamond. As is well known, sulphur can exist in two crystalline forms; rhombic sulphur, which is stable below 95.6 °C, and above that temperature the stable form belongs to the monoclinic system and exists up to the melting point. In the two forms the rather complicated S_8 molecules are arranged on different lattices and so have different energy levels and entropies. On cooling the high temperature form it does not immediately change into the rhombic form at 95.6 °C and supercooled thermodynamically unstable monoclinic crystals can be kept indefinitely at low temperatures. A more spectacular example in which the unstable form changes into the stable form even more slowly is tin. Below 18 °C α or gray tin is the stable form with a crystalline structure analagous to that of diamond. Above the transition temperature, β or white tin is the stable form with an octahedral arrangement of the atoms. Although tin has been known since the bronze age the existence of the α form was not discovered until 1851 when the collapse of a tin organ pipe in a cold winter drew the attention of Erdmann to the phenomenon. Setting aside the tiresome objection that no thermodynamic arguments can be correctly conducted with data from substances not in equilibrium, one can carry out the following cycle of measurements. Starting at absolute zero the stable (α) specimen is heated up to the transition point thereby increasing its entropy by $\int(C_{p\alpha}\mathrm{d}T)/T$. It is then allowed to change into the β modification stable above that temperature absorbing heat energy Q and gaining entropy Q/T_{trans} since the change is taking place through a series of equilibrium states. The metastable form is now cooled down to absolute zero losing entropy $= \int(C_{p\beta}\mathrm{d}T)/T$. If Nernst is right the resulting entropy change should be zero and Nernst's general statement has been turned into the definite relationship

$$\int_0^{T_{\mathrm{trans}}} \frac{C_{p\alpha} - C_{p\beta}}{T}\,\mathrm{d}T = \frac{Q_{\alpha\beta}}{T_{\mathrm{trans}}}$$

The measurements of Eastman and McGavock (1937) on the value of the heat capacities of the two forms of sulphur starting at 15 K in the temperature ranges for which ΔC_p is significant are shown in the abbreviated Table 8.3. From this data the value of $\int \Delta C_p/T\,\mathrm{d}T$ is 0.900 ± 0.21, this to be compared with $(397.7 \pm 42)/368.8 = 1.08 \pm 0.11$ obtained directly from the heat of transition 397.7 ± 42 J. The agreement is within the experimental errors and is often quoted as experimental evidence in favour of Nernst's law. Similar experiments with tin have been made by Lange. He found a predicted value of $Q/T_{\mathrm{trans}} = 7.322$ to be compared with the

Table 8.3 Heat capacity of sulphur

Temp. K	**60**	**80**	**100**	**160**	**200**	**300**	**360**	**368.5**
C_p(J/K) of monoclinic sulphur	8.755	11.062	13.87	17.73	20.09	23.71	25.41	25.65
ΔC_p	0.029	0.084	0.167	0.469	0.666	1.047	1.197	1.210

(From Eastman and McGavock, 1937. Published 1937 American Chemical Society.)

directly measured value of 7.494. The estimated error of the measurements was ±6 per cent so that the agreement is good. The equilibrium between diamond and graphite has intrigued the scientific and indeed the fashionable world ever since the Florentine Academy (1694) publicly confirmed Newton's remarkable prediction (1675) that diamond would be found to be an organic material. The change from diamond to graphite is not (as it is for sulphur and tin) a small distortion of the lattice but a complete rearrangement not only of the atoms but also of the orbits of the valency electrons: correspondingly it is accompanied by a large increase in entropy. Diamond and graphite in equilibrium will have the same Gibb's functions per unit mass and

$$\Delta G = U_D - U_G - T(S_D - S_G) + P(V_D - V_G)$$
$$= \Delta H - T\Delta S = 0$$

The value of ΔH at 25 °C is equal to the difference between the heats of combustion of diamond and graphite. This is a difficult quantity to measure as it is the difference between two large quantities. According to Prosen, Jessup, and Rossini (1944) $\Delta H = 1897 \pm 84$ J/mole, i.e. about 0.4 per cent of the heat of combustion. All the other values of ΔH must be calculated from the heat capacity data using the thermodynamic relationship $(\partial \Delta H/\partial T)_p = \Delta C_p$ analogous to that obtained for the vapour pressure equation. A selection of the values obtained in this way by Berman and Simon using the best available data is given in Table 8.4. It is obvious from this data that ΔG increases steadily with temperature because the entropy of graphite is so much larger than that of diamond, and that the two allotropic forms will never be in equilibrium at atmospheric pressure. However, the large increase in volume accompanying the change means that the high density form (diamond) is favoured at high pressures, when the

Table 8.4

K	ΔH	$-\Delta S$	ΔG	ΔV	Equilibrium pressure
298	453	0.804	694	1.7	16 100
400	403	0.954	785	1.6	18 200
700	330	1.09	1090	1.6	25 500
1100	310	1.11	1530	1.6	36 000
2000	310	1.11	2530	1.4	62 000
3000	310	1.11	3460	1.3	93 000

Thermal data in cal/mol, pressure in atmospheres, volume in cm³/mole. (Taken from Berman and Simon, 1965.)

difference between ΔH and ΔU, equal to $P(V_D - V_G)$, negligible at atmospheric pressure, becomes appreciable. A very crude calculation assuming that the thermodynamic data are independent of the pressure suggests that graphite and diamond will be in equilibrium at room temperature at a pressure given by

$$\Delta G = P(V_G - V_D)$$

so that

$$P = \frac{\Delta G}{V_G - V_D} = \frac{2906}{1.883 \times 10^{-6}} = 1.54 \times 10^9 \text{ Pa}$$

$$= 15\,400 \text{ atm.}$$

The corrected value obtained by Simon and Berman is shown in the last column of the Table 8.4. Data above 1200 K were obtained by linear extrapolation. The calculated values agree well with the experimental data of Bundy *et al.* (Fig. 8.7); as is well known, as the result of their pioneer work at Schenectedy, diamond synthesis is now carried out on a large scale by several corporations. Of course, achieving equilibrium conditions does not necessarily mean that the conversion takes place at a measurable rate. For successful diamond production a metallic catalyst (nickel) is used to reduce the activation energy. The influence of the existing solid phase is also important in view of the difference in structure of diamond and graphite. For example, it has been shown (Derjaguin and Fedoseev (1975)), that diamonds much below the equilibrium pressure exposed to carboniferous vapours, e.g. methane at 10 Pa, grow diamond and not graphite whiskers, the tendency for deposited carbon atoms to conform to the crystalline structure of the available seed exceeding that to form the most stable one.

A survey of measurements of many simple transitions disclosed an unanticipated difficulty. Many substances on careful measurement reveal anomalies in their heat capacity which are confined to quite small temperature ranges 5–10 °C which may be missed altogether if the experimental intervals are too large. Some of these anomalies can be attributed to two close energy levels and can be recognised by their shape (Schottky anomalies) as given in section 6.6 and illustrated in Fig. 8.8. All measurements of this type are subject to the possibility of anomalies below the lowest temperatures measured. For substances whose nuclei have magnetic moments these anomalies certainly exist at temperatures ~0.01 K but presumably both crystalline forms will contribute the same anomalous entropy and the experimental test of Nernst's theorem will not

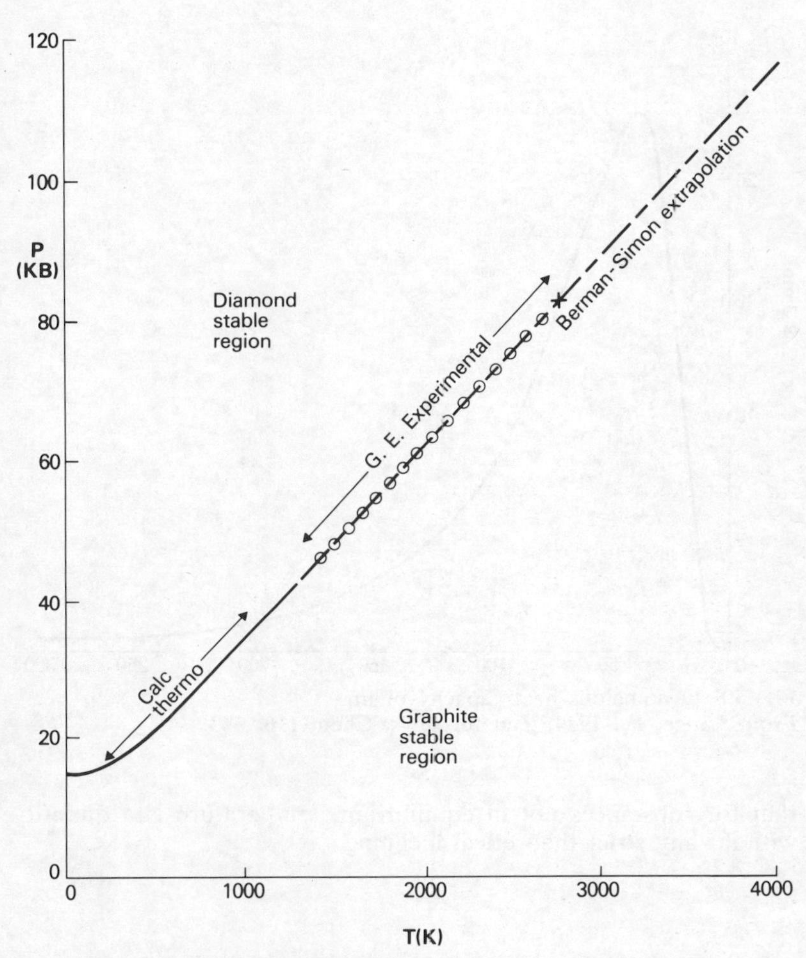

Fig. 8.7 Diamond – graphite equilibrium
ooo experimental points of Bundy F. P. et al. 1961 to match diagram extrapolated from the data of Simon and Berman 1965.

be affected. Another difficulty is the possibility of freezing-in crystalline imperfections. The outstanding example of this is supercooled liquids. Glycerine has been particularly well investigated. It solidifies at 17 °C but usually supercools and at low temperatures forms a glass, the entropy of which at absolute zero is 19.2 J/K per mole instead of zero. This can be looked upon as an illustration of the strict view that $\Delta Q/T$ is only the entropy change if the substance has the equilibrium distribution of energy over the energy levels and

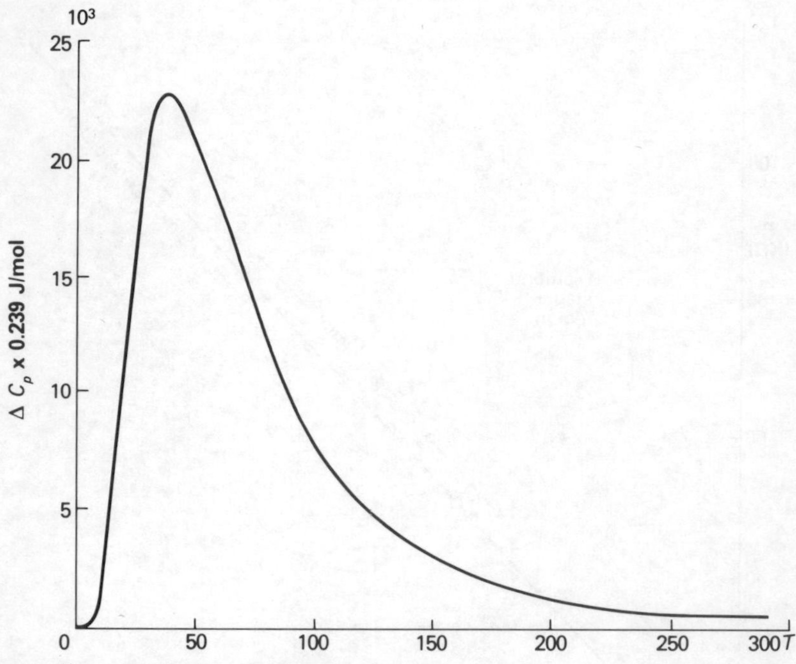

Fig. 8.8 Anomalous heat capacity of tin
From Lange, F., 1924, Zeit für Phys. Chem **110,** 343.

that for substances not in equilibrium, temperature is a quantity without any strict theoretical meaning.

8.7 Entropy of perfect gases

A perfect gas of measured entropy can be obtained by evaporating a pure crystalline solid into a vacuum at a temperature low enough for the vapour to behave like a perfect gas. As explained in (section 8.6), the entropy of the gas is

$$\int_0^T \frac{Cp}{T}\,\mathrm{d}T + \frac{\lambda}{T}$$

in which λ is the latent heat of evaporation at the temperature T.

The entropy of the gas has to be compared with the theoretical expression of section 6.10.

$$S = Nk\left(\frac{5}{2}\log T - \log p + \frac{3}{2}\log\frac{2\pi m k^{5/3}}{h^2} + \frac{5}{2}\right) \quad \text{(B.E.)}$$

in which p is the observed vapour pressure of the condensed phase at T. It does not matter at what temperature the comparison is made as long as it is low enough for the vapour to behave like a perfect gas. If the theoretical and experimental values of S agree at one temperature the Clausius–Clapeyron equation will ensure that they agree at all other temperatures. Some examples will now be given of the degree of agreement between theory and experiment.

The simplest measurements are made on a gas whose particles are electrons, usually referred to as an electron gas. It is a Fermi–Dirac undegenerate gas. The thermionic current which can be drawn from a hot filament in a vacuum is treated as the evaporation of electrons from the filament. As is well known in simple theory, as the voltage V across a diode is increased the thermionic current at first increases as $V^{3/2}$ and then reaches a constant value I_0 when all the electrons leaving the filament reach the anode. The rate of evaporation from a surface has already been treated in section (3.11) and can be expressed in terms of the static vapour pressure.

$$\text{Rate of evaporation} = \frac{n\bar{c}}{4}(1-r) = \frac{p(1-r)}{\sqrt{2\pi m k T}} = \frac{I_0}{e}$$

in which r is the reflexion coefficient, I_0 the saturation thermionic emission per unit area and e the electronic charge. This expression enables p the vapour pressure of the electrons, which they would exert on the wall of a cavity inside the filament, to be calculated from the experimental value of I_0. The theoretical expression for the entropy of an electron gas is thus

$$S = Nk\left(\frac{5}{2}\log T + \frac{3}{2}\log\frac{2\pi m k^{5/3}}{h^2} + \frac{5}{2} + \log 2 - \log\frac{I_0\sqrt{2\pi m k T}}{e(1-r)}\right) \quad \text{(F.D.)}$$

The experimental value of the entropy is easily expressed in terms of measurable quantities. In the hot metal the electrons form a highly degenerate Fermi–Dirac gas with a negligible heat capacity

so that

$$S = N\left[\int_0^T \frac{Cp}{T}\,\mathrm{d}T + \frac{\lambda}{T}\right] = \frac{N\lambda}{T}$$

in which λ is the latent heat of evaporation of an electron from the hot filament. Equating the two expressions for the entropy gives at once that

$$I_0 = \frac{4\pi mk^2e(1-r)}{h^3}T^2\,\mathrm{e}^{-\lambda/kT+5/2}$$

usually written as $I_0 = \mathrm{A}(1-r)T^2\,\mathrm{e}^{-e\,\phi/kT}$ in which A is a universal constant $= 1.2\times10^6\ \mathrm{A/m^2} = 120\ \mathrm{A/cm^2}$ and ϕ is the work function and represents the minimum energy needed to remove an electron from the metal. Many measurements of thermionic emission have been made. It is a field in which reliable measurements are difficult to achieve because one must ensure that the electrons are emitted from a surface uncontaminated by absorbed gas or impurities of different work function migrating from the body of the metal. The photoelectric threshold itself is rather an ill-defined quantity as it is demonstrably different for different crystallographic faces of a metallic crystal. For example, ϕ for a 111 face of a copper crystal is reported as 4.89 V and 5.64 V for a 100 face. Even the rather prosaic problem of measuring the temperature is not all that easy. When a set of satisfactory measurements have been made it is not practicable to ask what law they obey, because of the overwhelming influence of the exponential term. All that is feasible is to find out what are the values of the constants which give the best fit to a given theoretical formula. A selection from the more reliable results is given in Table 8.5. The experimental value of A($\sim$60) is in reasonable agreement with theory (120). Although the reflection coefficient r is not measured in these experiments, it is unlikely that it is large enough to bridge the gap. It is difficult to measure the reflexion coefficient of electrons at thermal energies. A measurement at higher energies by Davisson and Germer (1927), while it did not bridge the gap, led by a fortunate accident to the discovery of electron diffraction. The difference between theory and experiment is accounted for if ϕ has a small ($\sim5\times10^{-5}$ V/K) positive temperature coefficient. If $\alpha = \alpha_0 + \alpha T$ the emission constant A_α is $\mathrm{Ae}^{-\alpha/k}$ but remains independent of the temperature. That this is the probable explanation is shown by the work of Jain and Krishnan (1952). They avoided having to know the reflection coefficient by measuring the vapour pressure of electrons in a hot cavity by the

Table 8.5

Metal	Cr	Co	Mo	Mo	Ta	W	W	W
$A(1 - r)\left(\frac{A}{cm^2}\right)$	48	41	55	115	55 ± 5	75	72	60
$\phi(V)$	4.60	4.41 ± 0.10	4.20 ± 0.02	4.37 ± 0.02	4.05	4.55	4.52	4.52

(Taken from Conyers Herring and Nichols, 1949.)

effusion method. For nickel they found that A was 120 ± 5 A/cm^2 but for other metals it varied from 44(Mn) to 60(Cr) but did not depend on the temperature.

For a substance such as mercury, which can relatively easily be purified and for which the measurable vapour pressure occurs in a convenient temperature range, a direct comparison can be made using the very careful measurements of Busey and Giauque (1953). They found from measurement of the heat capacity of mercury from 15–330 K that the entropy of liquid mercury at 298.16 K was 76.10(7) J/K per mole. At this temperature the vapour pressure is 2.00×10^{-3} mmHg and the latent heat of evaporation is 61 303.(96). The entropy of mercury vapour is therefore

$$76.10(7) + \frac{61\,303.(96)}{298.16} = 281.7(1) \text{ J/K per mole.}$$

This has to be compared with the theoretical value with p = 0.266 644 Pa and $Nk = R = 8.3143$ J since the latent heat data refer to one mole of mercury, and the mass of a single atom is $M_r \times m_u$ in which M_r is the relative molecular weight (200.61 for Hg) and m_u is the unified mass number 1.660 56 10^{-27} kg. Thus

$$S = R\left(\frac{5}{2}\log 298.16 - \log 0.266\,644 + \frac{3}{2}\log M_r + \frac{3}{2}\log\frac{2\pi m_u k^{5/3}}{h^2} + \frac{5}{2}\right)$$
$$= 281.68$$

The agreement is good: it corresponds to predicting the vapour pressure of mercury to better than 0.4 per cent. Although the comparison of theoretical and experimental value of S at a selected temperature is in principle sufficient it is an uneconomical use of the data which do not usually have the extreme precision of Busey and Giauque's measurements for mercury. The value of the latent heat at the low pressures ($\sim 10^{-2}$ Pa) involved is not easily measured directly and is better obtained from the slope of the vapour pressure curve through the Clausius–Clapeyron equation. The raw data of the calculation is thus a set of vapour pressure measurements and a set of heat capacity measurements carried out over a wide range of temperatures. All the measurements are combined in the following way so as to make the maximum use of the data. The theoretical value of the entropy can be expressed in terms of the heat capacities in two ways.

Either
The theoretical value of the vapour pressure, obtained by integrating the Clausius–Clapeyron vapour pressure curve (with its integration constant i) (section 7.5) can be inserted into the theoretical formula for the entropy of a perfect gas (section 8.7 and section 6.10) to give

$$S = Nk\left[\frac{\lambda_0}{RT} + \frac{1}{R}\int_0^T \frac{C_p \mathrm{d}T}{T} - \frac{1}{RT}\int_0^T C_p \mathrm{d}T + \frac{5}{2} + \frac{3}{2}\log\frac{2\pi m k^{5/3}}{h^2} - i\right]$$

Or
the entropy can be calculated directly from

$$S = \int_0^T \frac{C_p}{T}\,\mathrm{d}T + \frac{\lambda}{T}$$

using the calculated (section 7.5) value of $\lambda = \lambda_0 + \int(C_p^g - C_p^s)\mathrm{d}T$ to give

$$S = Nk\left[\frac{\lambda_0}{RT} + \frac{5}{2} - \frac{1}{RT}\int_0^T C_p^s \mathrm{d}T + \frac{1}{R}\int_0^T \frac{C_p^s \mathrm{d}T}{T}\right]$$

Equating the two expressions shows that they are identical if

$$i = \frac{3}{2}\log\frac{2\pi m k^{5/3}}{h^2} = i_0 + \frac{3}{2}\log M_r$$

The experimental test of the theoretical expression for the entropy of a perfect gas is thus reduced to fitting the vapour pressure formula and seeing if the constant i has the theoretical value. This constant is usually referred to as the chemical constant. Its numerical value for metals like mercury which have two valency electrons is $7.86 + 3/2 \log M_r$ and $7.861 + 3/2 \log M_r + \log 2 = 8.554 + 3/2 \log M_r$ for metals like sodium and potassium which have a single valency electron as explained in section 6.11. The measured value of i_0 for potassium based on the vapour pressure data obtained by Edmundson and Egerton (1927) using the effusion method, and by Neumann and Volker (1933) using the reaction method is 8.56 ± 0.12 which is satisfactory agreement. Many other measurements of the chemical constant have been made and no outstanding discrepancies remain. Considering the great difficulty of the measurements involved one can say that the theoretical expression for the entropy of a perfect gas agrees well with the experimental value.

8.8 Separation of isotopes

The separation of isotopes is an important industrial process. Shortly after their discovery (1912) many attempts were made at separation with the object of confirming their existence. It was thought that separation would be easy: experience did not confirm this optimistic view although most of the methods now successfully applied were tried without success in the early days. This was usually due to ignorance of the rather exacting conditions which need to be fulfilled for a successful outcome. This can be illustrated by an attempt at a photochemical separation. It was well known that the reaction between hydrogen and chlorine proceeded faster when the gases were illuminated. It was argued that, if the isotopes of chlorine had sufficiently different absorption coefficients, light filtered through a long column of chlorine before admission to the reaction vessel would be preferentially absorbed by one of the isotopes. The hydrochloric acid formed in these circumstances would be formed from one isotope only and should have an isotopic composition different from the normal one. The experiment was unsuccessful. The idea was a good one and at the time of writing the possibility of a photochemical separation (Zare 1977) is being actively explored with high hopes of success. With hindsight it is obvious that the reaction between hydrogen and chlorine is an unsuitable choice because the absorbed quantum initiates a chain reaction

$$h\nu + Cl_2 \longrightarrow Cl + Cl$$
$$Cl + H_2 \longrightarrow HCl + H$$
$$H + Cl_2 \longrightarrow HCl + Cl \quad \text{etc.}$$

As the chain can be a long one it is clear that the composition of the photochemically formed hydrochloric acid is almost independent of which chlorine isotope initiated the chain.

A complete separation in one process can be obtained by using a mass spectograph, provided a suitable source of ions is available. Most atomic energy establishments have facilities for carrying out such a separation on a microgramme scale using a simple 180°

magnetic analysis of the beam. This meets the needs of nuclear physicists wishing to make targets of known nuclear composition. All other methods depend upon the repetition (cascading) of a process which in itself produces only a small separation. Thus the design of an isotope separation plant can be divided into two separate parts:

1 the choice of the separation process

2 the design of the cascade

The cascade presents a difficult mathematical and engineering problem. Owing to its importance in the Second World War (1939–1946) it has been exhaustively studied and reported on. All that is necessary here is to explain the nature of the problem. This is to find a compromise between maximum separation, maximum efficiency, and the time required for material to traverse the cascade. For large plants it is also necessary to consider the total cost including that of supporting research; however, if this is a serious factor it is almost certain that the plant will not be built. The interplay of these considerations can be illustrated by reference to the separation of the isotopes of uranium by diffusion (strictly speaking effusion) in gaseous UF_6. The naturally occuring isotopic ratio $^{235}U/^{238}U$ is 7.15×10^{-3}. After a single stage this ratio will be increased by the factor.

$$\sqrt{\frac{^{238}UF_6}{^{235}UF_6}} = \sqrt{\frac{352}{349}} = 1.004\ 29$$

To reach the desired ratio $^{235}U/^{238}U = 41.55$ will require a minimum of n stages given by

$$7.153 \times 10^{-3}(1.004\ 29)^n = 41.55$$

corresponding to about 2025 stages.
Such a plant would have a maximum separation factor and zero output since no allowance has been made for the change in composition of the stock as the enriched material is withdrawn. It should be compared with ~5000 stages of a plant with an output ~1 kg per day. To achieve a large output economically the cascade must be arranged as in a fractioning column so that the enriched material travels to the head of the cascade at the same time that impoverished material travels in the opposite direction to the tail (Fig. 8.9). For efficient working the input of the nth stage must have the same isotopic composition as the tailings of the (n + 1)th stage, otherwise separated material would be uselessly downgraded. The

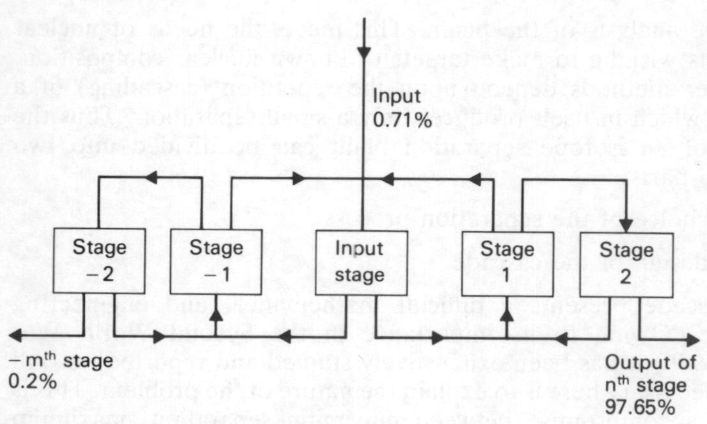

Fig. 8.9 Cascade for isotope separation.

stages will also differ in size, getting smaller as the rare isotope is concentrated. These conditions are not sufficient to determine the cascade since the separation depends upon the rate at which material is allowed to pass through each stage. Since material may pass more than once through each stage the overall rate of transfer from stage to stage is slow. If this time is too long a plant with many stages becomes unmanageable. For the Oakridge plant it takes several hours before a change in the controls is effective. The diffusion process is so well documented in principle (Cohen 1951) and the details are still so secret that it will not be further discussed.

8.9 Centrifugal separation

The simplest of all methods of separating the isotopes of a gas is to allow them to settle under gravity. Since the constituents of a perfect gas behave independently of each other a straightforward application of the barometer formula shows that the isotopic ratio $(\text{I.R.})_h$ at height h is related to the isotopic ratio $(\text{I.R.})_0$ at ground level by the equation

$$(\text{I.R.})_h = (\text{I.R.})_0 \, e^{\frac{-(m_1 - m_2)gh}{kT}}$$

The terrestrial value of 'g' is too small for this separation to be of any practical use but, as in a cream separator, use can be made of the large apparent value of 'g' which can be obtained in a centrifuge.

For a perfect gas rotating uniformly with angular velocity ω between cylinders of radii r_0 and R, simple analysis shows that

$$\text{Isotopic ratio at } R = (\text{Isotopic ratio at } r_0) \times e^{\frac{\omega^2 R^2 \Delta m(1 - r_0^2/R^2)}{2kT}}$$

the heavier molecules being concentrated at the outer periphery. The attractive feature of this result is that the degree of separation depends upon Δm and not on the much smaller factor $\Delta m/m$ as in the diffusion process. Thus at first sight centrifugal separation seems to be the right way of separating the isotopes of uranium for which $\Delta M = 3$ and $\Delta M/M = 1/117$. The degree of separation depends upon ωR, the peripheral velocity of the centrifuge and becomes appreciable when the energy of the molecules due to this circular motion approaches the mean squared energy of their thermal motion. The centrifugal stress at the base of a spinning rod (Fig. 8.10) is $\omega^2 R^2 \rho/2$ so that the maximum attainable peripheral velocity is $\sqrt{2T_B/\rho}$. The ratio of the breaking stress T_B to the density ρ of high tensile materials does not vary greatly; for piano wire it corresponds to a peripheral velocity of ~700 m/s. Some improvement can be obtained by shaping the rotor (Maxwell) and it is usually assumed that with modern materials (fibres of pure silica or carbon) there is no difficulty in obtaining a safe working figure of 500 m/s in a fabricated rotor. Adopting this figure as a basis for discussion shows that at 400 K the separation factor is $e^{0.0375\Delta M(1-r_0^2/R^2)} = e^{0.1125(1-r_0^2/R^2)}$

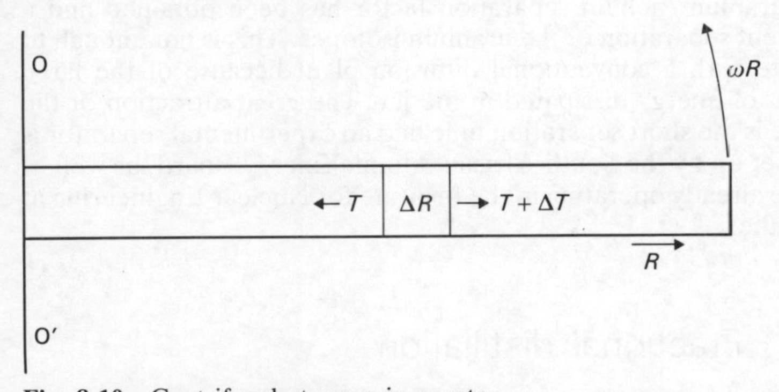

Fig. 8.10 Centrifugal stresses in a rotor.
If the stress in the rod rotating about the axis 00′ with angular velocity ω is T the centrifugal force on a section of unit area between r and $r + \mathrm{d}r$ is $-\mathrm{d}T/\mathrm{d}r\ \mathrm{d}r$ so that $\rho r \omega^2\ \mathrm{d}r = -\mathrm{d}T/\mathrm{d}r$, $T = -\omega^2 \rho r^2/2 + \mathrm{C}$ and on the axis $T_{\max} = \omega^2 \rho R^2/2$.

for uranium. The ratio r_0/R cannot be very small in a centrifuge designed for use with UF_6 because it condenses at about atmospheric pressure and the pressure at the base of rotor can hardly be allowed to fall below 1 Pa if the enriched gas is to be extracted at a reasonable rate. The pressure ratio is large because it depends upon $m(352)$ and not $\Delta m(3)$ in the exponential term. A pressure ratio of 10^5 corresponds to a value of 0.36 for r_0/R, and a corresponding separation factor of $e^{0.1125\times0.8726} = 1.103$, giving 87 for the minimum number of stages required. General considerations also indicate a design with a relatively narrow annular gap $R - r_0$ as this favours a more rapid flow of material through the centrifuge. It is no good drawing material out before it has had time to separate and the smaller the distance which the molecules have to diffuse the more rapidly the equilibrium is set up. The result of these primitive calculations suggests that the improvements in technology since the first diffusion plants were built 30 years ago have changed the balance of advantage in favour of centrifugal separation. At any rate a European consortium is building a centrifugal plant but no details have yet been published. The subject is now very secret. The last published account by Zippe (1960) described a small cylindrical centrifuge with a peripheral velocity of 350 m/s at 33°C capable of separation factors of 1.1 – 1.2. The large separation is due to internal cascading. Some further information is given in a general article by Whitley (1979) and Mezin (1979). A still simpler centrifugal separator which dispenses with the need for a mechanical centrifuge has been suggested by Professor Becker at Karlsruhe. A supersonic jet of helium containing 5 per cent of UF_6 is deflected through 180° and then divided by a knife edge. A large (5–10 per cent) uranium/helium separation factor has been obtained and a 1 per cent separation of the uranium isotopes. This is not enough to compete with a conventional diffusion plant because of the large amount of energy dissipated in the jet. The great attraction of the scheme is the short separation time and an experimental separator is being set up by the South African Atomic Energy Board, as well as the one already operating in the Institute for Nuclear Engineering at Karlsruhe.

8.10 Fractional distillation

The obvious way to attempt the separation of isotopes is by fractional distillation. It does not seem less obvious when one remembers that the vapour pressure constant is $i_0 + 3/2 \log M_r$, so that for a monatomic solid with isotopes of mass m and $m + \Delta m$, the difference in vapour pressure Δp due to the difference in the chemical constant is given by $\Delta p/p = 3/2\ \Delta m/m$. Such a large difference in vapour pressure would enable a separation by fractional distillation to be carried out quite easily. Experience shows that this is not always so. The explanation was given by Lindemann (1920), who calculated the vapour pressure of an ideal Einstein solid from the vapour pressure formula

$$\log p = -\frac{\lambda_0}{kT} - \int \frac{1}{kT^2} \int C_p^s \, \mathrm{d}T\mathrm{d}T + \frac{5}{2}\log T + i_0 + \frac{3}{2}\log m$$

The heat capacity of an Einstein solid is known in terms of the single frequency ν of vibration of the atoms in the lattice. The second integral is the energy added to the lattice and so can be replaced by $kT^2\,(\partial\,(\log Z)/\partial T) - 3h\nu/2$ to give (cf. section 6.86b).

$$\log p = -\frac{\lambda_0}{kT} - \log Z + i_0 + \frac{3}{2}\log m - \frac{3h\nu}{2kT} + \frac{5}{2}\log T.$$

Applying this to two Einstein solids differing only in the mass of their atoms gives that

$$\frac{\Delta p}{p} = -\frac{\Delta\lambda_0}{kT} - \Delta\log Z + \frac{3}{2}\frac{\Delta m}{m} - \frac{3h\Delta\nu}{2kT}$$

The difference between the two latent heats of evaporation at absolute zero $\Delta\lambda_0$ is not zero because although the forces between the isotopes are presumably the same the atoms have different kinetic energies due to their different zero point energies $3h\nu/2$. Thus $\Delta\lambda_0/kT = -3h\Delta\nu/2kT$. If the motion of the atoms in the lattice is strictly harmonic and the forces acting on them are the same because they are derived from the same electronic structure, then $\Delta\nu/\nu = \Delta m/2m$ and the R.H.S. of the expression for $\Delta p/p$ can be evaluated in terms of $\Delta m/m$ to give at temperatures for which $h\nu/kT \ll 1$.

$$\frac{\Delta p}{p} = \frac{3h\Delta\nu}{kT} - \frac{3h\Delta\nu}{kT} + \frac{3\Delta m}{2m} - \frac{3\Delta m}{2m}\left(1 - \frac{1}{12}\left(\frac{h\nu}{kT}\right)^2 + \text{higher terms}\right)$$

As the condition $(h\nu/kT) \ll 1$ is usually fulfilled when the vapour pressure is high enough for fractional distillation to be a practical process, the lack of success by this method is accounted for. For lead $(h\nu/kT) \sim 0.1$ at the melting point and the separation factor is about one thousandth of the originally expected value. At the time this explanation of the failure to separate isotopes by distillation was taken as good evidence for the existence of zero point energy (there was no Schrödinger's equation from which to derive it); Urey was the first to realise that Lindemann's demonstration applied to a limited class of substances in a particular temperature range. Indeed it has turned out that for some compounds distillation is quite a good way of separating isotopes. For example, the separation function for Boron trichloride is only $\sim 10^{-3}$ and separation by distillation is not feasible; for Boron trifluoride it is large enough for the separation to be carried out by fractional distillation in the laboratory on a kg scale. Indeed the best method separating the isotopes of the lighter elements is a combination of fractional distillation with chemical separation.

8.11 Chemical separation

Chemical separation depends upon the fact that compounds with different isotopic constitutions do not behave in exactly the same way when they take part in a chemical change. Gaseous chlorine for example contains three types of molecule, $^{35}Cl^{35}Cl$, $^{37}Cl^{37}Cl$ and $^{35}Cl^{37}Cl$. This is not a static mixture the exchange reaction $^{35}Cl^{35}Cl + {}^{37}Cl^{37}Cl \rightleftarrows 2\,{}^{35}Cl^{37}Cl$ taking place at an appreciable rate. If the proportions of the isotopes of chlorine were as α/β and the isotopes were distributed at random amongst the molecules the concentrations of the molecules $^{35}Cl_2$, $^{37}Cl_2$ and $^{35}Cl^{37}Cl$ would be

$$\frac{\alpha^2}{(\alpha+\beta)^2} : \frac{\beta^2}{(\alpha+\beta)^2} : \frac{2\alpha\beta}{(\alpha+\beta)^2}$$

and the reaction constant K given by

$$\frac{[^{35}Cl^{37}Cl]^2}{[^{35}Cl_2][^{37}Cl_2]} = \frac{4\alpha^2\beta^2}{\alpha^2\beta^2} = 4$$

would be precisely four. It will deviate from this value if the distribution of the atoms among the molecules depends upon their masses. The calculated value of K in this case is 3.999 7. It was pointed out by Urey (1947) that much larger deviations occur in selected compounds and that these deviations can be made the basis of a practical method for separating isotopes. Although the general theory is complicated in detail the principles upon which K is calculated are straightforward and can be illustrated by a simple example. Consider an exchange reaction between perfect gases

$$\mathrm{A} + \mathrm{B} \rightleftharpoons \mathrm{C} + \mathrm{D} + q$$

in which q is the heat energy given out per molecule when the reaction occurs at constant volume, and A differs from C and B from D by the exchange of a single isotope.

Let S_A, S_B, S_C, S_D be the absolute entropies of the molecules which are present in the reacting mixture at the four partial pressures p_A, p_B, p_C, p_D. The decrease in entropy which occurs when N molecules enter the forward reaction is $N(S_A + S_B - S_C - S_D)$. Since this change takes place through a state of equilibrium it is also equal to the calorimetrically determined decrease of entropy $= Nq/T$, so that

$$S_A + S_B - S_C - S_D = q/T$$

The whole heat change q comes (since there is no change of volume) from the decrease in the energy of the molecules and

$$q = E_A + E_B - E_C - E_D$$

in which E_A, E_B, E_C, E_D are the energies of the molecules made up in each case of kinetic energy of translation and internal energy of rotation and vibration. Eliminating q from these equations gives at once

$$\left(S_A - \frac{E_A}{T}\right) + \left(S_B - \frac{S_B}{T}\right) - \left(S_C - \frac{E_C}{T}\right) - \left(S_D - \frac{E_D}{T}\right) = 0$$

which on using the relationship $S = nk \log Z + E/T$ becomes

$$\log Z_A + \log Z_B - \log Z_C - \log Z_D = \log \frac{Z_A Z_B}{Z_C Z_D} = 0$$

and $$\frac{Z_A Z_B}{Z_C Z_D} = 1$$

The partition function of the gases can be calculated if the energy levels are known, either theoretically or from spectroscopic measurements. To make the calculation of the partition functions straightforward consider the particular case in which the four molecules A, B, C, D are all diatomic molecules as in the imagined reaction

$$^{12}C^{16}O + {}^{14}C^{18}O \rightleftharpoons {}^{12}C^{18}O + {}^{14}C^{16}O$$

The results so obtained can be generalised afterwards. The energy levels of a diatomic molecule are made up from three independent sets of energy levels all of which have already been used in the calculation of Z, section (6.8), namely the energy levels of a monatomic gas, the energy levels of a rigid rotator of moment of inertia I and the energy levels of a simple harmonic oscillator of reduced mass μ, and frequency ν. It is a fundamental theorem of quantum mechanics (Born and Oppenheimer 1927) that to a high degree of approximation the energy levels of a diatomic molecule $\varepsilon_{\mathrm{mol}}$ can be written as the sum of the energy levels of a free particle, a rigid rotator, and a simple harmonic oscillator in any linear combination. Accordingly the energy levels ε_{i} of a diatomic molecule in a box are any linear combination of $\varepsilon_{\mathrm{i}}^{\mathrm{t}}$ the translational levels of a monatomic gas, $\varepsilon_{\mathrm{j}}^{\mathrm{r}}$ the rotational levels of a rotator and $\varepsilon_{\mathrm{n}}^{\mathrm{v}}$ the energy levels of a vibrator. So that for a single molecule (using the notation of section 6.10).

$$z = \sum_{\substack{\text{all i}\\ \text{all j}\\ \text{all n}}} e^{-(\varepsilon_i^t + \varepsilon_j^r + \varepsilon_n^v)/kT} = \sum_{i} e^{-\varepsilon_i^t/kt} \sum_{j} e^{-\varepsilon_j^r/kT} \sum_{n} e^{-\varepsilon_n^v/kT}$$

$$= z_t z_r z_v$$

If there are q molecules in each member of the ensemble

$$z = \frac{z^q}{q!} = \frac{(z_t)^q}{q!} (z_r)^q (z_v)^q$$

so that

Z_{mol} = (Z for a monatomic gas) × (Z for a rigid rotator) × (Z for a harmonic oscillator) and the equation

$$\log \frac{Z_A Z_B}{Z_C Z_D} = 0$$

can be written as $\log(Z_A Z_B/Z_C Z_D)$ monatomic gas + $\log(Z_A Z_B/Z_C Z_D)$ for a rigid rotator + $\log(Z_A Z_B/Z_C Z_D)$ for a harmonic oscillator = 0
Taking these terms in order:

1 *monatomic gas*
The partition function for a molecule is given in section 6.10 and in terms of the pressure

$$\log Z = \log \frac{\mathrm{e}\, kT(2\pi mkT)^{3/2}}{h^3} - \log p$$

and the contribution of the translational term to Z_{mol} is

$$\log\left(\frac{m_A m_B}{m_C m_D}\right)^{3/2} + \log\frac{p_C p_D}{p_A p_B} = \log\left(\frac{m_A m_B}{m_C m_D}\right)^{3/2} + \log K$$

2 *rotation of a rigid diatomic molecule*

$$\log Z = \log\frac{8\pi^2 IkT}{h^2}\cdot\frac{1}{\sigma}$$

and the contribution to log

$$Z_{mol} = \log\left(\frac{I_A I_B}{I_C I_D}\cdot\frac{\sigma_C\sigma_D}{\sigma_A\sigma_B}\right)$$

in which σ (section 6.9c) is 2 for molecules with two identical atoms ($^{16}O^{16}O$) and 1 for molecules with two distinguishable atoms ($^{17}O^{16}O$)

3 *vibrational term*

$$\log Z = \log\frac{\mathrm{e}^{-h\nu/2kT}}{1-\mathrm{e}^{-h\nu/kT}} = \log \mathrm{E}(\nu) \quad \text{(E for Einstein)}$$

and the contribution to $\log Z_{mol}$ is $\log \mathrm{E}(\nu_A)\mathrm{E}(\nu_B)/\mathrm{E}(\nu_C)\mathrm{E}(\nu_D)$

Combining these three contributions gives at once that

$$K = \left(\frac{m_C m_D}{m_A m_B}\right)^{3/2} \times \frac{I_C I_D}{I_A I_B} \times \frac{\sigma_A\sigma_B}{\sigma_C\sigma_D} \times \frac{\mathrm{E}(\nu_C)\mathrm{E}(\nu_D)}{\mathrm{E}(\nu_A)\mathrm{E}(\nu_B)}$$

If the forces between the atoms are due to their electronic structure, which is the same for isotopes, the size of the molecules will not be changed by an isotopic exchange and the ratio of the moments of inertia will depend on the masses only. It is a straightforward exercise in elementary mechanics to show that

$$\frac{I_C I_D}{I_A I_B} \cdot \frac{m_C m_D}{m_A m_B} = 1$$

This leaves

$$K = \left(\frac{m_C m_D}{m_A m_B}\right)^{1/2} \frac{\sigma_A \sigma_B}{\sigma_C \sigma_D} \frac{\mathrm{E}(\nu_C)\mathrm{E}(\nu_D)}{\mathrm{E}(\nu_A)\mathrm{E}(\nu_B)}$$

This expression has an important limiting value at temperatures sufficiently high for $(h\nu/kT) \ll 1$ as is easily seen by expanding the Einstein functions.
For

$$\underset{T\to\infty}{\mathrm{Lt}} \frac{\mathrm{e}^{-h\nu/2kT}}{1 - \mathrm{e}^{-h\nu/kT}} = \frac{kT}{h\nu} + \text{higher terms}$$

and

$$\underset{T\to\infty}{\mathrm{Lt}}\ K = \left(\frac{m_C m_D}{m_A m_B}\right)^{1/2} \frac{\nu_C \nu_D}{\nu_A \nu_B} \cdot \frac{\sigma_A \sigma_B}{\sigma_C \sigma_D} = \frac{\sigma_A \sigma_B}{\sigma_C \sigma_D}$$

since the frequencies are all inversely proportional to the square roots of the reduced masses. In the particular example chosen all the symmetry terms (σ) are unity and $K = 1$. In the chlorine equilibrium previously considered in which $\sigma_A = \sigma_B = 2$ and $\sigma_C = \sigma_D = 1$, $K = 4$, as found by the simple probability argument. The conclusion that the reaction constant K for simple isotope exchange reactions depends at high temperatures only on the symmetry factors is an important one. Since the mass difference of the isotopes is not involved it means that when the temperature is high enough for $h\nu/kT \ll 1$ there can be no chemical separation. The condition that $h\nu/kT \ll 1$ is just the condition that the vibrational energy can be correctly calculated from the equipartition theory of classical mechanics. A more general treatment of the problem leads to the conclusion that isotope separation by chemical means is a quantum effect not shown by systems which obey the theorem of the equipartition of energy. It is a conclusion implicit in the pioneer work of Lindemann in this field and was first explicitly stated by Waldman and independently by Bigeleisen and Mayer (1947). The latter provided the understanding of the conditions necessary for chemical separation. These are that the greatest deviation of K from the classical value will occur when the isotope distributes itself between two compounds with very different binding energies. An approximate value of the magnitudes involved is given in terms of the reduced mass by the relationship

$$\Delta K = \frac{1}{24}\left[\left(\frac{h\nu_1}{kT}\right)^2 - \left(\frac{h\nu_2}{kT}\right)^2\right] \frac{d\mu}{\mu}$$

This result indicates that the light isotope is concentrated in the compound with the weak binding forces. In practice suitable compounds are complexes with many degrees of vibrational freedom so that accurate calculation is almost impossible and there is not very good agreement between experimental and calculated values of K. To make use of the small differences of the reaction constant to separate isotopes the equilibrium must take place between a liquid and a gaseous phase so that a cascade can be constructed as in a fractionating column with the liquid phase descending and the gaseous phase bubbling up through it. In these circumstances it is not profitable to discuss how much of the separation is due to distillation and how much to chemical exchange. To conclude two examples will be given in which these methods are used in practice to separate isotopes on a large scale.

Deuterium

As is well known, heavy hydrogen can be separated on a small scale by the electrolysis of water in which it occurs to about one part in five thousand. The separation factor is high; about 5–6 and relatively independent of the conditions. For large scale production the method is ruinously expensive because the free energy expended in electrolysing the 5000 parts of light water is not recovered. Electrolysis is a good method if one can start with a good concentration of D_2O, i.e. 30 per cent. Even in this case separation by fractional distillation is often preferred. On a large scale the complicated exchanges between hydrogen sulphide and water, of which the reaction

$$\text{HSH} + \text{DOH} \rightleftharpoons \text{HSD} + \text{HOH}$$

is a typical member, is used. The hydrogen sulphide at ~20 atm pressure bubbles up through a descending column of water at 30 °C.

This temperature is chosen in order to avoid the formation of a hydrate at 29 °C. The deuterium collects in the liquid phase so that the temperature has to be raised to 130 °C by flow down a second column to drive off the hydrogen sulphide. The towers are massive, 28 ft in diameter and 300 ft high and are some of the world's largest pressure vessels. The concentration is done in three stages which brings the deuterium content of the water up to 30 per cent. The final concentration to reactor grade (99.8 per cent) is carried out by distillation. The output is about 400 kg per hour. Other exchange reactions have been proposed but so far as is publicly known the H_2S exchange reaction is the only one actually being used.

Boron

The ratio of the boron isotopes in naturally occurring material is $^{10}B/^{11}B = 1/4.01$. They are separated on a large scale because ^{10}B has a large neutron capture cross-section and it is worth going to a great deal of trouble to obtain a five-fold increase in the sensitivity of neutron detectors. Also owing to its large capture cross-section it is used as a consumable poison to regulate the neutron economy of reactors. The separated ^{11}B is also useful because boron steels have valuable mechanical properties which cannot be used in the construction of reactors if they contain ^{10}B.

The separation is carried out in a fractionating column using the exchange reaction between boron trifluoride (BF_3) and its complex compound with ether:

($^{10}BF_3$ + $^{11}BF_3$ + ether + undissociated complex) is the gaseous phase and ($^{10}BF_3$ ether complex + $^{11}BF_3$ ether complex) the liquid phase. Originally ordinary ether (ethyl ether) was used but methyl ether is better because there is less thermal decomposition. The separation factor

$$\alpha = \frac{(^{10}B/^{11}B)\ \text{liquid}}{(^{10}B/^{11}B)\ \text{gas}} = 1.03$$

and a working figure 1.018 is usually adopted in designing the separation column. The light isotope concentrates in the liquid phase in agreement with simple theory. The apparatus consists of a boiler containing the liquid complex and a conventional fractionating column. The process can be made continuous by injecting the unseparated complex into the column at the appropriate place and drawing off the complex at the same rate from the head and tail fractions. A small British column (XXth Century Electronics) consists of a 3 in diameter rectifier 36 ft long surmounted by a further 18 ft of 4 in diameter stripping section. The pressure at the column head is 4×10^3 Pa and about 10^5 Pa at the boiler. The yield is about 8 g of separated ^{10}B per day at an expenditure of 500–600 W. A much larger American plant at Niagara Falls produced 20 kg per week.

The observed variation of the isotopic composition of boron minerals suggested that the isotopes might be separated by chemical exchange in a chromatographic column as suggested by Spedding. This technique has recently been applied on a large scale in a French plant designed for an output of 200 kg per year. A separation factor of 1.01 has been obtained. So far no data on the economics of the new method have been published.

8.12 Summary

It is clear that there is agreement, within the errors of measurement, between the deductions from the hypothesis that matter is a collection of molecules interacting according to the laws of mechanics and the observed behaviour of matter in bulk over a wide variety of observations. The mechanical properties of nearly perfect gases can be accounted for by a direct approach in terms of molecular collisions. For more strongly interacting molecules little progress can be made without introducing the concept of entropy as a function of state. We owe to Boltzmann the idea that entropy can be calculated from the molecular motion as determined by the laws of mechanics. The success of Boltzmann's ideas in dealing with such complicated phenomena as vapour pressure and chemical equilibria must have been surprising to its discoverers. For example even if it were feasible to make a direct calculation of the vapour pressure of sodium it would have to be based upon a more exact knowledge of the law of force between the atoms than we possess and upon the ability to predict the lattice structure of the metal from the law of force. In contrast to this formidable calculation the chemical constant $i = i_0 + 3/2 \log M_r$, which determines the vapour pressure, can be evaluated from the simple knowledge that for sodium $M_r = 23$. This remarkable simplification comes about because almost without noticing it, a semi-unified system of mechanics has been adopted in which only certain masses and laws of force are allowed. Problems are no longer worked out in terms of molecules of arbitrary mass m influencing each other through potential functions of assumed mathematical form and adjustable constants like the Lennard–Jones potential. Once it is decided that the problem is about sodium atoms characterised by $M_r = 23$ everything is settled: $m = 3.820 \times 10^{-26}$ kg and no other, all the forces and the crystalline structure of metallic sodium are determined by Schrödinger's equation. The circumstance that we cannot solve the equation doesn't matter. We can allow nature to act as a vast computer on our behalf and obtain the numerical values needed in the calculation from calorimetric measurements. The heat capacity and latent heat of evaporation of metallic sodium are in theory completely determined by the integer 23 and the general principles of mechanics embodied in Schrödinger's equation. This number is not in this context part of a continuum and obstinately proclaims the atomic nature of matter, the hypothesis from which we started (410 B.C.). Pythagoras who started it all (569–500 B.C.) can indeed rest quietly in the Elysian fields and chuckle at the attempts to undermine his philosophical position with the irrationality of $\sqrt{2}$.

Problems

1 Millikan observed the times of fall of an oil drop over a distance of 10.21 mm to be 11 s + 0.(848, 890, 908, 904, 882, 906, 838, 776, 840, 870, 952, 860, 846, 912, 910, 918, 870, 888, 894, 878). Pressure = 0.995 atm, density of oil = 919.9, temperature 22.82 °C, mean rate of fall $\bar{v} = 8.584.10^{-4}$ m/s, $\eta = 18.3$ μNsm^{-2}. Are these results compatible with the view that the fluctuations are due to the Brownian motion of the drop during its descent?

2 A capacitor (C) and an inductor (L) are connected to form a parallel LC circuit. If the system obeys the law of equipartition of energy show that $\overline{V^2}$ the fluctuating voltage across the capacitor is kT/C. Calculate its value if the capacitor consists of two concentric spheres of radii 20 and 30 mm. How would you attempt to measure this voltage in practice?

3 A flat surface aluminised mirror 2 mm diameter ($m = 1.6$ mg) is attached to the middle of a vertical 50 μm diameter quartz fibre 20 cm long anchored at both ends in a moderate vacuum. Calculate the period of torsional oscillations and the r.m.s. deflection due to the Brownian motion. Give a reasoned account of how you would set about showing that the Brownian motion was actually there.

Further reading

Avery, D.G. and **Davies, E.** (1973) *Uranium enrichment by gas centrifuge*. Mills and Boon, London.

Mitchell, A.C.G. and **Zemanski, M.W.** (1934) *Resonance radiation and excited atoms*, Cambridge University Press, London.

References

Berman R. (1965) *Physical Properties of Diamond*, Clarendon Press, London.

Bigeleisen J. and **Mayer, M.G.** (1947) *J. Chem. Phys.* **15,** 261.

Born, M and **Oppenheiner, J.R.** (1927) *Ann. d. Physik,* **84,** 457.

Born M. (1962) *Atomic Physics*, p. 333, Blackie and Son, London.

Bundy, F.P., Boverberk, H.P., Strong H.M., and **Wentorf, R.H. Jr.** (1961) *J. Chem. Phys.* **35,** 383.

Busey, R.H. and **Giauque, W.F.,** (1953) *J. Amer, Chem. Soc.* **75,** 806.

Cohen, K. (1951) *Theory of isotope separation*, McGraw-Hill N.Y.

Conyers Herrings and **Nichols, M.M.** (1949) *Rev. Mod. Phys,* **21,** 185.

Davisson C.J. and **Germer, L.S.** (1927) *Phys. Rev.*, **30,** 707.

Derjagiun, B.V. and **Fedoseov, D.B.** (1975) *Sci., Amer,* **223** (5), 102.

Eastman, E.D. and **Mc Gavock W.C.** (1937) *J. Amer, Chem. Soc.* **59,** 145.

Edmondson, W and **Egerton, A.** (1927) *Proc. Roy, Soc.* A CXIII 53.

Einstein, A. (1910) *Ann. der Physik* **XXV,** 205.

Fletcher, (1913) *Phys. Rev.* **24** 440.

Hardy, H.G. (1955) *Introduction to Pure Mathematics*, Clarendon Press, Oxford.

Hirschfelder, J.O. Curtiss, C.F. and **Bird, R.B.** (1964) *Molecular Theory of Gases and Liquids*, Wiley, N.Y.

Jain, S.C. and **Krishnan, K.S.** (1952) *Proc. Roy Soc.* **213A,** 143 and **215A** 437.

Jones, R.V. and **McCombie C.W.** (1952) *Trans. Roy Soc.*, **244,** 205.

Keesom, W.H. (1911) *Ann. der Physik,* **35,** 591.

Landolt–Bornstein. Tabellen, Springer, Berlin

Langevin P. (1908) *C.R.* **CXLVI,** 580.

Lindemann, F.A. (1919) *Phil. Mag.* 6 **38,** 173.

Mezin, M. (1979) *Revue Genérale Nuclaire, No. 2*, April, 124.

National Bureau of Standards, (1960) *Miscellaneous Publications*, **273** Washington, D.C.

Mendelssohn, K. (1973) *The World of Walther Nernst,* Macmillan, London.

Miller, R.C. and **Kusch P.** (1955) *Phys. Rev.* **99,** 1314.

Nernst, W. (1894) *Wied. Annalen*, **53,** 57.

Nernst, W. (1906) *Nachrichten von der Königlich*, Gesell der Wiss. XIII, 1.

Neumann, K. and **Volker, E,** (1932) *Zeit f. Physikalische Chemie,* **A164,** 33.

Nuckells, J. Emmelt, J. and **Wood, L.** (1973) *Physics Today* **26,** 8, 46.

Prosen, E.J., Jessup, R.S. and **Rossini, F.D.** (1944) *J. Res. Nat, Bur. Stand.* **33,** 47.

Simon F. (1930) *Fünfundzwanzig Jahre Nernstscher Warmesatz, Ergebnisse der exacten Wiss,* J. Springer, Berlin.

Smoluchowski, v, (1908) *Ann der Physik,* **XXV,** 205.

Urey, H.C. (1947) *J. Chem. Soc.*, 562.

Westgren, A. (1915) *Inaugral Dissertation*, Stockholm.

Whitley, S. (1979) *Physics in Technology,* **10,** 26.

Zare, R.N. (1977) *Sci. Amer.*, 23, 2, 80.

Zippe, G., (1960) *The development of short bowl centrifuges,* Research Laboratories for the Engineering Sciences, University of Virginia.

Appendix A

Quantum mechanics and energy levels

A.1 Early history

The study of atomic spectra by Rydberg and Ritz made it clear that the motion of electrons in the atom was not in accordance with the laws of classical mechanics. To account for the experimental results an entirely new form of mechanics, quantum mechanics, was invented. Quantum mechanics is presented as a complete system of mechanics capable of dealing both with matter in bulk and with the motion of the subatomic neutron, proton, and electron of which atoms are composed. Quantum mechanics is an entirely new way of solving mechanical problems and the only feature that it has in common with classical mechanics is that both systems assume that the fundamental unit is the massive point particle having an arbitrary mass m whose behaviour can be described in a space $(x, y, z,)$ and time (t) framework. Although the mathematical side of quantum mechanics is very abstract compared with Newton's system, its axioms are more directly connected with the physical world as revealed by experiment: this is because they contain Planck's constant h which has a value determined experimentally by measuring the photoelectric effect. Only the value of h is specified and quantum mechanics can be used to calculate the properties of non-existent hydrogen atoms with arbitrary values of e and m. Attempts to found a completely unified mechanics in which no solutions would be found for these arbitrary systems, notably by Eddington (1946) have not been generally accepted.

At the turn of the century the existence of h was not even suspected and all phenomena were explained either in terms of particles or waves; into which category they fell could be decided by a single universal experiment illustrated in Fig. A1. This is Young's two slit interference experiment with the slits replaced by some suitable periodic structure whose periodic distance has to be of the order of λ, the wave length of the radiation. The appearance of destructive interference used to be taken as conclusive evidence that

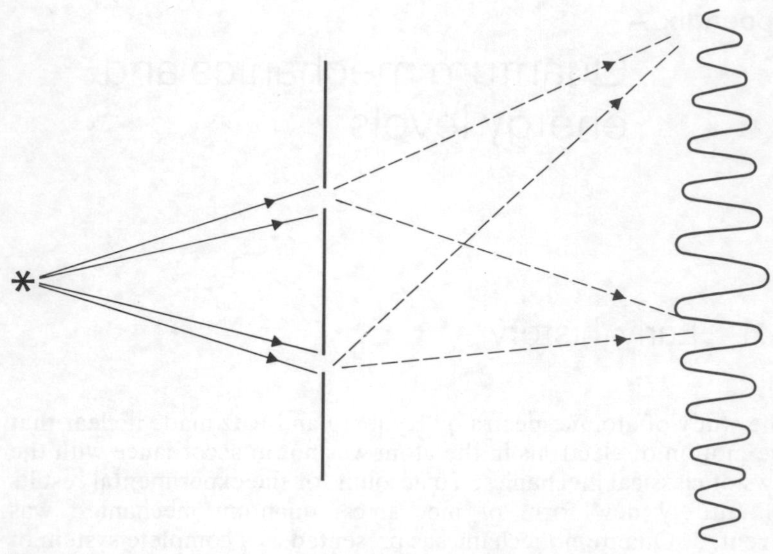

Fig. A.1 Young's two-slit interference experiment.

the radiation was a stream of waves. A good example of this type of investigation is provided by experiments with X-rays. These were at first thought to be a stream of neutral particles and showed no diffraction phenomena however fine the slits of the apparatus were made. However, when Friederich and Knipping substituted a crystal of blue vitriol for the slits the well-known diffraction pattern appeared and it was regarded as proved that X-rays were to be accounted for in terms of waves. If a radiation like γ-rays showed no destructive interference effects it could be either a stream of particles or a wavelike radiation for which no suitable periodic structure to act as a diffraction grating had yet been found. A positive experimental demonstration that a given beam of radiation was a stream of particles could only be obtained by making the detector so sensitive that it would register the arrival of the particles one by one. This was first done for α and β rays using a Wilson cloud chamber and later using a Geiger–Müller counter and other electronic devices. The photographic recording of α and β tracks in a cloud chamber was looked upon as conclusive proof that beams of α and β rays are streams of particles. This cosy classification of radiations into particles and waves was upset by Einstein's explanation of the photoelectric effect and the experimental verification of de Broglie's formula $\lambda = h/mv = h/p$ for the wave length λ of a particle of momentum $p = mv$. It turns out that, as far as we know, whether a beam of radiation is classified as a stream of particles or a beam of waves depends upon which experiment you do. This is illustrated by the work on cathode rays which when first discovered were thought to be waves. When J. J. Thomson isolated a parallel beam he passed it through electric and magnetic fields and showed that it behaved like a stream of charged particles, a conclusion later confirmed by the individual detection of β rays with a Geiger–Müller counter. If instead J. J. Thomson had carried out the experiment made years later by his son (using similar apparatus) and passed the cathode ray beam through a thin ZnO crystal he would have observed a magnificent diffraction pattern and concluded that he was dealing with a beam of wavelike radiation whose wave length λ was given by the relationship $\lambda = h/mv$. Similar accounts can be given of other types of radiation. The resolution of this paradox is achieved quite naturally by quantum mechanics which includes the numerical value of h in its axioms.

Quantum mechanics is difficult on the mathematical side and an adequate discussion of even so simple a problem as the motion of a single particle in a field free space needs a wide range of mathematical techniques and to show that an α particle will produce a track in a cloud chamber (obvious in classical mechanics) is really quite an advanced exercise. Fortunately the wave aspect of the problem can be made the basis of a simple graphical solution of the problems needed in the discussion of the kinetic theory. This is similar to the solution of acoustical problems given in most elementary text books of sound and can be illustrated by experiments on plucked strings and the vibrations of drums. This is because if the system has sharp energy levels the space dependence of the wave function obeys the same type of differential equation as the space part of a standing wave in acoustical resonance.

A.2 The calculation of energy levels

In this section it is proposed to calculate the energy levels of the systems needed in the elementary treatment of the kinetic theory. In the kinetic theory of a nearly perfect gas one represents a real situation in terms of a particle confined in a cubical box (side a) with plane, smooth, rigid, perfectly elastic walls. This simplification of the true state of the walls can be justified by detailed arguments and is usually taken for granted. The detailed justification of course depends upon the size of the box being very large compared with the range of the molecular forces. For smooth walls the change in the motion of the particle is normal to the walls and can be represented as a change in one variable (x say) if the coordinate system is suitably chosen. Similarly specular reflexion can be described in terms of two coordinates in a plane containing the normal to the wall at the point of impact. Such devices for simplifying the mathematics are commonplace in the treatment of particle motion and call for no comment. They can also be described as considering the motion of a particle along a line or confined to a plane. Such a particle can also be described as moving in a one-dimensional box (x only) or a two-dimensional box (x, y only); the constant motion in the other dimensions can be added afterwards to give the true motion in three dimensions. This way of describing the motion of a particle is useful if one wishes to consider the problem in quantum mechanics. In quantum mechanics the ideal smooth-walled box of classical mechanics is replaced by the rectangular potential well shown in Fig. A2, the rigid wall of the classical treatment being replaced by an abrupt and indefinitely large increase in potential. Such a well is described as a rectangular (angle at O is a right angle) potential well of depth V and width a. By analogy with one-, two- and three-dimensional boxes such potential wells can be one-, two- and three-dimensional ones. The one-dimensional well is correctly represented in Fig. A2a; a two-dimensional rectangular square well is represented by a deep well of square cross section in three dimensions as shown in Fig. A2b; the three-dimensional well cannot be represented graphically as it requires four dimensions (x, y, z, V) for a full representation. If the corners of the well are rounded off instead of being right angles, the well is named after the shape of the contour. The only well of this type that will be considered is the parabolic well ($V = Kx^2$) of Fig. A7 because of its importance in the theory of the s.h.m. oscillator.

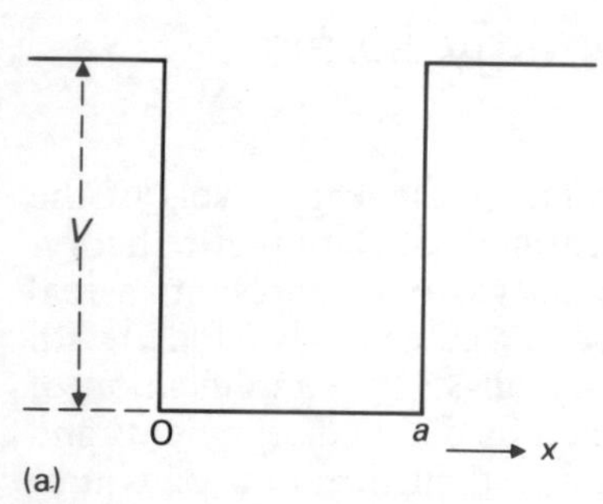

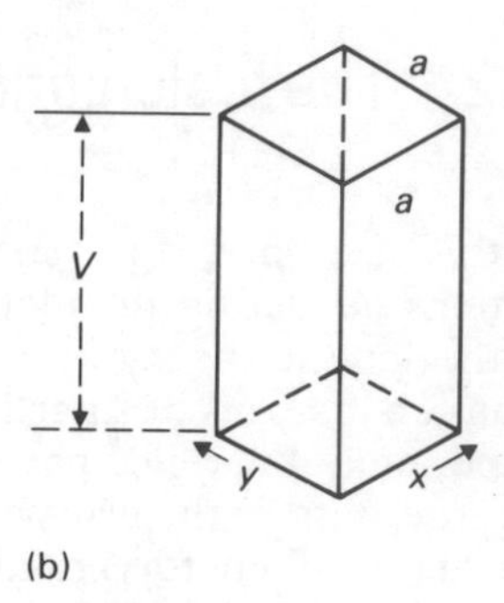

Fig. A.2 Potential wells.
(a) Square well of depth V in one dimension.
(b) Two-dimensional square well of depth V.

In quantum mechanics a particle in a one-dimensional square well has a wave function $\psi(x)$ which has the same mathematical form as a standing wave on a string stretched between the walls of the well and firmly anchored at the points $x = 0$, $x = a$. Note that the tension and density of the string which enter into the mechanical problem of the vibrations of the string do not enter into the quantum mechanical analogy which is concerned only with the shape of the standing wave. The wave lengths λ of the possible standing waves on the string are obviously $2a$, a, $2a/3, \ldots 2a/n$, as shown in Fig. A3.

The momentum of the particle can be calculated from the relationship

$\lambda = h/mv = h/p$, and the kinetic energy $= p^2/2m$ is $h^2/2m\lambda^2 = n^2h^2/8ma^2$. The total energy of the particle E is the sum of the potential and kinetic energies so that $E = (n^2h^2/8ma^2) + V$ in which V has a negative value for a well. This representation of the particle is shown in Fig. A3. In the language of wave mechanics the standing wave profile A $\sin(n\pi x/a) = \psi(x)$ is said to be the unnormalised (arbitrary A) wave function of a particle, mass m, confined in a one-dimensional rectangular potential well of depth V and width a. The horizontal lines at $E = (n^2h^2/8ma^2) + V$ are the negative energy levels of the particle. If a measurement is made of the energy of a particle in the box it will be found to have one of these energies and no other energy will ever be found. The spacing of the levels depends upon the numerical value of h. If h is allowed (in imagination) to become very small the spectrum of allowed

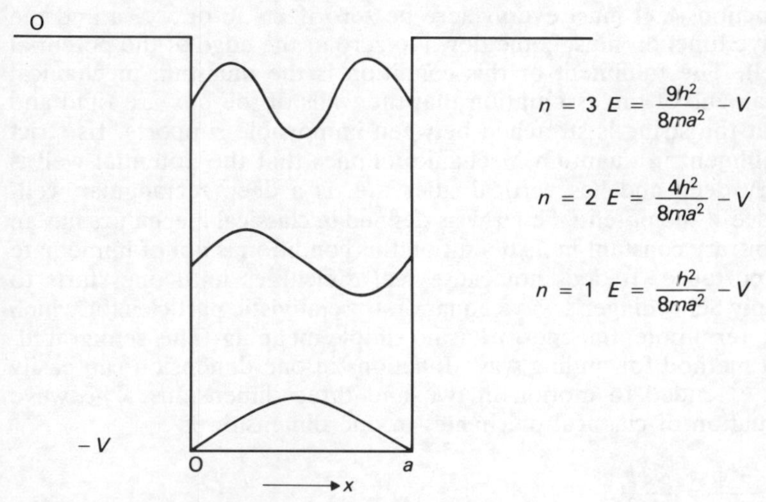

Fig. A.3 Wave functions in a square well.

energies tends to a continuum and the difference between classical and quantum mechanics disappears. The wave function itself A $\sin(n\pi x/a)$ has no direct physical interpretation and does not represent the amplitude of a wave in any real or even imagined (like the ether) physical medium: it is a mathematical symbol and is to be interpreted in the following sense. If initially the position of the particle in the box is unknown one can imagine an experiment to determine it. For example a narrow pulsed beam of light Δx in width could be directed perpendicular to the x axis and the light scattered by the particle, looked for and detected. The pulsed beam would not always illuminate the particle but there is a probability that the experiment would successfully locate the particle. The probability of finding the particle between x and $x + \Delta x$ is by definition

$$\left(\mathrm{A} \sin \frac{n\pi x}{a}\right)^2 \Delta x = (\psi(x))^2 \Delta x.$$

Clearly since the particle is confined to the potential well A must be chosen so that the sum of all the probabilities spread over all x between 0 and a is unity. This process of choosing A is called normalising the wave function, is not always mathematically trivial and is not necessary for determining the energy levels which depend only on the wavelength λ. The interpretation of $(\psi(x))^2$ as the probability of finding the particle at x brings out another important property of wave functions. Since the particle is assumed to be confined to the box and is never to be found outside it, the wave function $\psi(x)$ must everywhere be zero if $x < 0$ or $x > a$ and the wave function must come down to zero at the edge of the potential well. The fulfilment of this condition is the quantum mechanical analogue of the assumption that the walls of the box are rigid and that the string is stretched between immovable supports. Its strict fulfilment in quantum mechanics implies that the potential well is very deep and has vertical sides, i.e. is a deep rectangular well. Since V the potential energy as defined in classical mechanics has an arbitrary constant in its definition this condition is not of immediate importance. It does not cause real difficulties until one starts to apply Schrödinger's wave equation to relativistic particles for which the term potential energy has no simple meaning. This semigraphical method for finding wave functions in one dimension can easily be extended to motion in two and three dimensions. The wave equation of classical mechanics in one dimension is

$$\frac{1}{c^2}\frac{\partial^2 \phi}{\partial t^2} = \frac{\partial^2 \phi}{\partial x^2}$$

and has a universal solution

$$\phi = \text{A Sin}\left(\omega t \mp \frac{2\pi x}{\lambda} \pm \varepsilon\right)$$

which represents a plane wave (wave front normal to the x axis) travelling along the x axis. The parameter ϕ represents the disturbance which is being propagated with a constant velocity c which is independent of the wave length λ the amplitude A and the phase ε, of the harmonic wave. If the actual physical wave is travelling not along the x axis but in some arbitrary direction whose direction cosines are l, m, n, the equation of the wave is

$$\phi = \text{A sin}\left(\omega t \mp \frac{2\pi}{\lambda}(lx + my + nz) + \varepsilon\right)$$

which (as is easily verified by substitution) is a solution of the three-dimensional wave equation

$$\frac{1}{c^2}\frac{\partial^2\phi}{\partial t^2} = \frac{\partial^2\phi}{\partial x^2} + \frac{\partial^2\phi}{\partial y^2} + \frac{\partial^2\phi}{\partial z^2} = \nabla^2\phi$$

by definition of the symbol ∇^2. Pairs of waves moving in opposite directions may, by a suitable choice of ε, be combined to give standing waves $\phi = 2\text{A} \sin \omega t \sin(2\pi/\lambda)\ (lx + my + nz)$ in which the motion (as with standing waves on a string) is the product of a function of the time only with a function of position only. The part which depends on the spatial coordinates only is a solution of the differential equation $\nabla^2\phi = -\text{K}^2\phi$ in which the constant K is determined by the mechanical properties of the isotropic medium in which the wave is being propagated. This equation is seen formally to be the same as Schrödinger's time independent wave equation which is effectively by definition

$$\nabla^2\psi = -\frac{8\pi^2 m}{h^2}(E - V)\psi = -\frac{4\pi^2}{h^2}p^2\psi$$

in which V is the potential energy and p the momentum of the particle. For rectangular wells with a flat bottom such as are being used to represent a box, the analogy is exact since E, V, and p are all constants independent of the position of the particle in the well.

Now consider the motion of a particle in a two-dimensional box. This is like that of a billiard ball set in motion on a smooth square (side a) billiard table with perfectly elastic cushions and represented in a cartesian system whose origin is at a corner of the square and whose axes are parallel to the sides. In this case there is an exact analogy between the wave function representing the particle and the standing waves in a membrane stretched over a rigid square frame, i.e. a 'square tambourine' of side a. The analogy is exact because not only are the two time-independent wave equations of the same form but the mechanical waves and the Schrödinger waves both have to come down to zero amplitude over the sides of the square.

As is easily seen by substitution the profile of the standing wave is given by $\psi = \mathrm{A}\sin(q\pi x/a)\sin(r\pi y/a)$ in which q and r are the positive integers 1, 2, 3 . . . The energy levels are easily found from the differential equation. Since

$$\nabla^2\psi = -\frac{\pi^2}{a^2}(q^2 + r^2)\psi = -\frac{4\pi^2}{h^2}p^2\psi$$

so that

$$p^2 = \frac{h^2}{4a^2}(q^2 + r^2)$$

and the kinetic energy

$$E - V = \frac{h^2}{8a^2m}(q^2 + r^2).$$

As in the one-dimensional example the levels are discrete and can be written down by assigning successive values to the integers q and r. It will be noticed that the energy levels are degenerate because the quite different wave functions, $\mathrm{A}\sin(q\pi x/a)\sin(r\pi y/a)$ and $\mathrm{A}\sin(r\pi x/a)\sin(q\pi y/a)$ have the same energy level with kinetic energy

$$\frac{h^2}{8a^2m}(q^2 + r^2) = \frac{h^2}{8a^2m}R^2$$

in which R^2 is necessarily an integer. For large values of R^2 the number of degenerate levels is large and the point of interest is the density of levels. By density of levels is meant the number of levels between two energies E and $E + \mathrm{d}E$ usually expressed in terms of the corresponding momenta p and $p + \mathrm{d}p$. For large R this density is readily approximated to as shown in Fig. A4 in which the pairs of integers q and r are plotted to give a square lattice. The quarter circle of radius R contains all the pairs of integers for which

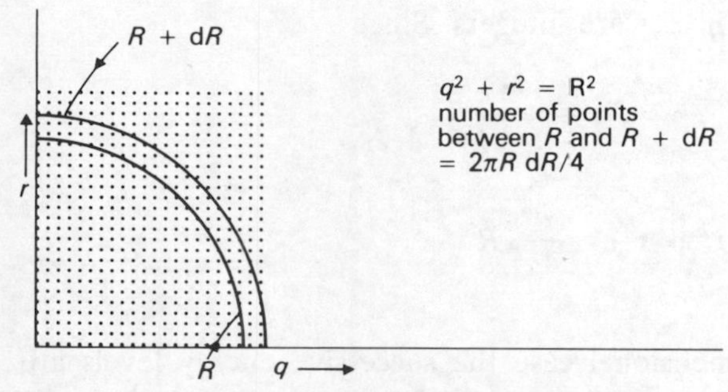

Fig. A.4 Density of states in two dimensions.

$q^2 + r^2 < R^2$. For large values of R the number n of such pairs is $\pi R^2/4$ because the density of lattice points is one per unit area and $n = \pi/4 \cdot 4a^2/h^2 \cdot p^2$ so that the density of states

$$\frac{\mathrm{d}n}{\mathrm{d}p} = \frac{2a^2\pi}{h^2}\,p\mathrm{d}p.$$

Note that the expression is linear in p and proportional to the area of the square in which the particle is contained. This particular result is not of any great physical interest but serves as an introduction to the calculation of the energy levels of a particle in a cubical box of side a and volume a^3 upon which the quantum theory of a perfect gas is founded.

In three dimensions the box is represented by a three-dimensional rectangular well of depth V and width a and one has to find a solution of Schrödinger's equation in three dimensions

$$\frac{\partial^2\psi}{\partial x^2} + \frac{\partial^2\psi}{\partial y^2} + \frac{\partial^2\psi}{\partial z^2} = -k^2(E - V)\psi = -\frac{4\pi^2p^2}{h^2}\,\psi$$

which is zero over the six faces of a cube of side a. This is essentially the same problem as finding the acoustical resonances of a cubical box resonator with rigid walls. By analogy with the two-dimensional case one can write down the solution as

$$\psi = \mathrm{A}\sin\frac{q\pi x}{a}\sin\frac{r\pi y}{a}\sin\frac{s\pi z}{a}$$

which is readily found by substitution to solve the differential equation, and to be zero over the six planes $x = y = z = 0$,

$x = y = z = a$ if q, r, s are integers. Since

$$\nabla^2\psi = \frac{-\pi^2}{a^2}(q^2 + r^2 + s^2) = -\frac{4\pi^2 p^2}{h^2}\psi$$

$$p^2 = \frac{h^2}{4a^2}(q^2 + r^2 + s^2) = \frac{h^2}{4a^2}R^2$$

As in the two-dimensional case the successive energy levels are obtained by assigning positive integral values to q, r and s. The energy of the molecules of a perfect gas which are to occupy these levels is $3kT/2$ and a little arithmetic shows that this energy corresponds to very large values of R^2. For large R the density of states can be calculated in the same way as was used in the two-dimensional case. The triplets of integers q, r, s are set out in a three-dimensional cubic lattice and a sphere of radius R is constructed with the origin as centre. Then all the points for which $q^2 + r^2 + s^2 \leqq R^2$ lie in the positive octant of this sphere, and this is the number of energy levels n whose kinetic energy is less than corresponds to a momentum given by $p^2 = h^2/4a^2 R^2$. Since the density of lattice points is one per unit volume

$$n = \frac{1}{8}\cdot\frac{4\pi}{3}R^3 = \frac{\pi}{6}\cdot\frac{8a^3p^3}{h^3}$$

and the density of states $\mathrm{d}n/\mathrm{d}p$ is given by

$$\frac{\mathrm{d}n}{\mathrm{d}p} = \frac{4\pi a^3 p^2}{h^3} = \frac{4\pi p^2}{h^3} \text{ per unit volume}$$

This is the result used in section 6.10. An important feature of this expression is that when expressed in this way the dimensions of the box do not appear and it turns out to have been a mathematical dummy introduced to make the calculations easy and the result is true for any large box not necessarily cubical. If thought desirable the derivation of these formulae can be cast in a form which avoids mentioning a definite volume a^3 at all. If one applies this expression to particles which are light quanta for which $p = h\nu/c$ one must include a factor 2 to give $\mathrm{d}n = 8\pi\nu^2\mathrm{d}\nu/c^3$ as the number of quanta per unit volume whose frequency lies between the frequencies ν and $\nu + \mathrm{d}\nu$. The factor 2 is justified by saying that light waves are vector waves and have two independent planes of polarisation.

The energy levels of a particle moving in a circle can be obtained by an extension of the method used in calculating the levels in a one-dimensional well. The mechanical analogue is now the standing radial waves in a loop of wire bent into a circle of radius r. The condition of resonance is now that the circumference of the loop is a whole number of wavelengths so that $n\lambda = 2\pi r = nh/mv$ and $mvr = nh/2\pi$. This is equivalent to quantising the angular momentum of the particle in units of $h/2\pi$, a result originally obtained by Bohr in his pioneer study of the hydrogen atom. It leads at once to the expression

$$E = \frac{n^2h^2}{8\pi^2mr^2} = \frac{n^2h^2}{8\pi^2I}$$

for the energy levels of a particle rotating in a plane. To extend this result to the more realistic case of a particle rotating in three dimensions is mathematically more difficult and corresponds to investigating the standing waves in the vibration of a thin spherical shell. The result is not very different from that obtained in two dimensions and the energy levels are given by $K(K + 1)h^2/8\pi^2I$ in which K is the integral rotational quantum number 0, 1, 2 . . . and I the moment of inertia of the rotator. For a detailed treatment see Schiff (1949) or Pauling & Wilson (1935). If the moment of inertia of the rotating body cannot be represented by a single moment of inertia I the expressions for E become complicated and a simple treatment is confined to molecules like N_2 and O_2 which can be looked upon as two point masses m_1 and m_2 a fixed distance R apart whose moment of inertia is $(R^2m_1m_2)/(m_1 + m_2)$ about any axis perpendicular to the line of centres and passing through the centre of mass.

Finally the energy levels of a particle oscillating with simple harmonic motion must be considered because of the important part they play in the elementary theory of simple solids. The motion of a particle in a one-dimensional box is the same as that of a bead on a smooth horizontal wire moving between fixed perfectly elastic stops. The motion would be described as that of a symmetrical saw toothed oscillator (Fig. A5). This oscillatory motion can be modified by bending the wire into a concave curve in a vertical plane (Fig. A6) and it is well known (Huyghens, 1673) that if the wire is bent into a cycloid the motion becomes simple harmonic with a period independent of the amplitude: it requires more than trivial mathematics to show this however. Similarly the simple sinusoidal wave functions of a particle in a one-dimensional rectangular potential well are modified by rounding off the corners of the potential well and the spacing between the energy levels is changed.

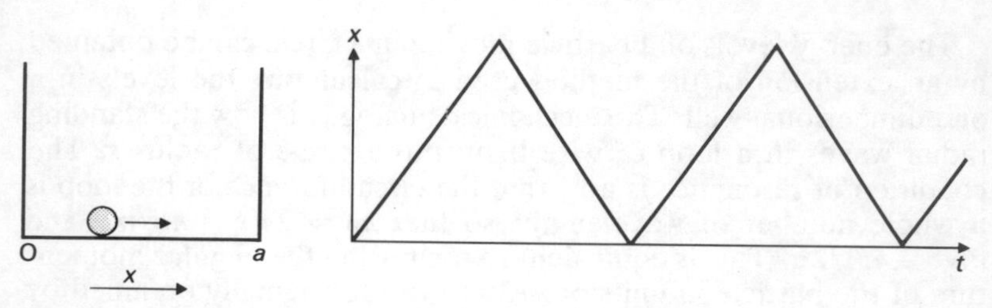

Fig. A.5 Oscillations of a ball in a smooth box.

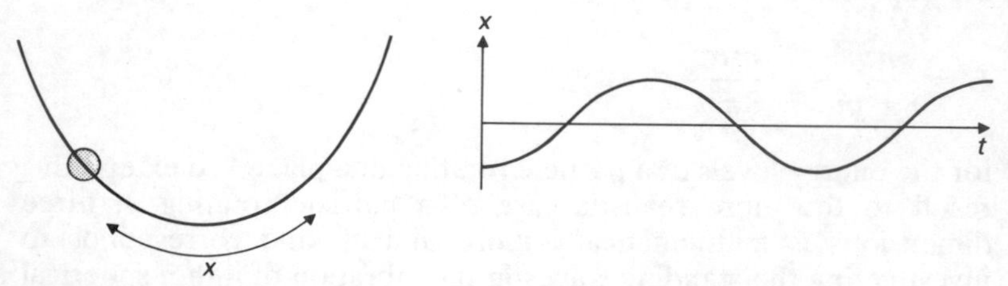

Fig. A.6 Oscillations of a smooth bead on a cycloidal wire.

An important shape for the potential well (corresponding to the cycloidal wire) is the parabolic potential well in which $V = \mathrm{K}x^2$. The dependence of the potential upon position is the same as that in a classical harmonic oscillator ($\mathrm{K} = \omega^2 m/2$ in the usual notation) and the wave functions and energy levels are those of the quantum mechanical harmonic oscillator. These wave functions and energy levels are to be obtained by solving the time independent Schrödinger equation with variable potential energy $V = V(x)$,

$$\frac{\mathrm{d}^2\psi}{\mathrm{d}x^2} = k^2(V - E)\psi = \frac{8\pi^2 m}{h^2}\left(\frac{m\omega^2 x^2}{2} - E\right)\psi$$

This is not an easy equation to solve but the resultant energy levels form a particularly simple system, $E = (h\nu/2; + nh\nu) = (n + \frac{1}{2})h\omega$ in which ν is the frequency, ω the angular frequency of the corresponding classical oscillator and n is one of the positive integers $0, 1, 2, \ldots$.

The first two wave functions and energy levels are shown in Fig. A7. The simple ladder structure of the energy levels was discovered by Planck (1905) and is the starting point from which the whole subject has been built up. The term $\frac{1}{2}h\nu$ which prevents the oscillating particle ever coming to rest is called the zero point energy. It is characteristic of a quantum mechanical treatment and is well justified by experiment.

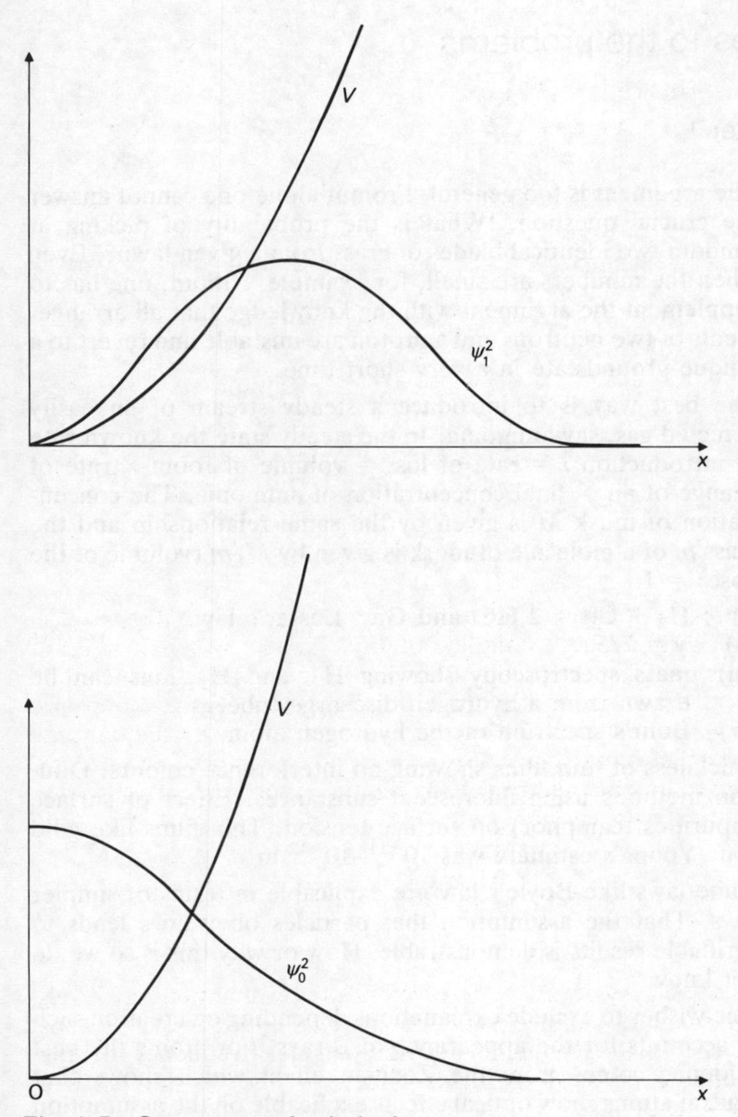

Fig. A.7 Quantum mechanical s.h.m.
The square wave function Ψ^2 shows the probability of finding a particle executing S.H.M. in a parabolic potential well $V = \frac{1}{2}\omega^2 mx^2$.

Notes to the problems

Chapter 1

1 The argument is too general. From it alone one cannot answer the crucial question, 'What is the probability of picking at random two identical blades of grass from a given lawn?' Even when the numbers are small, for example, tritium, one has to supplement the argument with the knowledge that all arrangements of two neutrons and a proton are unstable and revert to a unique groundstate in a very short time.

2 The best way is to introduce a steady stream of an easily detected gas, say ammonia. In the steady state the known rate of introduction r = rate of loss = volume of room × rate of change of air × final concentration of ammonia. The concentration of musk M is given by the same relationship and the mass m of a molecule of musk is given by M/m (volume of the nose) = 1.

3 (i) $H_2 + Cl_2 = 2\,HCl$ and Gay–Lussac's law.
(ii) $\gamma = 7/5$.
(iii) mass spectroscopy showing H^+ and H_2^+ ions can be drawn from a hydrogen discharge tube.
(iv) Bohr's spectrum of the hydrogen atom.

4 Thickness of thin films showing no interference colours. Dilution methods using fluorescent substances. Effect of surface impurities (camphor) on surface tension. Thin films like gold leaf. Young's estimate was 10^{-11}–10^{-12} m.

5 Some laws like Boyle's law are explicable in terms of simpler laws. That the assumption that particles obey laws leads to verifiable results is demonstrable. How or why this is so we do not know.

6 One wishes to exclude explanations depending on creation such as accounts for the appearance of β rays. For atoms the best evidence comes from the Zeeman effect which shows that neutral atoms show optical effects explicable on the assumption that they contain particles with same e/m as that of the electron.

7 For a simple pendulum of 'honest mass' m and added non-gravitational mass Δm

$$T = 2\pi\sqrt{\frac{l(m + \Delta m)}{mg}} = T_0\left(1 + \frac{\Delta m}{m}\right)^{1/2}$$

An easily obtained accuracy in T of one part in ten thousand requires $\Delta m/m \approx 2 \times 10^{-4}$ small compared with the $\Delta m/m \approx 10^{-3}$ obtained from the known masses. The experiment has of course been tried, with negative results.

8 Arrange for the reaction to take place in a sealed vessel and weigh before and after. The chief difficulty is in ensuring that the external surface does not change its layer of absorbed water vapour and that the liberated heat energy does not expand the balance arm. An easily achieved accuracy is 10^{-5} per cent. For the case quoted the loss was traced to the formation of nitric oxide which expanded the glass containing vessel and upset the buoyancy correction.

9 The data only suffice for a rough calculation neglecting end effects, thickness of walls etc. The number of neutrons involved = (volume of gas leaving) × (number of atoms per unit volume) × (neutrons per argon atom) = approximately 7.0×10^{28}.
The total charge = $VC = 3 \times 10^{-6}$ × capacity of an ideal cylindrical condenser = $3 \times 10^{-6} \times 2\pi r l \varepsilon_0/\Delta r = 6.6 \times 10^{-15}$ Coulombs = 4.1×10^4 electron charges. Therefore each neutron carries not more than 6×10^{-25} electron charges.

10 Neither Van Helmont's figures or conclusions are disputed. However, air is now known to contain 3–4 parts in ten thousand of carbon dioxide which is the source of the increase in weight. If this is removed the plant will not grow.

11 Since the field outside a spherical distribution of charge is the same as that of an equal point charge at the centre the field outside the atom is zero and there is no interaction if $R > 2a$. If the atoms are a little closer than this the distributed charges overlap. The intruding negative charge is now in the field of a net positive charge and there is an attractive force. When the two atoms approach to a distance $R < a$ the positive point charges are no longer screened and the force changes sign. Heitler and London's treatment of the interaction between two hydrogen atoms is of course quite different.

12 The total negative charge is

$$\int_0^\infty 4\pi r^2 \rho \, dr = \int_0^\infty 4\pi A e^{-r/a} \, r^2 dr$$

which integrates by parts to give $A = Ze/8\pi a^3$. The total charge within a radius r is $Ze/2a^3 \cdot e^{-r/a}\,(r^2 a + 2ra^2 + 2a^3)$

and the corresponding potential on the surface of this charge is

$$\frac{Ze}{8\pi\varepsilon_0 a^3} e^{-r/a} \left(\frac{2a^3}{r} + a^2\right).$$

13 Applying $PV = \tilde{n}RT$ to 6.7×10^{-8} moles of a perfect gas gives $V = 1.66 \times 10^{-7}$ m^3 = 166 cubic mm at 1/10 atm pressure. It is always a surprise to find that absorbed layers amount to quite large volumes of gas measurable with quite crude equipment. There are $1/\sigma^2$ molecules per m^2 in a monomolecular layer and the figures correspond to layers 1, 2 and 3–4 molecules thick if $\sigma = 9.6 \times 10^{-11}$ m. This is about a quarter of the expected molecular diameter so that the layers must have been several molecules thick.

14 See Fig. 1.1b. $V = 0$ if $r = \sigma$, $\mathrm{d}V/\mathrm{d}r = 0$ if $r = 1.12\sigma$. The attractive force is greatest when $\mathrm{d}^2V/\mathrm{d}r^2 = 0$ and $r = 1.25\sigma$ and falls to 1 per cent of this value when $r = 2.68\sigma$.

15 If r_1 and r_2 are measured from the C of G

$$m_1 r_1 = m_2 r_2, \; r_1 + r_2 = r \quad \text{and} \quad I = \Sigma m r_i^2$$

$$r_1 = \frac{m_2 r}{m_1 + m_2}, \quad \text{and} \quad I = \frac{(m_1 m_2^2 + m_2 m_1^2) r^2}{(m_1 + m_2)^2} \quad \text{etc.}$$

Chapter 2

2 Use $p = \tilde{n}kT$ to give $p = 10^{21} \times 10^6 \times 1.38 \times 10^{-23} \times 273 = 3.77 \times 10^6$ Pa = 37.2 atm.
Closely packed spheres occupy $\pi/3\sqrt{2} = 0.74$ of the available volume (this should be deduced from the geometry of the tetrahedron) leaving 0.26 available for the helium. The pressure will be greater by at least a factor of $1/0.26 = 3.84$.

3 The pressure is infinite when $v = b$ and when the spheres touch. For cubical packing $b = 6/\pi = 1.92 \times$ volume of the spheres. For close packing the factor is 1.35. These factors are to be compared with van der Waals' estimate of b = four times the volume of the molecules.

4 Differentiate twice and then put the differentials equal to zero. Taking Dieterici's equation as an example this gives at once that

$$0 = e^{-a/RT_cV_c} \left(\frac{a^2}{RT_c V_c^4} - \frac{2a}{V_c^3}\right)$$

so that $V_c = a/2RT_c$ and at the critical point the exponential factor is e^{-2}. From the first differentiation when $(\partial p/\partial V)_T = 0$ it follows at once that $P_c = a/V_c^2 \, e^{-2}$ and $RT_c/P_cV_c = \frac{1}{2}e^2 = 3.69$.

5 The vapour pressure equation $p = p_0 \, e^{-\lambda/RT}$ is looked upon as giving the relationship between the internal pressure p_0 and the external pressure p in general. Writing the equation for a gas as $p_{internal} \, (V - b) = RT$ gives $p_{external} \, e^{\lambda/RT} \, (V - b) = RT$. Since λ is proportional to the work done in removing a molecule from the attractive forces exerted on it by the molecules within the small range of the attractive forces it is reasonable in the case of a gas to replace λ by a/v in which a represents the attractive forces and $1/v$ the number of molecules interacting with each molecule in the body of the gas and so obtain Dieterici's equation of state.

6 From the text $T_2 = T_1 \, (p_1/p_2)^{-2/5}$ and $\log T_2 = \log 300 - 0.4 \log 120$ giving $T_2 = 44$ K. The error due to assuming helium to be a perfect gas will be small but the fall in temperature will be less because the cylinder walls will not cool down. In the second example the effect of the walls will be small but owing to the low temperature reached (~5 K) the deviations from the perfect gas formulae should be taken into account.

7 Work done on the gas $= W = -\int_{p_1}^{p_2} p \, dv$. For an isothermal change $W = RT \log_e(p_2/p_1)$ per mole and the power is $8.31 \times 300 \times \log_e 100 = 11.48$ kW. For an adiabatic change $W = -\int_{p_1}^{p_2} p \, dv$ with $pv^\gamma =$ constant. This integrates at once but it requires a little reduction to eliminate k and give

$$W = \frac{RT_1}{\gamma - 1}\left(\left(\frac{p_2}{p_1}\right)^{\gamma - 1/\gamma} - 1\right) \text{ per mole}$$

The required power is

$$\frac{3 \times 8.31 \times 300}{2} \cdot ((100)^{2/5} - 1) = 19.9 \text{ kW}.$$

It is straightforward to introduce an intermediate pressure p and differentiate to show that the most economical value is given by $p = \sqrt{p_1 p_2}$.

8 Equate the values of RT given by the virial expansion and van der Waals' equation; substitute van der Waals' value for RT (expanding when necessary by the binomial theorem) and equate coefficients to find

$$A_v = b - \frac{a}{RT} \quad \text{and} \quad B_v = b^2.$$

9 Differentiate to find $(\partial p/\partial T)_v = R/(v - b) =$ constant.

10 From elementary mechanics K.E. $= mgh$ in any consistent set of units. Hence $h = 0.97$ m.

11 As the composition of the gas in the bulbs will change express the gas laws in the general form $PV = \tilde{N}RT$. Initially the helium will expand into the argon leaving $\tilde{n}_1$ moles of pure helium in the small bulb in an obvious notation. Then $pv = \tilde{n}_1 RT_1$ and $pV = (\tilde{N} + \tilde{n} - \tilde{n}_1)RT_2$. Equating the values of p gives

$$\tilde{n}_1 = \frac{Pv(V + v)T_2}{RT(T_1V + T_2v)}$$

$$\text{and } p = \frac{\tilde{n}_1 RT_1}{v} = \frac{PT_1T_2(V + v)}{T(T_1V + T_2v)} = \frac{14(V + v)}{15V + 7v} \text{ atm.}$$

The initial ratio of He/A in the large bulb is

$$\frac{\tilde{n} - \tilde{n}_1}{\tilde{N}} = \frac{v}{V}\left(1 - \frac{(V + v)T_2}{T_1V + T_2v}\right) = \frac{8v}{15V + 7v}$$

Diffusion will not change the pressure but the composition will finally have the uniform value $\tilde{n}/\tilde{N} = v/V$.

13 For any gas $\gamma = (C_v + R)/C_v$ in which R represents the work done on expansion. For nearly perfect gases the C_v's add: so for the mixture $C_v = \frac{1}{2}(3R/2 + 5R/2) = 2R$ and $\gamma = 3/2$. The change in temperature is calculated from

$$\frac{T_1}{T_2} = \left(\frac{p_1}{p_2}\right)^{\gamma - 1/\gamma} \quad \text{to give } T_2 = 1105 \text{ K}.$$

Chapter 3

1 Since the formula weight of $PbSO_4$ is 303.3 the rate at which SO_2 is being absorbed is

$$\frac{1 \times 10^{-3} \times 6.022 \times 10^{23} \times 10^4}{303.3 \times 30 \times 8.64 \times 10^4} = 7.66 \times 10^{15} \text{ mol/m}^2 \text{ s}$$

At s.t.p. $n\bar{c}/4$ for SO_2 is $2.687 \times 10^{25} \times 300/4 = 2.01 \times 10^{27}$ so that the amount of pollution is $7.66 \times 10^{14}/2.01 \times 10^{27} = 3.8 \times 10^{-6}$ ppm. (The normal amount in industrially polluted air is 0.1 ppm.)

2 The radiometer pressure is approximately, cf section 4.7, $p\Delta T/4T$ with $\lambda > r$ the radius of the spherical vessel. In terms of the molecular diameter σ this becomes $k\Delta T/4\sqrt{2}\pi\sigma^2 r$ so that a gas like helium with a small value of σ^2 is best. The existence of the radiometer effect shows that the effect of light pressure is only important when the pressure is very low and conduction through the gas is small. If conduction can be neglected ΔT is determined by radiation. Assuming the vane to be a non-conducting black body the heat energy radiated is $4\sigma T^3 \Delta T = 6.12\ \Delta T$ W/m^2. (σ = Stefan's constant, T = room temperature). Thus for an incident flux F, $\Delta T = F/6.12$ and the radiometer pressure $\approx pF/24.48T$. The corresponding radiation pressure is F/c for normal incidence and the two pressures are equal if $p \approx 24.48T/c = 2.4 \times 10^{-5}$ Pa. This is about 10^{-4} of the pressure at which the radiometer effect becomes noticeable. If required a second approximation can be made allowing for conduction in the gas and through the vane.

3 The value of $n\bar{c}/4$ for O_2 at 513 °C and 0.01 atm is 1.68×10^{25} molecules m^{-2}s^{-1} = 0.893 kg m^{-2}s^{-1}. The max rate of absorption by a 1 μm diameter particle is 2.8×10^{-12} kg/s. The Arhenius factor $\exp(-E/RT) = \exp(-24.33) = 2.714 \times 10^{-11}$ and the maximum rate if gain of weight $\Delta w = 7.6 \times 10^{-23}$ kg/s. Taking the density of B_4C to be approximately 2.25×10^3 one obtains 30×10^{-4} hour for $\Delta w/w$ the quantity measured by Dominey to be 2×10^{-3} h. The difference between the maximum possible calculated rate and the observed rate is no great cause for embarrassment. The process of oxidation is a complex one whose detailed mechanism is unknown as is shown by the observation that the rate is proportional to (pressure)$^{0.62}$ and not to the pressure as assumed in the calculation.

4 No great accuracy can be expected from the measurement of so small a diagram but a change of 5 in $\log_{10}$(Rate) is caused by a change in $1/T$ of approximately 5.02×10^{-4}. If Rate = $Ae^{-E/RT}$, $\log_e$(Rate) = $\log_e A - E/RT$, $E/R = 2.29 \times 10^4$ and $E = 1.90 \times 10^5$ J.

5 The explanation is that the capillaries in the walls of the pot are so narrow that the molecules in them collide with the walls rather than each other even at atmospheric pressure. The conditions are therefore those of molecular flow. The rates of flow are proportional to $\bar{c}$ and the hydrogen flows in $\sqrt{14.6} = 3.82$ times as fast as the air flows out. The porous material must therefore have capillaries $\sim 10^{-7}$ m in diameter, which is very fine. There must be no large capillaries through which the mixed gases will flow out owing to the excess mechanical pressure and the walls should be as thin as is consistent with mechanical strength. Assume a wall thickness of 1 mm and capillaries spaced 10 diameters apart. The flow of H_2 per capillary is $2\pi n\bar{c}r^3/3l = 4.25 \times 10^{-16}\, n$. So the *net* flow is $4.25\, n \times 10^{-4} \times 2.82/3.82 = 3.14\, n \times 10^{-4}$ mol/m²s into the pot. For a pot radius R containing $4\pi R^3 n/3$ molecules the rate of rise of pressure is $9.42 \times 10^{-4}/R$ atm/s. For $R = 2.5 \times 10^{-2}$ this corresponds to 38 cm of water pressure per second so that an effective lecture experiment can be based upon a porous pot which departs considerably from the ideal specification.

6 The air pumped is 1000 l $\times 10^{-6} = 1$ cm³ per second at s.t.p. and can conveniently be collected and measured at the outlet of the forepump. The vacuum pressure is adjusted to 0.1 Pa with a variable leak.

7 The apparatus must be on rather a large scale for laboratory work and one has to devise economical ways of avoiding end corrections.

8 The same calculation as that of section 3.3.

9 (a) The total number of diffusing molecules is $\int_0^\infty 4\pi r^2 n dr$. Change the variable from r^2 to $y^2 = r^2/4Dt$. The integral then becomes $K \int_0^\infty y^2 e^{-y^2} dy$ which does not depend upon t.

(b) Differentiate with respect to t at constant r to give

$$\left(\frac{\partial n}{\partial t}\right)_r = \frac{1}{(4\pi D)^{3/2}} e^{-r^2/4Dt} \left(\frac{r^2}{4D} t^{-7/2} - \frac{3}{2} t^{-5/2}\right)$$

This is zero and n is a maximum when $t = r^2/6D$.

10 Preliminary calculations: radius of the bulb $R = 7.81 \times 10^{-2}$ m, area $= 7.67 \times 10^{-2}$ m², $\bar{c}$ at 400 °C = 889.5 m/s. Take 1 atm = 10^5 Pa. Number of molecules to be pumped out $= 20 \times \text{area}/\sigma^2 = 20 \times \text{area} \times \sqrt{2}\pi n\lambda = 3.74n \times 10^{-7}$. Rate of pumping $= 2\pi n^*\bar{c}r^3/3l$ with n^* = molecules/m³ at 0.01 Pa and 400 °C $= 4.05n \times 10^{-8}$ giving a pumping rate of

7.5×10^{-12} n molecules/s. The pumping time is therefore $3.74n \times 10^{-7}/7.5 \times 10^{-12}$ $n = 4.98 \times 10^4$ s $= 0.57$ days. From the relationship $n_1\lambda_1 = n_2\lambda_2$ the mean free path is 156.2 m when $n_2 = 3.521n \times 10^{-10} = 8.69 \times 10^{-3}$ n^* and the pressure has to be reduced by this factor. Allowance must now be made for the diminishing pressure as the residual gas is pumped out. If $\mathrm{d}n/\mathrm{d}t = -\mathrm{A}n$ then $n_t/n_0 = \mathrm{e}^{-\mathrm{A}t}$. In this case $n_t/n_0 = 8.69 \times 10^{-3}$ and $\mathrm{A} = 1.86 \times 10^{-1}$ to give the correct initial pumping speed of 1.86×10^{-4} n^* mol/s. Thus $\mathrm{A}t = \log_e n_0/n_t$ and $t \approx 25$ s.

11 To escape from an earth of radius a and mass M the total energy of the molecule $= \frac{1}{2}mc^2 - mMG/a$ must be positive and $c^2 > 2MG/a = 2ag = 1.25 \times 10^8$ and $c > 1.12 \times 10^4$ m/s. This is about ten times the mean velocity of helium molecules at s.t.p. The fraction of molecules having 100 times the mean energy is so minute ($\sim\mathrm{e}^{-100}$) that escape from the earth can be ruled out except in astronomical times. If, as is shown in the tables, the temperature of the upper atmosphere is some 7 times that at the tropopause where the isothermal atmosphere starts the position is quite different and simple considerations can no longer be used to give a definite answer.

12 From Boyle's law $\rho = mp/kT$ so that $1/p \;\; \mathrm{d}p/\mathrm{d}h = -(mg/kT)$. This equation integrates at once to give

$$\log p - \log p_0 = -\int_0^h \frac{mg\mathrm{d}h}{kT}$$

in which g the local value of 'g' is a function of h. From Newton's law $g_h = MG/(a + h)^2$; thus

$$\log p - \log p_0 = -\frac{mMG}{kT}\int_0^h \frac{\mathrm{d}h}{(a + h)^2} = -\frac{mMG}{kT}\cdot\frac{h}{a(a + h)}$$

which is the required result since $MG = a^2 \times$ (surface value of g_h). This result applies independently to the molecules of a mixture so that

$$\log_e p_{\mathrm{He}} - \log_e p_{\mathrm{N_2}} = \log_e p_{0\mathrm{He}} - \log_e p_{0\mathrm{N_2}} + \frac{g}{kT}\cdot\frac{ah}{a + h}(m_{\mathrm{N_2}} - m_{\mathrm{He}}).$$

Using the known composition of the air, this becomes

$$\log_e 100 = \log_e 5.2/7.8 \times 10^5 + \frac{24g \times 10^{-3}}{RT} \cdot \frac{ah}{a+h}.$$

Taking $T = 236$ K its value at the tropopause where the isothermal atmosphere starts gives $ah/(a+h) \approx 137$ km and $h \approx 140$ km. It will be noted that the variation in gravity is not an important factor. The result is not in agreement with observation because the upper atmosphere is not isothermal.

Chapter 4

1 Mass flow $= F = -p\mathrm{d}p/\mathrm{d}l = -\frac{1}{2}\mathrm{d}p^2/\mathrm{d}l$ so that $p_2^2 - p_1^2 = 2\,Fl$ as stated.

2 If a unit cube is sheared by a force F the angle of shear $\theta = F/G$ by definition. If there is a relaxation time τ this shear will disappear so that $\theta = \theta_0\,\mathrm{e}^{-t/\tau}$ and $\mathrm{d}\theta/\mathrm{d}t = -\theta/\tau$. If this relaxation is looked upon as due to a viscous force $= \eta \times$ (velocity gradient) $= -\eta\,\mathrm{d}\theta/\mathrm{d}t$ one has $F = \eta\theta/\tau = G$ and $\tau = \eta/G$.

3 The viscous forces acting on a unit square of glass dx thick are a retarding force $-\eta\,\mathrm{d}v/\mathrm{d}x$ on the bottom surface and a force in the direction of the flow

$$\left(\frac{\mathrm{d}v}{\mathrm{d}x} + \frac{\mathrm{d}}{\mathrm{d}x}\left(\frac{\mathrm{d}v}{\mathrm{d}x}\right)\mathrm{d}x\right)$$

on the top surface. The viscous forces will be in equilibrium with the gravitational force $g\rho \sin\theta\,\mathrm{d}x$ acting down the plane perpendicular to x if

$$\eta\,\frac{\mathrm{d}^2v}{\mathrm{d}x^2} = -g\rho \sin\theta = \text{constant}.$$

Accordingly

$$v = \frac{g\rho \sin\theta}{\eta}\left(\mathrm{d}x - \frac{x^2}{2}\right)$$

a form which satisfies the two boundary conditions $v = 0$ if $x = 0$ and $\mathrm{d}v/\mathrm{d}x = 0$ if $x = d$.

$$\text{The net flow} = \int_0^d v\mathrm{d}x = \frac{g\rho \sin \theta d^3}{3\eta}$$

$$= 9.81 \times 2.5 \times 10^3 \times 0.342 \times (5 \times 10^{-3})^3/3\eta$$

$$= \frac{3.49 \times 10^{-4}}{\eta} \text{ m}^3/\text{s per meter of front.}$$

The velocity of the surface of the glass is $g\rho \sin\theta \, d^2/2\eta = 1.05 \times 10^{-1}/\eta$ m/s which becomes 7×10^{-122} 3.4×10^{-16} and 3.11×10^{-5} m/s at the three specified temperatures. Only the last velocity is actually observable. It is from such measurements that the numerical value of η is obtained.

4 Partial differentiation gives $\partial\theta/\partial t = -(4\pi^2 a/\lambda^2)\theta$, a $\partial^2\theta/\partial x^2 = -4\pi^2 a\theta/\lambda^2$ which justifies the solution.

5 Direct substitution (remembering that c_p = heat capacity per kg) gives $\alpha = \nu^2 \times 1.6 \times 10^{-10}\,\text{m}^{-1}$.

(a) Taking the criterion for loss of balance to be

$$\frac{\text{Intensity at 100 Hz}}{\text{Intensity at } 10^4 \text{ Hz}}$$

to change by a factor 2 gives $\exp(-2\alpha(10^4 - 10^8)x) = 2$ and $x \approx 21.6$ m.

(b) for 1 cm sonar $\nu^2 \approx 1.1 \times 10^9$ and $I = I_0 \exp(-0.352x)$ corresponding to $I = 8.76 \times 10^{-4}\, I_0$ at 20 m.

6 About 5 per cent more due to the molecules with long free paths carrying a greater proportion of the drift momentum. On Knudsen's hypothesis the drift momentum of a molecule is $mvc/\bar{c}$ instead of a uniform mv. The increased momentum transport can be approximately calculated by using a velocity weighted free path $1/\bar{c}\int_0^\infty c\lambda_c f(c)\mathrm{d}c$ instead of λ. It so happens that this difficult integral was evaluated long ago by Tait to be 1.05; this is the basis of the statement that the viscosity will be increased by 5 per cent.

7 The calculation is made in the same way as in the worked example of section 4.7. The expression for the pressure on either plate must be modified because the molecules leave the plates with temperatures $T_{\text{reflected}}$, conveniently written as t. The pressure on the hot plate is accordingly $n_1kt_1/2 + n_2kt_2/2$.

Using the defined value of

$$\alpha = \frac{T_{\text{gas}} - T_{\text{reflected}}}{T_{\text{gas}} - T_{\text{plate}}}$$

one finds that

$$t_1 = \frac{\alpha_1 T_1 + \alpha_2(1 - \alpha_1)T_2}{\alpha_1 + \alpha_2 - \alpha_1\alpha_2}$$

and a similar expression for t_2. The expressions for n_1 and n_2 are also different. Accordingly the pressure on the inner surface of the hot plate is

$$\frac{p\left[\left(\alpha_1 + \alpha_2(1 - \alpha_1)\dfrac{T_2}{T_1}\right)^{1/2} + \left(\alpha_2\dfrac{T_2}{T_1} + \alpha_1(1 - \alpha_2)\right)^{1/2}\right]}{2(\alpha_1 + \alpha_2 - \alpha_1\alpha_2)^{1/2}}$$

This rather untidy expression can be made more intelligible by assuming that $(T_2 - T_1)/T_1 = \Delta T/T_1$ is small enough to justify neglecting $(\Delta T/T_1)^2$.
The expression for the net outward pressure on the plate then becomes

$$\frac{p}{2}\left(\frac{1}{2}\cdot\frac{\Delta T}{T} + \frac{1}{2}\frac{\Delta T}{T}\left(\frac{\alpha_2 - \alpha_1}{\alpha_1 + \alpha_2 - \alpha_1\alpha_2}\right)\right)$$
$$= \frac{p}{2}\left(\sqrt{\frac{T_2}{T_1}} - 1\right)\left(1 + \frac{\alpha_2 - \alpha_1}{\alpha_1 + \alpha_2 - \alpha_1\alpha_2}\right)$$

This interesting result shows (rather surprisingly) that if the plates have the same accommodation coefficient no correction is needed to the elementary theory of the Knudsen gauge.

Chapter 5

1 Work done $\approx \Delta p(v_2 - v_1)$: heat taken in $= \lambda$ the latent heat of fusion; so $\eta = \Delta p(v_2 - v_1)/\lambda$. This efficiency can be made to approach 1 by raising p. This should be compared with a better founded calculation: $\eta = \Delta p(v_2 - v_1)/\lambda$; but from the Clausius–Clapeyron equation $\Delta p/\Delta T = \lambda/T(v_2 - v_1)$ and $\eta \approx \Delta T/T$.

2 Go from A → B by a reversible path so that

$$\int_{\mathrm{A}}^{\mathrm{B}} \frac{\mathrm{d}Q}{T} = \int_{\mathrm{A}}^{\mathrm{B}} \mathrm{d}S = S_{\mathrm{B}} - S_{\mathrm{A}}$$

which does not depend upon the path. Return to A by a different path so that $\int_{\mathrm{B}}^{\mathrm{A}} \mathrm{d}S = S_{\mathrm{A}} - S_{\mathrm{B}}$ and $\oint \mathrm{d}S$ round the closed cycle is zero.

3 Heat flow into the house

$$= kA\,\frac{\mathrm{d}T}{\mathrm{d}x}$$

$$= \frac{6 \times 10^2 \times 4.2 \times 10^{-2} \times 52}{20 \times 10^{-3}}$$

$$= 65.52 \text{ kw.}$$

The heat pump takes in $-Q_1$ at T_1 to heat the house and Q_2 at T_2 and gains no entropy so that

$$\frac{Q_2}{T_2} - \frac{Q_1}{T_1} = 0$$

and $$Q_1 - Q_2 = Q_1\left(1 - \frac{T_2}{T_1}\right)$$

$$= \frac{20Q_1}{295} \text{ J/s}$$

$$= 65.52 \times 20/295 = 4.4 \text{ kW.}$$

4 In cylindrical coordinates, $x = r\cos\theta,\ y = r\sin\theta,\ z = z$; $\mathrm{d}x = \cos\theta\,\mathrm{d}r - r\sin\theta\,\mathrm{d}\theta$ etc. Substituting into the Pfaffian gives at once that $\mathrm{d}Q = r^2\mathrm{d}\theta + k\mathrm{d}z$. For radial paths $\mathrm{d}\theta = \mathrm{d}z = 0$ and all radial paths are adiabatic. Any two points r, θ, z and $r + \mathrm{d}r,\ \theta + \mathrm{d}\theta,\ z + \mathrm{d}z$ can be connected by an adiabatic path. The simplest case is when $\mathrm{d}\theta$ and $\mathrm{d}z$ have opposite signs. Move radially to a point at which $r_1^2\mathrm{d}\theta = -k\mathrm{d}z$; move tangentially to $r_1,\ \theta + \mathrm{d}\theta,\ z + \mathrm{d}z$. This change is adiabatic since $\mathrm{d}Q = r_1^2\mathrm{d}\theta + k\mathrm{d}z = 0$ by the choice of r_1. Return radially to $r + \mathrm{d}r;\ \theta + \mathrm{d}\theta,\ z + \mathrm{d}z$; a similar scheme can be devised if $\mathrm{d}\theta$ and $\mathrm{d}z$ have the same sign.

5 The analytical proof to which reference has been made follows fairly simply from the definition of Curl **R**. The following intuitive arguments may help to explain the formal mathematics. The previous example which goes to the root of the matter showed a system in which many radial adiabatic vectors radiated from an axis like the spokes of a wheel. It was always possible to join neighbouring points by adiabatic paths because one could always find a loop in the plane of the spokes traversing which always contributed to dQ. The expression $\mathbf{R}$. Curl $\mathbf{R} = 0$ says that this is not possible because all the components of Curl $\mathbf{R}$ are perpendicular to the plane of the spokes. The plain loop corresponds to $\oint dQ \neq 0$ at constant T.

6 The difficulties are (a) finding a vessel not permeable to the gases at 1000°C. This means mastering the technology of making a platinum iridium vessel and joining it to a glass system. (b) Calibrating the manometer of a constant volume gas thermometer. This involves columns of mercury at least 100 mm in diameter and several meters high. These are difficulties which patience, money and good workmanship can overcome. (c) The greatest difficulty and the final source of the experimental error is to ensure that the gas in the hot bulb has a temperature. Thermocouples welded to the bulb have always disclosed that it has not got a uniform temperature however carefully the furnace windings are graded.

7 As in the worked example; $C_1 \log T_1 + C_2 \log T_2 =$ constant.

8 On working through, the only difference will be that one must write $du_1 = T_1 ds_1 + p_1 dv_1$, $du_2 = T_2 ds_2 + p_2 dv_2$. Proceeding as before and equating the coefficients of the differentials to zero one gets

$$\lambda T_1 + \mu = 0, \; \nu - \lambda p_1 = 0$$
$$\lambda T_2 + \mu = 0, \; \nu - \lambda p_2 = 0$$

so that $p_1 = p_2$ and $T_1 = T_2$ as previously assumed.

9 The following outline shows how a demonstration can be constructed. Starting from the obviously true statement that $dQ = dQ_1 + dQ_2$ convert it by the given relationships into

$$t(s_1, s_2, \theta)ds = t_1(s_1, \theta)ds_1 + t_2(s_2, \theta)ds_2$$

and convert this into

$$h(s)ds = h_1(s_1)ds_1 + h_2(s_2)ds_2$$

Eliminate $\mathrm{d}s$ through the relationship

$$\mathrm{d}s = \left(\frac{\partial s}{\partial s_1}\right)_{s_2} \mathrm{d}s_1 + \left(\frac{\partial s}{\partial s_2}\right)_{s_1} \mathrm{d}s_2$$

Note that since $\mathrm{d}s_1$ and $\mathrm{d}s_2$ are independent variables that their coefficients must be zero to give

$$h_1(s_1) = h(s)\left(\frac{\partial s}{\partial s_1}\right)_{s_2} = \text{function of } s_1 \text{ only}$$

$$h_2(s_2) = h(s)\left(\frac{\partial s}{\partial s_2}\right)_{s_1} = \text{function of } s_2 \text{ only}$$

Form the partial derivatives of these two equations with reference to s_2 and s_1 respectively, noting that the l.h.s. will in both cases be zero. Use the relationship $\partial^2 s/\partial s_1 \partial s_2 = \partial^2 s/\partial s_2 \partial s_1$ to obtain the desired result.

10 To find Q_2 draw an adiabatic through the final state volume $2V$ internal energy U_1 the same as the internal energy of the initial state. This adiabatic will cut the original isothermal at the point at which the isothermal compression must be terminated. This must always be at some point such that $Q_2 < Q_1$ because an adiabatic compression always increases the internal energy.

11 Preliminary calculations; $a = 1.06 \times 10^{-1}$, $b = 2.507 = 10^{-5}$ $V_1 = 47.5 \times 10^{-5}$ about 1 per cent less than the measured value. Using the results of the worked example of section 5.6 one gets

$$T_2 - T_1 = -8.95 \text{ K}$$
$$S_2 - S_1 = 5.61 \text{ J/K}$$

Chapter 6

1 Yes. The first drawer's chance of winning is $1/n$. The second drawer's chance is $1/(n-1)$ (chance that the first drawer does not win) $= 1/(n-1)(1-1/n) = 1/n$ and so on.

2 The multinomial theorem states that the general term of $(a+b+c+d\ldots)^n$ is $n!/(r!\ s!\ t!\ldots)\ a^r b^s c^t \ldots$ with $r+s+t\ldots = n$.
Each pig can be placed in one of four sties so that there are $(4)^5$ ways of placing the five pigs; of these $5!/(l!\ m!\ n!\ o!)$ correspond to the specified arrangement. The probability = number

of ways of fulfilling the conditions/total number of ways of placing pigs in the sties $= n!/(l!\ m!\ n!\ o!)\ (\frac{1}{4})^5$ which is the coefficient in the multinomial expansion.

3 25.188, 64.48, 157.969.

4 Differentiate to obtain

$$\frac{\partial^2 A}{\partial t^2} = -\omega^2 A, \frac{\partial^2 A}{\partial x^2} = -\frac{4\pi^2 l^2 A}{\lambda^2}$$

etc. and substitute. Since $\lambda\omega = 2\pi c$ for a wave of constant velocity c the equation is satisfied if $l^2 + m^2 = 1$.
The boundary conditions imply that $\sin 2\pi la/\lambda = \sin n_1\pi$, $\sin 2\pi ma/\lambda = \sin n_2\pi$ with n_1, n_2 positive integers, so that $l^2 + m^2 = 1 = \lambda^2/4a^2\ (n_1^2 + n_2^2)$. The fundamental mode occurs when $n_1 = n_2 = 1$ and $\lambda = \sqrt{2}a$. The second harmonic has $\lambda = \sqrt{2}a/3$ which corresponds to $n_1 = n_2 = 3$. The total number of standing waves is therefore 11, nine from the values of n below 4 and two from the combination $n_1^2 + n_2^2 = 17$ with a wave length $2a/\sqrt{17}$.

5 (a) The two level system has not had the degrees of freedom specified. The mean energy is easily shown to the kT by the same mathematical device as is used for the linear oscillator.

(b) The linear oscillator
$E = h\nu/2 + h\nu/(h\nu/kT - 1)$ for two degrees of freedom x and $\dot{x}$. For large T expand the denominator to give $h\nu/kT\ (1 + h\nu/2kT \ldots)$ so that

$$E \approx \frac{h\nu}{2} + kT\left(1 - \frac{h\nu}{2kT}\right) = kT$$

corresponding to $kT/2$ per degree of freedom. Note that the zero point energy has been absorbed into the high temperature limit.

(c) The rotator
The energy is $\frac{1}{2}I(\omega_1^2 + \omega_2^2)$ so that there are two degrees of freedom

$$Z = \frac{2IkT}{\hbar^2\sigma}, \ E = nkT^2\,\frac{\partial \log Z}{\partial T} = \frac{nk^2}{Z}\frac{\partial Z}{\partial T} = nkT$$

and each degree of freedom has energy $= kT/2$.

6 For a particle mass m single degree of freedom p and kinetic energy $p^2/2m$ the average energy is

$$\frac{\text{total energy}}{\text{number of particles}} = \frac{A\Sigma\varepsilon_i e^{-\varepsilon_i/kT}}{A\Sigma e^{-\varepsilon_i/kT}}$$

$$= \frac{\dfrac{1}{2m}\displaystyle\int_{-\infty}^{+\infty} p^2 e^{-p^2/2mkT}\,dp}{\displaystyle\int_{-\infty}^{+\infty} e^{-p^2/2mkT}\,dp}$$

$$= \frac{kT\displaystyle\int_{-\infty}^{+\infty} y^2 e^{-y^2}\,dy}{\displaystyle\int_{-\infty}^{+\infty} e^{-y^2}\,dy} = kT/2$$

for any degree of freedom whose square represents energy.

7 For a Poisson distribution near the mean value $n/2$

$$P = \left(\frac{1}{2}\right)^n \frac{n!}{\left(\dfrac{n}{2}+a\right)!\left(\dfrac{n}{2}-a\right)!}$$

to be evaluated by Stirling's theorem. This is rather long; a typical term is

$$\log\left(\frac{n}{2}+a\right)! = \left(\frac{n}{2}+a\right)\log\frac{n}{2}\left(1+\frac{2a}{n}\right) - \frac{n}{2} - a$$

which on expanding the logarithmic term gives $n/2 \log n - n/2 \log 2 + a \log n/2 - n/2 + 2a^2/n$. On summing all the terms one finds

$$\log P = \log\frac{2}{\sqrt{2\pi n}} - \frac{4a^2}{n} \quad \text{and} \quad P = \frac{2}{\sqrt{2\pi n}}\, e^{-4a^2/n}.$$

On comparison with the Gaussian expression one finds that $n = 2\sigma^2$ and that σ the mean square deviation $= \sqrt{\bar{n}}$.

8 The o-p change at $\sim$ 20 K consists of three-quarters of the molecules changing from a state with $J = 1$ to a state $J = 0$. The energy released is thus $3N/4 \cdot h^2/4\pi^2 I$ J/mol. The moment of inertia μr^2 can be calculated from the known distance apart $r = 7.46 \times 10^{-11}$ m of the protons in H_2 and is

4.60×10^{-48}. The heat of conversion is accordingly 1.092×10^3 J/mol in fair agreement with the 895 J/mol measured by Farkas.

9 The value of $\varepsilon = mgh$ so that $n_h = n_o\, e^{-mgh/kT}$ by direct substitution. Note that this is an example which does not lead to the equipartition of energy.

10 $S = 8.314\ (\frac{5}{2}\log_e\ 273.15 - \log_e 1.01325 \times 10^5 + \frac{3}{2}\ \log_e 754.45 + \frac{5}{2})$
$= 8.314\,(14.025 - 11.526 + 9.939 + 2.5) = 124.2$ J/mol.

11 The whole curve is determined by a single parameter ε. Verify that $\varepsilon \approx 2.4\ kT_{max} = 1.22 \times 10^{-21}$ J. It requires advanced statistical treatment to go beyond a subjective opinion that the fit is or is not satisfactory in the circumstances.

Chapter 7

1 By definition

$$\left(\frac{\partial S}{\partial T}\right)_p = \frac{C_p}{T};\ \left(\frac{\partial S}{\partial p}\right)_T = -\left(\frac{\partial v}{\partial T}\right)_p$$

by Maxwell; but S is a function of state so that

$$\left(\frac{\partial^2 S}{\partial T \partial p}\right) = \left(\frac{\partial^2 S}{\partial p \partial T}\right) \text{ and } \frac{1}{T}\left(\frac{\partial C_p}{\partial p}\right)_T = -\left(\frac{\partial^2 v}{\partial T^2}\right)_p.$$

2 $T\mathrm{d}S = \mathrm{d}Q = \mathrm{d}u + p\mathrm{d}v$; for an isothermal change $\mathrm{d}u = (\partial u/\partial v)_T \mathrm{d}v = (T(\partial p/\partial T)_v - p)\mathrm{d}v$ (from section 7.4) so that $\mathrm{d}S = (\partial p/\partial T)_v \mathrm{d}v = R\mathrm{d}v/(v - b)$ which integrates at once to the stated relationship.

3 $PV = RT$; $U = RT/2\,(3 + d)$ with d the number of degrees of freedom of internal molecular motion;

$$\gamma = \frac{5 + d}{3 + d},\ \gamma - 1 = \frac{2}{3 + d}$$

so that $U = \dfrac{RT}{\gamma - 1} = \dfrac{MP}{(\gamma - 1)\rho}$.

4 As in problem 2,

$$\left(\frac{\partial u}{\partial v}\right)_T = T\left(\frac{\partial p}{\partial T}\right)_v - p = \frac{RT}{v-b} - \left(\frac{RT}{v-b} - \frac{a}{v^2}\right)$$

and $u = -a/v +$ constant. Work done against internal forces $= p_{\text{Forces}} \times v = a/v$ and $P_{\text{int}} = P_{\text{ext}} + a/v^2$ as assumed by van der Waals.

5 In general $\gamma = (\partial v/\partial p)_T(\partial p/\partial v)_s$: if $p(v-b) = RT$ this becomes $(\partial p/\partial v)_s = -\gamma(p/(v-b))$ so that $\mathrm{d}p/p + \gamma\ \mathrm{d}v/(v-b) = 0$ equivalent to $p(v-b)^\gamma =$ constant; also $(\partial u/\partial v)_T = 0$ so that $u = \mathrm{K} + F(T)$.

6 If $PV = RT, \beta = 1/T, \rho c_p = C_p/V$ and

$$\left(\frac{\partial T}{\partial P}\right)_s = \frac{RT}{PC_p} = \frac{\gamma - 1}{\gamma} \cdot \frac{T}{P}$$

equivalent to $\log T = (\gamma - 1)/\gamma \log P +$ constant the well-known result for a perfect gas.

7 One has to evaluate $(\partial l/\partial T)_F$ which from $l = l_0 + l_0\alpha T + Fl_0/AE$ is $l_0\alpha - Fl_0/AE^2\ \mathrm{d}E/\mathrm{d}T$. Substituting in the general result (section 7.11) this gives

$$\Delta T = \frac{-T}{c_p\rho}\left(\alpha\left(\frac{F}{A}\right) - \left(\frac{F}{A}\right)^2 \frac{1}{2E^2}\frac{\mathrm{d}E}{\mathrm{d}T}\right).$$

In numerical calculations for substances like steel the second term can be neglected. Using tabulated constants at 300 K $\Delta T = -2.75 \times 10^{-3}$ K. This result is uncertain because 'steel' is too vague a specification. The calculated cooling agrees to about 30 per cent with some measurements made by Joule. In a heat engine with an invar wire ($\alpha = 0$) as a working substance the work done on an extension $\mathrm{d}l$ is $-\frac{1}{2}\ F\mathrm{d}l$. In a cycle of contraction at T and extension at $T - \Delta T$ the work done per cycle is $\frac{1}{2}\ \mathrm{d}F/\mathrm{d}T\ \mathrm{d}l\ \mathrm{d}T = Q\mathrm{d}T/T$ with Q the heat taken in during the contraction. The heating on a sudden extension $\mathrm{d}l$ is accordingly

$$\frac{T}{2c_p\rho Al}\frac{\mathrm{d}F}{\mathrm{d}T}\,\mathrm{d}l = \frac{T}{2c_p\rho}\left(\frac{F}{A}\right)^2\frac{1}{E^2}\frac{\mathrm{d}E}{\mathrm{d}T}$$

as previously found.

8 Use the Clausius–Clapeyron equation. All the data are available in the tables.

9 If $p = p_0\,\mathrm{e}^{-\lambda/RT}$ and $p' = p_0'\,\mathrm{e}^{-\lambda'/RT}$ and the pressures are the same at T_0, $p_0' = p_0\,\mathrm{e}^{-(\lambda-\lambda')/RT_0}$. Forming $(p - p')$ and differentiating for a maximum gives at once that

$$\frac{\lambda}{\lambda'} = \mathrm{e}^{(\lambda - \lambda')\left(\frac{1}{RT} - \frac{1}{RT_0}\right)}$$

which is the exponential form of the given expression. The maximum occurs at about −12 °C.

10 Such a drawing shows that the number of molecules whose attractive forces are preventing a molecule in the surface from leaving are less for a convex surface.

11 If one puts $p_r/p = 1.38$ one finds $r \approx 1.1 \times 10^{-9}$ m which is 4–5 molecular diameters.

12 Taking the change of melting point with pressure to be 7.4×10^{-8} K/Pa the rate of decent is

$$7.4 \times 10^{-8}\, gK \frac{\rho_{\mathrm{Cu}}}{\rho_{\mathrm{Ice}}} = 8.5 \times 10^{-9}\ \mathrm{m/s}.$$

13 The problem can be discussed in terms of: stopping distance, the possibility of skating at −30 °C, whether a film of water is involved and the area of contact on the two hypotheses.

14 $\mathrm{d}Q = \mathrm{d}u + p\mathrm{d}v = np\mathrm{d}v + nv\mathrm{d}p + p\mathrm{d}v$

$$C_p = \left(\frac{\partial Q}{\partial T}\right)_p = (n+1)p\left(\frac{\partial v}{\partial T}\right)_p.$$

Form $\mathrm{d}u + p\mathrm{d}v = C_v\mathrm{d}T + T\left(\frac{\partial p}{\partial T}\right)_\sigma \mathrm{d}v = 0$ for an adiabatic change

16 $\mathrm{d}T = \dfrac{T\varepsilon_0 E^2}{2c_p\rho}\dfrac{\mathrm{d}\varepsilon}{\mathrm{d}T} = -6.9 \times 10^{-7}$ K

17 Form $-T(\mathrm{d}v/\mathrm{d}T)_p + v$. Plot the data to give IJT coefficient $-33.4 + 7.64 \times 10^{-2}\ p$, giving at 167 K $\mathrm{d}A/\mathrm{d}T \approx 1.4 \times 10^{-4}$ atm^{-1} and $-\dfrac{\mathrm{d}B}{\mathrm{d}T} \approx 3.3 \times 10^{-7}$ atm^{-2}. Interested students should find the best straight line and estimate the mean square error of the results.

Chapter 8

1 From Stokes' formula calculate $r \approx 2.8 \times 10^{-6}$ m. Calculate $\overline{\Delta x^2}$ from Einstein's formula and put

$$\frac{\overline{\Delta x^2}}{v^2} = \overline{(\Delta t)^2} \approx 1.4 \times 10^{-3}$$

which is less than the observed value (1.47×10^{-3}) of $\overline{\Delta t^2}$. This is to be expected since some of the variance must be due to errors of measurement.

2 The stored energy of a condenser $= \frac{1}{2}C\overline{V^2} = kT/2$ and for equipartition $\overline{V^2} = kT/C$. For $C = \frac{2}{3} \times 10^{-11}$ F $\overline{V^2} = 1.24 \times 10^{-9}$. Amplify the voltage with an amplifier of known gain and band width. Compare the amplified and rectified output with that from a calibrated noise generator, usually a saturated diode; or plot the output of the amplifier against the temperature of the condenser.

3 $T^2 = 4\pi^2 I/\text{couple per unit twist} = 4\pi^2 MR^2/4 \times l/\pi r^4 G$ and $T = 1.28 \times 10^{-2}$ s ·

$$\overline{\theta^2} = \frac{2kTl}{\pi r^4 G} = 1.08 \times 10^{-14} \text{ at } 300 \text{ K}$$

and R.M.S. $\theta = 1.04 \times 10^{-7}$ radians. The difficulties inherent in the measurement of such small angles can in principle be avoided by the use of a divided photocell; those of a stable mounting free from vibration remain.

Index